Construction Technology

Third Edition

Mark W. Huth

Thomson Learning® **TOOLS**

For more information, contact:

Thomson Learning® Tools
5101 Madison Road
Cincinnati, OH 45227

International Thomson Publishing
Berkshire House
168-173 High Holborn
London, WC1V7AA
England

Thomas Nelson Australia
102 Dodds Street
South Melbourne 3205
Victoria, Australia

Nelson Canada
120 Birchmont Road
Scarborough, Ontario
M1K 5G4, Canada

International Thomson Publishing GmbH
Konigswinterer Str. 418
53227 Bonn
Germany

International Thomson Publishing Asia
221 Henderson Bldg. #05-10
Singapore 0315

International Thomson Publishing Japan
Kyowa Building, 3F
2-2-1 Hirakawa-cho
Chiyoda-ku, Tokyo 102
Japan

1 2 3 4 5 6 7 8 9 0 KI 04 03 02 01 00 99 98 97 96 95

Printed in the United States of America

Cover Photo courtesy of CTA Architects
Cover Design: Sandy Weinstein

Publisher: Brian Taylor
Sponsoring Editor: Suzanne F. Knapic
Production Editor: Melanie Blair-Dillion

Library of Congress Cataloging-in-Publication Data

Huth, Mark W.
   Construction technology/Mark W. Huth—3rd ed.
      p. cm.
   Includes text.
   ISBN 0-538-64471-0
   1. Building—Juvenile literature. [1. Building.]
      I. Title.
   TH149.H88 1994   94-2210
   690--dc20   CIP

CONTENTS

PREFACE

Who Should Read This Textbook?

In the mid 1980s the educational community recognized that all students should study technology. While there are different views on which technology to study and how it should be studied, almost all experts agree that construction technology is one of the major topics. Most schools include a course in construction technology. This is a textbook for those courses. The purpose of CONSTRUCTION TECHNOLOGY, 3rd edition is to help you understand all of construction—how it is organized, how it is controlled, its impacts on society, and the opportunities it offers. This is not a text on architectural design or carpentry, but future architects and carpenters will find this a good starting place.

Problem Solving and the Systems Approach

The construction industry is one of the biggest sectors of our economy. Unlike other sectors, the construction industry is made up mainly of relatively small companies working on projects in the field. Because of the decentralized nature of the construction industry, there are almost as many solutions to construction problems as there are construction projects. This makes the study of construction a great way to develop problem-solving skills.

In recent years there has been a trend toward viewing all technology as following a system. This system-oriented approach works very well for construction technology. There are many types of problems encountered in construction and their solutions are equally varied, but all of construction can be understood if one understands the common systems of technology. The systems approach is explained in Chapter 1. Each section begins with a systems diagram to indicate where the content of that section fits the system.

Organization

The book is divided into seven major sections. Each section is further divided into short chapters. Section One explains the construction industry in broad terms. The topics covered include the systems approach, occupations, the major categories of construction, designing for construction, and the business of construction. Section Two explains how to use the tools of the trades. Because safety is so important in the use of tools, safety precautions are especially frequent in this section. Section Three covers the most common materials used in construction. You will learn the most important properties of the materials and how to work with them. Section Four describes the construction of the structural parts of small buildings. Section Five describes the construction of heavy structures. Section Six covers the nonstructural parts of a building. Some courses do not include the study of heavy construction, so your teacher might not give you assignments from Section Six. It follows heavy construction in the text because the nonstructural systems for light construction and heavy construction are very similar. Section Seven describes advanced construction systems—buildings assembled in factories and construction in space.

Features

CONSTRUCTION TECHNOLOGY is packed with features to help make learning fun and easy.

▼ **Short Chapters.** Each section is divided into short chapters, so you will be able to study one chapter thoroughly before going on to the next. A chapter is a complete learning package with objectives, information and procedures, activities, and review questions.

▼ **Activities.** Technology should be learned by "doing." At the end of each chapter there are several hands-on activities that will help you experience construction technology. These activities can be done with the tools and materials found in most technology laboratories. You probably will not do every activity in the book. Your teacher will help you decide which activities can be done in the time you have available.

▼ **Applying Construction Across the Curriculum.** There are several new activities that help the student apply mathematics, science, social studies, and communications to their study of construction in each chapter. These activities are less prescriptive than the first set of activities in each chapter. Like the more detailed activities, many of the *Applying Construction Across the Curriculum* activities require students to work cooperatively in small teams.

▼ **Objectives.** Each chapter begins with a list of the major objectives to be accomplished. Read the objectives before the rest of the chapter so that you will know what to concentrate on in the chapter. When you have completed the chapter, reread the objectives to see if you have met them.

▼ **Key Terms.** After the objectives at the beginning of each chapter is a list of the most important terms used in that chapter. The list of key terms will help you in two ways. By paying particular attention to how they are used in the chapter, you will be focusing your attention on the most important topics. Second, these are the vocabulary words you will need to know to study construction. Learning the meanings of the key terms will make the chapter easier to read. The key terms are highlighted in boldface when they first appear in the text.

▼ **Special Interest Topics.** There are several articles on topics of special interest located throughout the text. Although it is not necessary to study these articles to use the text, they are particularly interesting to students of construction.

▼ **Review Questions.** Each chapter ends with a series of objective-type questions. If you have learned the material in the chapter, the questions should be easy. If you have not learned the material, the questions may not be so easy.

▼ **Glossary.** After the last chapter there is a complete glossary. Although each key term is defined as it is introduced in the text, when you hear a construction term used outside the text, the glossary will be a handy reference. Don't let a new term go unexplored. When you hear a new term, look it up and learn to use it. Expanding your vocabulary makes learning fun and easy.

▼ **Mathematics.** To work well with construction systems, you will need to apply some basic mathematics. Although you have probably already studied more math than you need for this course, you may find problems for which a quick math review would be helpful. The math review in the back of the text will explain all of the math you will need.

▼ **Student Associations.** If your school has a club for technology students, join it. It will be fun and will help you learn technology. If your school does not have a technology club, you and some of your friends may want to organize one. The appendix on student associations will tell you how.

Teacher's Resource Guide

A complete Teacher's Resource Guide (TRG) is available. The Teacher's Resource Guide includes several items that will be of use to the teacher who is starting a new program and the teacher who is using CONSTRUCTION TECHNOLOGY in an established program. The Teacher's Resource Guide includes teaching outlines to help the instructor prepare for each chapter, supplemental questions for students to demonstrate their understanding of the material, quizzes, and transparency masters. It also includes two types of software. One disk contains a student version of Timberline Software's best-selling *Precision Estimating Light*. There is a complete tutorial for this software in the Teacher's Resource Guide. The other disk contains a program for tracking the position of the sun for use in planning energy efficient construction.

Acknowledgments

I am fortunate that many of my friends are recognized as the leading experts in technology education. From time to time we have individual and group discussions about what should be learned and how it should be taught. Although space does not permit me to thank each of the experts individually who helped shape this edition, I'd like to thank them collectively for their guidance and encouragement. I would like to thank two professionals at Delmar Publishers who were particularly important in making this book and its supplements the best they could be. Wendy Troeger, my production editor, shaped the design of this third edition and helped ensure that the final product was what we had planned. Sandy Clark, my editor has been a constant source of inspiration and creativity. There are also many experts in the construction industry who helped make this edition what it is. The companies and individuals who contributed to the contents and illustrations are named in the credits throughout the book, but I would like to give special thanks to two who gave extensive help. Jim Besha, President of Besha Associates, contributed many valuable hours of his and his associates' time, air transportation to and from the project described in Chapter 22, and a great deal of expertise. John O'Sullivan, P.E., taught me most of what I know about highway construction. I would also like to express my appreciation to the many teachers who helped shape this edition; both those who used the earlier editions and provided feedback for this edition, and the following reviewers who helped develop the manuscript for this edition:

▼ Rob Campbell,
Tacoma Public Schools, Washington
▼ Charles Hining,
Prairies High School, Iowa

- ▼ Jeffrey Krynen,
  Don Bosco Technical Institute, California
- ▼ Terry Smith,
  Dorchester High School, Nebraska
- ▼ Larry Stiggins,
  J.T. Hutchinson Junior High, Texas

- ▼ Richard Thompson,
  Winterset High School, Iowa
- ▼ Al VanQuekelberg,
  Apollo High School, Minnesota
- ▼ John Wozniak,
  Edgerton Middle School, Wisconsin

SAFETY

SAFETY is an especially important concern on construction sites and in school construction laboratories. In the industry, safety is so important that people are designated to watch over the safety of construction workers.

Safety Coordinator

Companies often have safety coordinators who manage special safety programs. The safety coordinator is responsible for safety education. Safety education may include classes on the safe use of tools or handling materials, safety posters, and incentive programs to get employees thinking about safety. The safety coordinator also ensures that tools, structures, and the work environment are free of unnecessary safety hazards. The coordinator may personally inspect potential hazards and write recommendations to management for correcting the hazards. On very large projects the safety coordinator may plan a program that includes safety inspections by other employees.

OSHA

In 1970, the U.S. government established the Occupational Safety and Health Administration. OSHA (pronounced *oh sha*) has written regulations that require employers to meet certain safety standards. OSHA regulates the safety conditions in all areas of employment. Many of the regulations of OSHA pertain to the construction industry.

Personal Responsibility

Although OSHA requires the employer to maintain a safe place to work, the worker has as much responsibility for safe work habits. Just as the construction worker on a site is responsible for avoiding accidents, the student in the school laboratory is also responsible for safe conduct. School laboratories are generally as safe as a construction environment can be, but if the students in that laboratory don't practice good safety habits, the chance of serious injury is high.

Throughout this text you will find frequent safety cautions. These pertain to particular operations or tell how to avoid specific hazards. There are many general safety rules that apply to the construction laboratory in general. Most of these rules will seem very natural to you; some may not. Read these rules carefully. Ask your teacher for help if you do not understand all of the rules. Then, be sure to practice them. It is not enough to know all the safety rules; you should practice them so they become habits.

▼ Do not operate any machine or use any tool until you have been instructed on its use.

- ▼ Use tools and machines only in the manner they were intended to be used.
- ▼ Do not distract others while they are working.
- ▼ Running and horseplay are not permitted in the laboratory or on the construction site.
- ▼ Wear proper safety attire in the laboratory and on the construction site (e.g., safety glasses, hard hats, sturdy shoes, etc.).
- ▼ Use electrical equipment only with proper grounding and in dry areas.
- ▼ Operate a power tool only when all safety guards are in place.
- ▼ Do not work with dull or damaged tools.
- ▼ Know the locations of emergency devices (e.g., fire extinguishers, electrical disconnects, first aid kits, etc.), and how to use them.
- ▼ Do not attempt to lift or carry excessive loads—get help.
- ▼ Do not leave hazards for others (e.g., hot metal, exposed nail or screw points, open pits or shafts, etc.).
- ▼ Report any injury, no matter how slight, to your teacher.

The rules above are some common sense guidelines, but safety should be more than a list of special precautions you observe when you are working around dangerous machines. Mature people in construction make safety an attitude. They are so conscious of safe work habits and a safe environment that to do something unsafe causes an uncomfortable feeling. You may be surprised at how quickly and easily you, too, can develop a safety attitude. Every time you see a possible safety hazard, think about the danger it can present and make a point of avoiding that hazard. In practically no time, you will be using good safety habits without stopping to think about them.

Many accidents are caused by bad housekeeping. When clutter accumulates, it becomes more difficult to focus on the task at hand and to do it safely. Tools and equipment are most enjoyable to use when they work properly. They are also safest when they work properly. An instrument left in the corner or a tool shoved out of the way on a work table is more apt to become damaged than one that has been returned to its proper storage space.

In the heat of designing or constructing a structure, scraps or waste are created. It is not realistic to discard every piece of scrap material as it is created, but it is easy to clean the area periodically. Your teacher will allow time at the end of each class period to clean the lab and put away your equipment. If you see scraps accumulating around your work area, you might need to take a minute to clean up during the class. You'll get more accomplished in the class and there will be less chance of someone being injured, if you are not working around clutter.

Electrical Safety

Electricity is an important element in most construction work. Even the planners and designers who work primarily in offices use electricity. There can be enough electricity stored in an unplugged computer or television to kill a person who touches certain parts with a screw driver. Many deaths have been caused by frayed lamp cords. When large amounts of electric current are brought to a wet construction site to run power tools, the potential hazard is far greater.

Improperly used electricity can also be a serious fire hazard. In 1892, the first large-scale electric lighting project was done at the World's Fair in Chicago. The plan was that two hundred fifty thousand electric light bulbs would illuminate the fair before the eyes of the world. At the key moment, the switch was closed and the lights came on. Then the lights went off, as the overloaded wires burned. It has been said that the electrical fires gave more light than the light bulbs at the 1892 World's Fair.

Since 1892 electrical codes have been written to provide rules to ensure that electrical wiring is safe. When used according to electrical codes, there is very little chance that an electrical device will cause serious shock or start a fire. Electrical codes are designed to protect a person who does not know anything about the code. If you use electrical devices only in the way they are intended to be used, you will generally be safe. Never use electrical equipment that is in poor condition or that has been tampered with. Never open any power panel, switch box, or other electrical enclosure until you have been trained to do so.

Hazardous Materials

The United States Department of Transportation (D.O.T.) classifies hazardous materials. They are classified according to the type and amount of danger they present. Some hazardous materials can be found on construction sites or even in school laboratories. Some chemicals that have not been classified as hazardous are found in construction labs and can present some danger. For example, uncured concrete can cause serious chemical burns if left in contact with your skin. The fumes from chlorine bleach (the kind used to launder clothing) can burn the inside of your nose or lungs.

OSHA requires that every worker who comes in contact with any potentially dangerous material must have access to certain information about that material. A *Material Safety Data Sheet* (MSDS) must be available for every such material. For example an MSDS must be on file for most plywood used on construction sites, because the glue used to make the plywood is a hazardous material and wood dust can cause severe fire and health risks if not handled properly.

An MSDS for contact cement is reprinted on pages 53–57 of this text. Read this sample MSDS carefully. You'll find some scientific information that you might not understand, but you will also find that you do understand most of the information. This type of contact cement is extremely flammable and the fumes are dangerous to breathe. Look for this information. You will also see information about how to handle spills, fires, and health hazards.

A few simple rules can eliminate most of the danger from common hazardous materials.

▼ Always keep chemicals and solvents in their original containers, with the manufacturer's label in place. It is very dangerous to use chemicals without knowing what they are.

▼ Read labels. Before you use even the most ordinary supplies, read the manufacturer's label completely. If there is any danger, the manufacturer usually explains the danger, how to avoid it, and what to do if an accident occurs.

▼ Read the MSDS for each hazardous or questionable product with which you come in contact.

▼ Use chemicals and other supplies only for their intended purposes. There is a solvent, chemical, or other substance designed for just about anything you might want to do. Why try to force another substance to do the job?

Section One

Construction Planning and Drawing

Section One explains the planning, designing, and managing of a construction enterprise. If the whole construction industry is viewed as a technological system, the units in this section touch on all parts of the system.

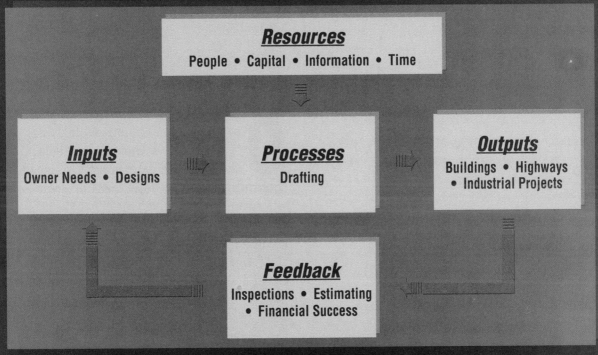

Chapter 1
The Construction Industry

provides an overview of the construction industry and its relationship to all industrial technology. This unit also explores the types of careers in the construction industry and the training needed for those careers.

Chapter 2
Designing for Construction

explains the design process, including designing for aesthetics and designing for function. It also discusses how good design relates to societal needs.

Chapter 3
Architectural Drawings

explains drafting and computer-aided drafting. It covers drafting tools, symbols, and abbreviations so that the student can read basic drawings.

Chapter 4
Specifications and Contracts

covers the remaining construction contract documents and further explains the design processes.

Chapter 5
Business of Construction

explores the major issues of the management of a construction company: forms of ownership, estimating, and scheduling.

CHAPTER 1

The Construction Industry

OBJECTIVES

After completing this chapter, you should be able to:

▼ discuss the importance of technological advancement in areas related to the construction industry;

▼ recognize how the economy affects the construction industry;

▼ describe the various occupations in the construction industry; and

▼ explain the differences between light, heavy, industrial, and civil construction.

KEY TERMS

technology

system

input

process

resource

output

feedback

light construction

commercial construction

industrial construction

civil construction

laborer

skilled trades

apprenticeship program

professions

journeyman

technician

The Development of Technology

One of the most noticeable traits of civilization is the use of tools and materials to create things. This is called **technology**. One of the first uses of technology was to improve shelter. This may have been as simple as using a pole as a lever to pry large stones out of a cave.

From these early beginnings, civilization continued to develop better tools and discover new ways of using available materials. Technology fulfilled a wide range of needs. Among the most ancient archeological finds are constructions for shelter, recreation, worship, and even for the advancement of technology, **Figure 1–1**

For at least 2-1/2 million years, technology was limited to the raw materials that were readily available. Tools were primitive devices made of wood and stone. The materials of construction were grass-like plants, wood, clay, and stone. About 4000 B.C., it was discovered that copper could be extracted from the earth. After copper, tin and zinc were found. These metals produced better tools and materials that, in turn, made possible more complex and permanent construction.

FIGURE 1–1 Stonehenge is a well known example of ancient construction. *(Courtesy of Susan Warren)*

Technology advanced with each new discovery. Metal fittings, for example, produced sturdier animal-drawn carts and wagons that could move material more efficiently. With better vehicles, heavier loads could be transported over longer distances. Larger and more sophisticated structures were easier to build. From the first uses of the earth's raw materials to the Industrial Revolution (circa 1800 A.D.), each new discovery led to others and civilization expanded.

During the Industrial Revolution, technology advanced at a rapid rate. Soon new methods were discovered for extracting raw materials from the earth. New machinery modernized the industrial world and new industries were created. Better sources of power and materials with more desirable properties produced even more advances and the whole technological chain advanced. This rapid advancement continues today.

Interdependence of Industries

All areas of technology have a great impact on one another. In the preceding examples of the development of tools, tool manufacturing and the discovery of metals played an important part in the advancement of construction practices.

In the modern world this interdependence of one industry on another is even more striking. When the automobile was developed around the turn of the century, better roads had to be built. These roads were built by the construction industry. With better roads, transportation became faster and safer. People began to depend on automobiles and highway travel. To support the growing demand for motor vehicles, the manufacturing industry had to open new factories and offices, **Figure 1–2**.

The construction industry affects every other sector of industry and the other sectors affect construction. All manufactured goods are produced in factories built by construction workers, stored in warehouses built by construction workers, transported over roads built by construction workers, and sold in stores built by construction workers. In return, the construction industry relies on other modern industries. For example, construction materials are transported in trucks, trains, and ships built by the manufacturing industry.

When all industries have the needed resources for new projects, a country can

FIGURE 1–2 This automobile manufacturing plant is a product of the construction industry. *(Courtesy of Nissan Motor Manufacturing Corp., USA)*

The Tallest Building

The 100-story Sears Tower in Chicago is the world's tallest building. It contains 4-1/2 million square feet of space, making it the largest private office building in the world. Designed by the architectural company of Skidmore, Owings & Merrill, the tower reaches 1,454 feet above the ground. Twin antenna towers atop the building bring its total height to 1,707 feet (nearly one third of a mile). Sears Tower has some spectacular statistics:

▼ Construction of Sears Tower took three years. During peak times, 1,600 people worked on the project.

▼ The framework of the tower consists of 76,000 tons of steel. The building contains enough concrete to build an eight-lane highway, five miles long. It has 16,000 bronze-tinted windows and 28 acres of black aluminum skin.

▼ 114 concrete piles support the 222,500-ton building and each is securely socketed into the bedrock.

▼ The building's 104-cab elevator system, including 16 double-decker elevators, divides the building into three separate zones, with lobbies between each of them. A special elevator for baby strollers and wheelchairs serves the public areas of the tower.

▼ Six automatic window-washing machines clean the building exterior eight times a year.

▼ There is room to accommodate up to 12,000 office workers in Sears Tower.

▼ Sears occupies the lower half of the building; the remainder is leased to tenants.

▼ The tower contains 30 stores featuring everything from souvenirs to women's and men's clothing. Seven restaurants offer everything from a quick bite to a full dinner. ■

Courtesy of Sears, Roebuck & Company

develop and advance. This is only possible when the economy is healthy. When the economy is failing, all industries suffer. Because of the interdependence of industries, when one sector of industry is doing poorly, it affects all other sectors. Without money to build roads, there is no reason to make more cars or find new methods of transporting materials. When the economy is healthy, resources are available for every industry. In this way, the economy greatly affects the development of a country.

Systems and Construction Technology

Technology is changing at a fast pace. More technological developments have taken place in this century than in all of history before this century. In the last decade, more technological development has occurred than in the first three-fourths of this century. Since most of the readers of this book were born, we have seen the introduction of microcomputers, robots have gone to work in industry, and hundreds of new products have been developed for construction.

All technology, whether it is new this year, coming next year, or has been around for twenty years, can be studied and controlled as a system. The **systems** of technology all have the same basic parts: inputs, processes, outputs, and feedback, **Figure 1–3**. The **input** for a system

FIGURE 1–4   This construction estimator is getting feedback from his computer about the costs of construction. *(Courtesy of Brown & Roote, Inc.)*

of technology is generally the desired output. For example, if the desired output is a four-story office building, the input for the system is the design for the building. The **process** that creates the building requires **resources**. The resources of technology are people, information, materials, tools and machines, energy, capital (money), and time. The **output** in this building example is the finished building. Throughout the process there are ways of observing the results and adjusting the input. This is called **feedback**, **Figure 1–4**.

Each of the sections of this book covers a segment of the construction system. Some units cover inputs, some cover resources, and others cover the processes. Feedback, such as estimating and inspecting, is discussed throughout the book. As you study construction technology, you will learn to recognize the parts of the system. You will also begin to understand what effect changing one of the components will have on the output.

Types of Construction

Construction can be broadly defined as the assembly or erecting of structures that cannot

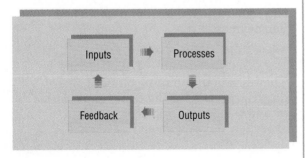

FIGURE 1–3   All systems of technology have the same basic components.

Kathy Sutphin

Occupation:

Carpenter Apprentice Instructor

How long:

3 Years

Typical day on the job:

Kathy's local has 260 apprentices at different stages of training. The apprenticeship requires 160 hours of instruction per year for 4 years. Apprentices spend two days per month in school. Usually the first 1 1/2 hours of the two-day session are spent in the classroom, then the rest of the time is spent in the shop. Because this is a self-paced program, apprentices are working at various levels and on various skill blocks at the same time. The instructor has to be able to jump from one topic to another as the apprentices need attention. All training is graded on the basis of what would be acceptable in the field. Either the apprentice's work is acceptable or it must be redone. When the apprentices are not in school, they are employed in the field as apprentice carpenters. As they complete each of the seven skill blocks of the apprenticeship, their pay is increased.

Education or training:

Kathy is a product of the system: she went through the same apprentice training. She has also completed a training program administered by Associated General Contractors (AGC) for superintendents and taken college classes in drafting and estimating.

Previous jobs in construction:

She had several jobs as a carpenter, especially in heavy construction. Just before becoming an instructor, Kathy was foreman of an 8-person framing crew, then an 18-person drywall crew.

Future opportunities:

Kathy says she is very happy where she is, but she is nearly finished earning a bachelor's degree in college and might become a contractor some day.

Working conditions:

Compared with working in the field as a carpenter or foreman, the working conditions for an apprentice instructor are great. The work is mostly indoor. There is a lot of stress involved in this job, however, because all of her students are just learning and safety is a constant issue.

Best aspects of the job:

She likes watching people grow. Many apprentices come in sort of rough around the edges, then blossom into polished journeymen.

Disadvantages or drawbacks of the occupation:

The work is in the same building every day. Carpenters in the field get to work at a new location, with new projects every few days or weeks.

FIGURE 1–5 Light construction refers to houses and other small buildings. *(Courtesy of the California Redwood Association)*

be readily moved once completed. This definition includes structures that fit into four categories: light, commercial, industrial, and civil.

Light construction is the building of homes and other small buildings, **Figure 1–5**. These may be single-family dwellings, small apartment buildings, offices, stores, etc. Larger structures are not normally included in this category because the materials and techniques are quite different from those used in light construction.

Light construction, perhaps more than other categories, often follows traditions or conventions. For example, many designers still think of 2x4 lumber as the basic building unit for light construction, even though other materials could do the same job. Very recently, however, many very comfortable and useful small buildings have broken away from these traditions. The designer of light structures (small buildings) considers the basic needs of the occupants, the surrounding community, and the owner to design the best possible structure to satisfy those needs. Many of the new generation of light buildings use large areas of glass, earthen embankments, and prefabricated sections where they will help satisfy the requirements of the building.

Commercial construction includes the construction of large buildings that are used for commercial purposes, **Figure 1–6**. Commercial building construction includes larger

FIGURE 1–6 Office buildings are examples of commercial construction. *(Courtesy of PPG Industries)*

apartment buildings, schools, churches, office buildings, and warehouses. This type of construction involves larger equipment, a greater financial investment, and more time than light construction.

Commercial construction generally allows for the greatest amount of creativity on the part of the designers. Commercial construction is usually larger than light construction, so a bigger design team may be involved. Also, commercial buildings serve a wider variety of needs, so there is more room to change the design to satisfy the need. For example, a shopping mall has very different requirements than does an office building. In the shopping mall, customers need to be able to move around freely while stores need to be able to display their goods. In an office building the occupants need to have a comfortable work area, but there will probably be fewer non-employees in the building.

Industrial construction includes structures other than buildings that are erected for industrial purposes, **Figure 1–7**. Such things as dry docks for shipbuilding, nuclear power plants, and steel mills cannot be considered buildings. Many construction companies specialize in one type of industrial construction, such as oil refineries.

Function is the most important concern in most industrial construction. Community planners will want to protect quiet residential neighborhoods from noisy industrial operations, and architects will make the work areas comfortable and safe for workers, but the main reason for the construction is to perform an industrial function. Engineers who have special knowledge about industrial functions and industrial equipment will be deeply involved in the design of industrial projects.

Civil construction is more closely linked with the land than other areas of construction, **Figure 1–8**. Civil construction generally benefits the public as a whole more directly than other types of construction. Examples of

FIGURE 1–7 Industrial construction includes structures other than buildings that are used for industrial purposes. *(Courtesy of Niagara Mohawk Power Corporation)*

FIGURE 1–8 Highways are examples of civil construction. *(Courtesy of Conrail)*

civil construction are airport runways, highways, dams, and bridges.

Civil construction, like industrial construction, must be very functional. Whether the project is an airport, a bridge, or a dam, it is designed to perform a function. For an airport to be functional it must involve more than just a place for planes to take off and land. There must be plenty of parking for departing passengers, the noise of planes cannot be bothersome to nearby homes, and the terminal buildings must provide a variety of services for both passengers and employees. Bridge designers have to consider what is under the bridge—river, trains, highway, etc. The bridge has to be strong enough to support the expected traffic and it has to be located where that traffic can approach and leave the bridge.

Construction Personnel

The construction industry employs about one-sixth of the working people in North America. The occupations in the construction industry can be divided into four categories.

▼ unskilled labor
▼ skilled trades or crafts
▼ technical
▼ professional

Unskilled Labor

A construction project requires a large amount of labor. Unskilled construction workers are usually called **laborers**. Laborers are sometimes assigned the tasks of moving materials, running errands, and working under the close supervision of a skilled worker, **Figure 1–9**. Because their work is strenuous, construction laborers must be in excellent physical condition.

Construction laborers are workers who have not reached a high level of skill in a

FIGURE 1–9 Laborers usually work under the supervision of workers who are more skilled. *(Courtesy of the United Brotherhood of Carpenters and Joiners of America)*

construction trade and are not in an apprentice program. These laborers often specialize in working with a particular trade, such as mason tenders. Although a mason tender does not have the skill of a bricklayer, the mason tender knows how to mix mortar and erect scaffolding, and is familiar with the bricklayer's tools. Laborers who specialize in a particular trade are generally paid slightly more than general construction laborers.

There are many opportunities for unskilled laborers. General laborers do whatever work is assigned. General unskilled laborers are normally the lowest paid workers on a construction site.

Skilled Trades

A craft is a career that is directly involved with working with tools and materials. Crafts require a high level of skill. The *building trades* are the crafts which deal most directly with building construction. There are over twenty **skilled trades**, **Figure 1–10**

The skill needed to be employed in the construction trades is usually learned in an **apprenticeship program**. Apprentices attend classes a few hours a week to learn the necessary theory. The rest of the week they

Michael Malone

Occupation:
Electrician

How long:
14 Years

Typical day on the job:
The day starts as Michael looks at the prints for a job and gets materials and equipment together for the day's work. The materials are ordered a week before they are needed, then stored in a job-site trailer or storage area. Most of the day is spent installing electrical systems: bending and installing conduit, pulling wires, and installing electrical devices. On large jobs, the design of the system is done by the engineer before the electrician begins installing it, but there is still some need to calculate wire sizes.

Education or training:
Michael attended college for aerospace engineering, but chose to change careers and entered a union electrician's apprenticeship program. Half way through the apprenticeship, Michael took and passed the journeyman's exam, so he was already a journeyman when he completed his apprenticeship. He has continued his studies, with seminars on the *National Electrical Code,* and preparing for the master electrician's exam.

Previous jobs in construction:
Michael was the electrical foreman on the Joe Robie Stadium and the Miami Arena. He has also been superintendent on several other jobs.

Future opportunities:
Michael might someday start his own electrical contracting company.

Working conditions:
All of the work is on job sites, which vary somewhat.

Best aspects of the job:
This is not a repetitious job. There is something new every day.

Disadvantages or drawbacks of the occupation:
On very large jobs, you are in the same location day after day.

Structural Trades	Finish Trades	Mechanical Trades
Carpenter	Plasterer	Plumber
Mason	Terrazzo Worker	Electrician
Iron Worker	Tile Setter	Sheet Metal Worker
Operating Engineer	Painter	Millwright
	Floor Covering Installer	
	Cabinet Maker	

FIGURE 1–10 Some of the most common building trades.

FIGURE 1–11 Carpentry is a skilled trade. *(Courtesy of the United Brotherhood of Carpenters and Joiners of America)*

work on a construction project under the supervision of a **journeyman** (skilled laborer who has completed the apprenticeship). The term *journeyman* has been used for decades and probably will be for many more, but it is worthy to note than many supervisory journeymen in construction are women. Apprentices receive a percentage of the regular salary for the trade during their apprenticeship. Some skilled laborers receive their training through vocational school and informal on-the-job training. Classroom training and construction experience are still required to attain the necessary skill.

The construction trades are among the highest paying of all skilled occupations. However, work in the construction trades depends on the weather. This type of work may move from one location to another, **Figure 1–11** During the winter months in the north, many construction workers are unemployed. They must rely on income earned during the summer months. When a major construction project is completed, construction workers may have to look in other geographical areas for work. This should not be too much of a threat to the person interested in a career in construction. The construction industry is growing at a high rate nationwide. Generally, plenty of

FIGURE 1–12 Many construction technicians work with computers. *(Courtesy of Hewlett-Packard Company)*

work is available to provide a comfortable living for a good worker.

Technicians

Technicians provide a link between the professions and the trades. Technicians work with professionals as part of a team. The work of a construction technician is very similar to the work of a professional. Both of them use mathematics, computer skills, and knowledge of construction principles to solve construction problems, **Figure 1–12**

Technicians spend much of their time in an office. They also make frequent visits to the

Technical Career	Some Common Tasks
Surveyor	Measure land, draws map, lays out highways
Estimator	Calculates time and materials necessary for project
Expeditor	Ensures that labor and materials are properly scheduled
Drafter	Draws plans and designs some details
Inspector	Inspects project at various stages
Planner	Plans for best land or community development

FIGURE 1–13 A few common categories of technicians

construction site. **Figure 1–13** lists some kinds of technical careers in the construction industry. Technicians usually have a broad education so they can do a variety of jobs.

Technicians get their training in community colleges, technical institutes, and senior colleges. The program of study for construction technology usually focuses on drafting, architecture, estimating, surveying, building construction materials and methods, or con-

struction management, **Figure 1–14** Upon completion of the program of study, the technician is employed by an architect, engineer, government agency, construction company, or construction materials company. The starting salary is about the same as the salary of a skilled worker. On an hourly basis, the technician's salary may even be a little lower than the salary of the skilled worker. However, the technician is more certain of regular work and has greater opportunity for advancement.

Professionals

Architects, engineers, building code officials, and land-use planners are involved at the beginning of construction projects. These are the design **professions**. Architects consider the surrounding environment, and the needs of the people who will use the building, to design a useful, attractive building. Architects often have a strong background in art, so they are well prepared to design an attractive building within the environment,**Figure 1–15**

CIVIL/CONSTRUCTION TECHNOLOGY (CIV)
Associate in Applied Science Degree Program

SPECIALIZED COURSES (50 HOURS)

CIV 1603	Engineering Properties of Construction Materials
CIV 1613	Soils and Foundations
CIV 2603	Construction Estimating
CIV 2613	Concrete and Asphalt
CIV 2623	Structural Analysis and Detailing
CIV 2633	Civil Drafting
CIV 2643	Civil Computer Drawings; or CON 2642
ARC 1613	Introduction to Construction Drawings
ARC 1633	Professional Office Practices
ARC 1643	Mechanical and Electrical Systems
CON 2613	Introduction to Computer Drawings
CON 2623	Construction Management I

ENR 1613	Engineering Graphics I
ENR 1623	Engineering Graphics II
SRT 1603	Basic Surveying Approved Electives (6 hours)

GENERAL EDUCATION AND RELATED COURSES (20 HOURS)

ENG 1613	English Composition I; or ENG 1303
GOV 2623	Texas State and Local Government; or GOV 2613
MTH 1323	Technical Algebra and Trigonometry I; or MTH 1633
MTH 1333	Technical Algebra and Trigonometry II; or MTH 1643
PHY 1614	General College Physics I
PSY 1611	Orientation to College
SPE 1633	Business and Professional Communication

FIGURE 1–14 Typical program of study in construction technology *(Courtesy of Tarrant County Junior College)*

FIGURE 1–15 The use of natural wood siding and its low profile make this building especially well suited to the natural environment. *(Courtesy of Pella Windows and Doors)*

The architect acts as the owner's representative throughout the construction process. The architect, for example, is the person who ensures that the contractor is building to the owner's expectations.

When structures or mechanical systems are needed in buildings, the architect relies on an engineer. Engineers specialize in mechanical systems and electrical, structural, and civil engineering. Civil engineers design projects such as highways, dams, and bridges, which involve extensive earthwork. Land-use planners work with state and local governments to plan for the most desirable use of land. These planners are concerned with such things as traffic flow, community services, recreational areas, and population density.

Construction design professionals use computers widely in their work. They keep track of huge amounts of information with computerized data base programs. Because so much of their work is communicating ideas and solutions to others, word processing is an important computer use for design professionals. *Computer-aided design (CAD)* and *computer-aided engineering (CAE)* have completely changed the way construction design work is done, **Figure 1–16** Using these computerized tools, the professional can solve problems in minutes that used to take days. Solutions can be communicated quickly through computer-aided drafting. Because so many of the time-consuming tasks can be done accurately and quickly on a computer, the professional can turn more of this work over to the technician. This frees the professional to concentrate on more advanced design issues.

In addition to the design professions, technology education employs many construction professionals. In the early 1980s, there was a decline in the number of technology teachers. However, a recent trend is toward teaching more technology subjects, like construction, so the number of teaching positions has increased.

The James Bay Hydro Power Project

Each year we consume more and more electricity to light our cities, run our subways, refine aluminum, cool our homes, and operate many of our other products of technology. A very large part of that electricity is generated by burning fossil fuels—coal, oil, and natural gas. These fossil fuel power plants have been linked to acid rain, smog, and other environmental problems.

One alternative to burning fossil fuels is to construct dams in rivers and use the force of the water to turn turbines that power electric generators. This is called hydro-electric power. The largest hydro-electric power project ever planned is in northern Quebec, Canada. The developers of this dam plan for the James Bay hydro-electric project to take place in three phases. Phase one includes 8 dams on and near the LaGrande River, most of which have already been constructed. Phase two will flood a large portion of the Great Whale River basin. The third phase will divert the Nottaway and Rupert Rivers into the Broadback River. The three dozen dams that are planned for twenty rivers will flood dry land (not counting hundreds of existing lakes) with an area equaling the State of New Jersey. The total area affected by changes in river flows will be as large as all of France.

The James Bay Project has changed many times since plans for it were first announced in 1971, but it still seems likely that most of the planned

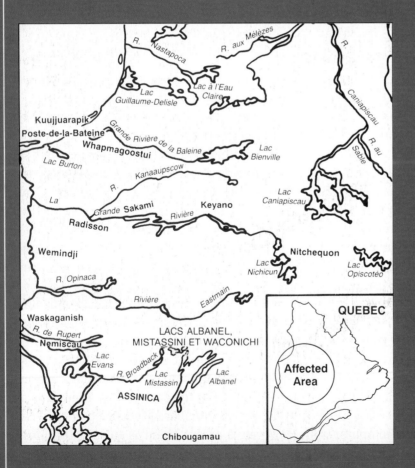

construction will be completed shortly after the turn of the century. Tens of thousands of construction workers will be employed by the three phases. This project is important to Americans as well as Canadians. For example, part of the consideration for planning the project was the hope of a contract with New York State to purchase 15% of New York's total electrical power from Hydro-Quebec, the developers of the James Bay project. Supporters of the project say that without a large source of hydro power, more fossil fuel plants will be necessary. They also point out that there are few people, only about 12,000 Cree and 6,000 Inuit Eskimo people, living in the area.

There are also several negative views on the project. What will become of the unique culture of the Cree and the Inuit? As vegetation decomposes underwater, bacteria transforms mercury, a harmless component of soil, into a toxic form that is absorbed by fish. The fish, which are a staple in the diets of the Cree and the Inuit, become dangerous to eat. Methyl mercury is believed to cause birth defects.

The change in the river currents also is blamed for destroying wildlife. Fish, migratory birds, and mammals are affected. In 1984, the first year that the LaGrande River reservoirs were full enough to require controlled water releases, 10,000 caribou drowned in the swift current.

Throughout this century humans have struggled with the basic question: Does the land and the natural environment belong to us, to be developed as a great natural resource, or are we just using it during our lifetimes, with a responsibility to preserve the natural environment at all cost? ■

FIGURE 1–16 Architects and engineers use CAD to design construction projects. *(Courtesy of Hewlett-Packard Company)*

The construction professions require at least four years of college education. Some of the professions, such as architecture, require five years of college. Technology teaching requires four years of college, and a fifth year may be needed to continue teaching. The salary is higher in the design professions than in teaching, but many enjoy the experience of helping others learn.

ACTIVITIES

Types of Construction

One of the most effective ways to learn about technology is to observe how experts have used it. In this activity you should begin to notice the types of construction in your community and how construction planners have designed structures to satisfy the needs of the structures.

Equipment and Materials

Cardboard
Balsa wood
Tape
Library paste
Scissors
X-Acto knife
Glue gun
Other model making materials

Procedure

1. List several examples of the four types of construction in your community.
2. Choose one example of each type and list the major functions or needs for that example.
3. Construct a model of one of the examples you analyzed in step 2, above. Your model does not need to look anything like the example you chose from the community. In fact, this should be your chance to be a real problem-solver. Concentrate on finding better ways to satisfy the needs you listed in step 2.

Influence of Other Industries

1. Name one important industry in your community.
2. Explain how this industry has affected the local construction industry.

Construction Planning

1. List three new construction projects that would benefit your community.
2. Compare your list with other students. From all the projects listed, choose the three that would be most beneficial.
3. Discuss the resources of the community that would be used to complete and operate each of these construction projects.

Occupations

Name one occupation in the construction industry that interests you and describe the education and training necessary to enter that field.

Applying Construction Across the Curriculum

Social Studies

How has the growth of the construction industry affected other modern industries?

History

Find information about an important historical structure and explain why the structure was built or how it was used by the people of its time.

Communications

Choose one category of construction occupations or personnel that interests you. Use the library to find a trade or professional organization that serves that category. Write a letter to that organization to obtain information about the occupational category.

REVIEW

A. Matching—Match the kind of construction in Column II with the correct structure in Column I.

Column I

1. Bus terminal
2. Eighty-unit apartment building
3. Single-family home
4. Hospital
5. Water tower
6. Housing development
7. Grain elevator
8. Subway

Column II

a. Civil construction
b. Industrial construction
c. Light construction
d. Heavy building construction

B. Matching—Match the occupation in Column II with the duty in Column I.

Column I

1. Works with a skilled worker to learn a trade
2. Designs machinery for operating a drawbridge
3. Designs the approach to a major highway
4. Handles materials at a construction site
5. Owner's representative during construction
6. Inspects new homes for compliance with city ordinances
7. Does drafting in an architect's office
8. Lays out the route for a highway

Column II

a. Mechanical engineer
b. Laborer
c. Architect
d. Building inspector
e. Apprentice
f. Civil engineer
g. Technician
h. Surveyor

CHAPTER 2

Designing for Construction

OBJECTIVES

After completing this chapter you should be able to:

▼ list the needs to be served by a construction project,

▼ explore alternative ways to serve those needs, explain elements of a structure's design that serve specific needs,

▼ explain the impacts of a particular design on the environment and on society.

KEY TERMS

need

function

aesthetic

brainstorming

impact

building code

zoning law

Designing to Fulfill Needs

Technology can be a powerful force to serve humans. Throughout the history of humanity we have used the resources available to us to satisfy our wants and needs. When we do that by erecting permanent structures, we are using construction technology. The structures we build are designed to satisfy our needs. The professionals who design whatever is to be constructed start by determining what **needs** are to be fulfilled. The most obvious needs for any structure are its functional needs. The **functions** of a structure are the purposes it must serve. For example, the functions of the bridge in **Figure 2–1** are to allow cars to cross the river and to allow shipping traffic to use the river. **Figure 2–2** lists some of the functional needs to be served by a public gym. The needs to be served by the gym will be quite different from those to be served by a store, although both might be in similar sized buildings.

Aesthetic Design

Most construction projects are designed to fulfill aesthetic needs as well as functional needs.

FIGURE 2–1 This functional bridge allows traffic to pass below uninterrupted.

Public Gymnasium Needs
Weather protection
Strong to support activity and equipment
Men's and Women's shower rooms
Storage for unused supplies and equipment
Offices
Meeting rooms for consultations
Parking
Safe entrance/exit for patrons at night
Fire exits
Service entrance for supplies
Trash removal
Climate control

FIGURE 2–2 Public gymnasium needs.

Aesthetic means having to do with how pleasing something is to the senses. A building that looks good, as if it belongs in its surroundings, is aesthetically pleasing. The building in **Figure 2–3** is visually pleasing. Of course, if it lacks good ventilation and there is an unpleasant odor inside the building, it has not fulfilled all of its aesthetic needs.

Architecture is often considered a form of art. In fact many architects begin their college education as art majors and later switch to architecture. The exterior design of an attractive building might be developed as much from artistic sketches as from rigid technical drawings. The technical drawings come later, when the construction details are worked out. If you compare the exteriors (outsides) of buildings designed in different areas, you can see the changes in artistic styles, **Figure 2–4**. Of course the design of the exterior has to accommodate the layout of the interior (inside). All of the parts of the design must work together. Consider an office building designed to take advantage of a beautiful view. The offices inside the building should be laid out so that as many office spaces as possible are on that side of the building. Perhaps the design will include an all-glass wall on that side. The exterior design should take advantage of the glass wall.

FIGURE 2–3 A well designed building is a work of art.

FIGURE 2–4 The style of a building reflects the era in which it was designed.

Design Ideas

The design of a building actually starts when the new owner first decides that the building is needed. The owner usually has some very definite needs in mind before any design professionals are even brought into the job. At their first meeting, the architect asks questions to learn what those needs are. At this point the architect tries to remain very objective. Some very creative designs have come as a result of listening to the needs without being too quick to decide on a final design, **Figure 2–5**.

A very popular method for generating ideas is called **brainstorming**. Brainstorming is usually done in groups. One person in the group is chosen to record ideas. Everyone tells the group any ideas they have, without worrying about how good it is or how it might sound to the group. The recordkeeper lists all of the ideas on a chalkboard or flip chart, **Figure 2–6**. The key to a successful brainstorming session is that everyone must feel free to contribute as many ideas as possible, without worrying about how they sound to the rest of the group. Criticism of or laughing at another person's ideas is not allowed in brainstorming.

FIGURE 2–5   The New York State auditorium is a creative solution to the client's needs.

FIGURE 2–6   Brainstorming involves listing as many ideas as possible.

Once as many ideas as possible have been listed, the group can discuss each. Again, no one is allowed to laugh at or criticize another's ideas. As the ideas were being listed, the intent was to get as many as possible on the list. The intent was not to screen out the ideas that wouldn't work. Now that the list is complete, it is time to think about and discuss each idea.

Look for creative ways to solve problems and fulfill the needs of the project. As each idea is discussed, allow the person who contributed the idea to explain it. If they don't have an explanation, that's okay. Maybe someone else will see some possibilities. Eventually you will probably narrow the list to two or three ideas that should be studied further. Many of the ideas that were listed originally can probably be eliminated.

Brainstorming is most often a group activity. For example, the design team, consisting of an architect, an engineer, a drafter, and a real estate developer might brainstorm ideas for a recreational area in a new housing development. It is also possible to brainstorm in pairs. A designer and a future building owner might brainstorm ideas for exterior building materials.

Functional Design

As the general style and functional design of the project begins to take shape, attention begins to focus on the next level of details. In the design of a bridge such things as height, span (distance from one side to the other), and approaches begin to take shape. If the project is a building, the basic arrangement of the rooms is developed, **Figure 2–7**. Sometimes the arrangement of the rooms and the structural requirements of a building are very closely related. For example, in the design of our gym, we have included a large room with an indoor track. Without some support in the center of the room, the span of the ceiling would make it impractical to put other rooms on a second floor, above the track room. The weight of the materials and occupants of the second floor rooms would put a very heavy load on the ceiling of the track room. It is more practical to put a second floor above the smaller rooms,

Anthony Clemente

Occupation:
Model and Mock-up Technician

How long:
7 Years

Typical day on the job:
The job involves making three-dimensional models of proposed satellites and satellite systems for the space division of General Electric. These models are used as training aids for engineers and in demonstrations for government buyers and private merchants. The models are also used for research and development. The job includes tooling work, frame making, and plaque making. In addition, a day's work may include cleaning and waxing machines and ordering supplies.

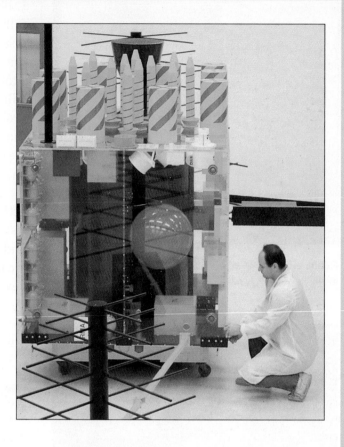

Education or training:
Anthony has a bachelor's degree in industrial arts education. His studies included metalworking, woodworking, drafting, electricity, graphic arts, and photography.

Previous jobs in construction:
Anthony did residential remodelling and spent five years teaching high school industrial arts classes.

Future opportunities:
Anthony hopes either to return to teaching or open a cabinetmaking business.

Working conditions:
The building is air-conditioned, but there is very little natural light.

Best aspects of the job:
Anthony's pay and benefits are good. The job involves a lot of creative problem solving. He also enjoys working with his hands.

Disadvantages or drawbacks of the occupation:
Anthony sees no major disadvantages in his job.

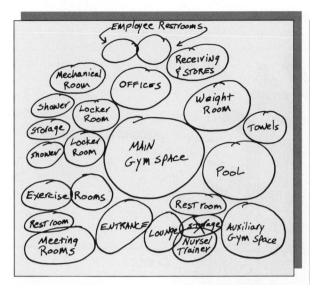

FIGURE 2–7 The first step in planning a building is often a rough bubble diagram.

because the walls of the smaller rooms give structural support to the second floor.

As the design takes shape, the structural requirements and construction details become a bigger part of the design activity. Eventually, materials must be selected for the structural components of the floors, walls, roof, etc. The choice of structural materials will depend on cost, structural requirements, and aesthetic requirements. If a building is intended for temporary use, and cost is more important

FIGURE 2–8 An air structure is an efficient way to create a temporary building quickly and at moderate cost. (*Courtesy of Industrial Fabric Association International*)

than a long life or aesthetics, an air-supported structure might be a good choice, **Figure 2–8**. Many of the homes and smaller buildings constructed today use a standard system of structural components, **Figure 2–9**. This system is based on hundreds of years of trial and error, refining the materials that are readily available. New materials and new uses of existing materials are introduced by creative designers every year, but it is rare to see a completely new system of building construction. Throughout most of this textbook, you will study the materials and methods that are widely used today. You should also be constantly searching for new ways to use existing materials and new materials to satisfy the needs of construction.

Community Planning

Another important aspect of construction design is the planning and designing for construction which will serve the needs of the community. Whenever we erect a structure, it affects others. At the very least, everyone who lives nearby will have to look at the structure. The way it fits its surroundings can have a big impact on the value of neighboring property. Consider, for example, what the **impact** (effect) would be on your neighbors if you started a kennel to keep up to 100 dogs in a quiet residential neighborhood. Some construction projects have impacts on whole communities or even entire regions, **Figure 2–10**.

Nearly all towns and cities have Building Departments. The municipal Building Department is responsible for ensuring that buildings in the community serve the needs of the entire community. Usually a **building code** sets the requirements for any building erected in the community. Building codes are laws that describe the minimum requirements

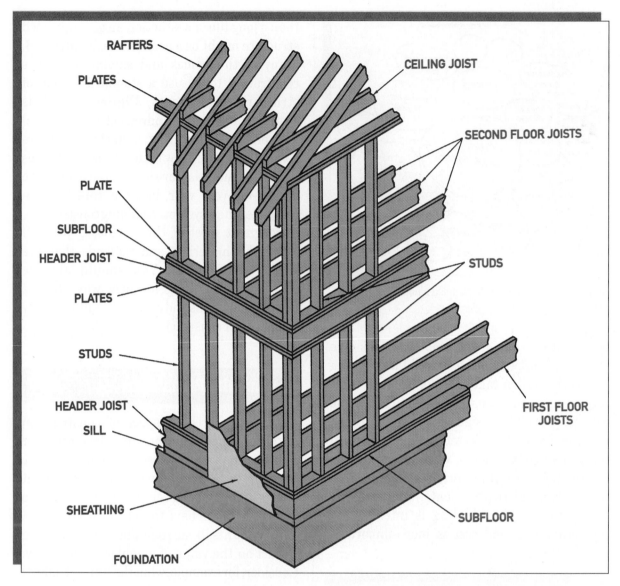

FIGURE 2-9 This is the style of framing system that has been used in most small buildings constructed in the twentieth century.

to make buildings safe for their intended use. For example, most building codes require that any public meeting space must have two exit routes in case of fire. Some building codes are concerned with a specific aspect of construction. Most building codes are based on suggested codes that are published by national organizations, **Figure 2-11**

Towns and cities also have **zoning laws** which regulate the kinds of structures that can be constructed in specific parts of the com-

munity (zones). Zoning laws ensure that residential neighborhoods remain as quiet residential neighborhoods, by requiring factories to be in industrial zones and farming to be in agricultural zones. Zoning laws also regulate such things as how far a new building must be placed away from property lines and how tall a building may be.

To regulate construction that affects more than a single community, states and the federal government have land use and environmental

FIGURE 2–10 When the dam failed in Johnstown, Pennsylvania, hundreds of people were killed. *(Courtesy of The Tribune-Democrat, Johnstown, PA.)*

FIGURE 2–11 Building codes cover all phases of construction and building uses.

protection laws. Architects, civil engineers, and land use planners are expected to be familiar with all of the municipal, state, and federal laws affecting the structures they design. Before any construction work begins on the site, the necessary permits must be obtained from these agencies. For a small building in the proper zone of a city, a city building permit may be all that is required. For a large hydroelectric power plant that will create a new lake, permits will be required from several state and federal offices.

ACTIVITIES

Community Planning

Zoning officials and other community planners determine where construction for various purposes will best serve the needs of the community. They have to allow for new construction to take place, so the community can continue to grow, but they need to control development so that it is acceptable to all. They spend a great deal of time reviewing the zoning maps of the community.

Equipment and Materials

Zoning map or street map of your community

Magazines or brochures with photos of various types of buildings

Procedure

1. If you cannot obtain a copy of a local zoning map, draw your own recommendations for zoning boundaries on a street map of your town or city. Plan zones that you think will be good for your community. The following are a few kinds of zones often described on zoning maps:

    Residential (1- and 2-family homes only)
    Residential & light commercial (includes apartments and small commercial buildings)
    Heavy commercial (small and large commercial buildings)
    Industrial (factories and processing plants)
    Agricultural (farming and homes on very large lots)
    Recreational (no buildings, except those associated with parks and outdoor recreation)

2. Tape or paste pictures of each type of construction into each of the zones on your map.
3. Write a short description of what is or is not allowed in each zone.

Design a Teen Center

Your class has just been given a grant from the Parent Teacher Organization to build a recreational center for teens in your community. In this activity you will design the center. Your center should be designed for up to 100 teens and one adult for every 25 teens. You should decide what hours it will be open. Your teacher might have the whole class work together on this, divide the class into smaller groups, or have you design the center individually.

Procedure

1. If you are working in a group, use brainstorming to decide what kinds of activities will take place in your center. If you are working individually, interview at least 4 other members of your class to get their ideas about what activities to include in your plan.
2. Make a list of the spaces or rooms that you will need in order to accommodate all of the activities you expect. Do not plan for many extra rooms, because you might have to cut back on your design if it is not within your budget.
3. Decide where in the community the center should be built. Be prepared to explain why you chose that location.
4. Sketch the layout of the rooms, arranged for good supervision by adults, yet plenty of fun for young people. Be sure to include things like restrooms, storage areas, and fire exits.
5. Sketch the exterior design of the center. Be as artistic as possible and be sure to match your exterior to fit the interior.

Applying Construction Across the Curriculum

Social Studies

Visit your City Hall or Town Hall and make a list of the functions or needs that it serves. You might want to make an appointment to talk with a town or city official (town clerk or city clerk, councilman or councilwoman, mayor, supervisor, etc.) to learn more about what functions are carried out in the hall.

Social Studies

Contact your local building department and find out about the building code in effect in your area. When was the code established? How can it be changed? Is it based on some model code or was it written from scratch, without reference to any model? (Very few building codes are written without reference to any other

models.) Who is responsible for enforcing the code in your community? List a few specific topics that are covered in your local building code. (If your town does not have a local building code, find information about the county or state code.)

Mathematics

If it has been decided that an auditorium must have 12 square feet of floor space for each person it can seat, plus 20% additional space for service areas like offices, hallways, and restrooms, how much total floor space is needed for an auditorium to seat 800 people? 1,000 people?

REVIEW

1. Which of the following is an example of a functional requirement of a house?
   a) a place to prepare food
   b) attractive to the neighbors
   c) can be built within the budget
   d) must be in a residential zone
2. Which of the following is an example of an aesthetic requirement for a house?
   a) protection from the weather
   b) large enough for all expected uses
   c) materials can be bought locally
   d) the choice of interior materials creates a cozy atmosphere
3. What is the first thing an architect does in designing a commercial building?
   a) decides on the style of the outside
   b) plans the basic arrangement of rooms
   c) interviews the owner to determine his needs
   d) finds a suitable building site
4. Which of the following is apt to affect the exterior design of an office building?
   a) the arrangement of interior rooms
   b) the surroundings in the neighborhood
   c) the owner's budget
   d) all of the above
5. Which of the following is a rule of brainstorming?
   a) try not to list ideas that you know won't work
   b) do not rule out any ideas as the list is being developed
   c) list ideas in order of their value
   d) each person should contribute one idea

6. Which of the following might be decided by brainstorming?
   a) choice of flooring in the entrance
   b) the total budget for landscaping
   c) the part of the community in which to build
   d) all of the above

7. Which of the following would be most affected by the structural requirements of an office building?
   a) the size of the parking lot
   b) the choice of materials in the building frame
   c) the size of the windows
   d) the height of the ceilings

8. What is the purpose of building codes?
   a) to prevent too much building in a town or city
   b) to regulate the number of people who can live in one house
   c) to see that all buildings are kept in good shape
   d) to see that all new buildings are constructed safely

9. What is the purpose of a zoning law?
   a) to see that all new buildings are constructed safely
   b) to keep buildings of each type in the same areas
   c) to prepare for emergencies
   d) all of the above

10. Which of the following government levels might have regulations controlling the construction of a large dam?
    a) federal   c) local
    b) state   d) all of the above

CHAPTER 3

Architectural Drawings

OBJECTIVES

After completing this chapter, you should be able to:

▼ demonstrate how to draw architectural plans and make models from those drawings.
▼ locate dimensions of components and overall sizes on architectural drawings.
▼ identify common symbols used on architectural drawings.
▼ measure scale drawings with an architect's scale.

KEY TERMS

scale

pictorial drawing

multi-view drawing

elevation

plan view

section view

detail drawing

topographical drawing

contour lines

standard

CAD

data

CRT

input device

mouse

digitizer

output device

plotter

binary code

software

Scales

It is not possible to make architectural drawings actual size, so they are drawn to **scale**. The dimensions of all parts are reduced to a size that can be drawn on a sheet of paper. For example, floor plans for most residential construction are drawn 1/48th of actual size. At this scale, 1/4 inch on the drawing represents 1 foot on the actual construction site. This scale is written 1/4″ = 1′-0″. When it is necessary to draw a large object, small-scale drawings are used. Smaller objects and detail drawing use a larger scale. The detail drawing in **Figure 3–1** is drawn to a scale of 1/2″= 1′-0″.

Reading the Architect's Scale

An *architect's scale* is used to work with these scale drawings. The triangular scale, shown in **Figure 3–2**, combines eleven frequently used scales. The architect's scale is *open divided*. This means that the main units of the scales are undivided and a fully subdivided extra unit is placed at the zero end

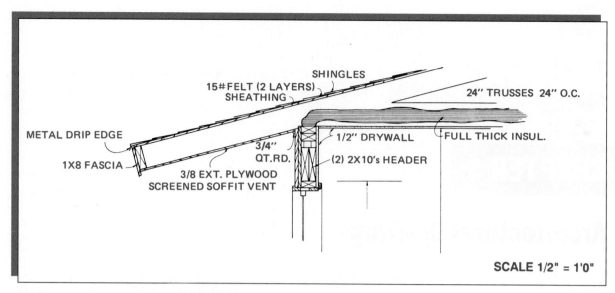

FIGURE 3–1 This detail drawing is done in a scale of 1/2″ = 1′-0″. *(Courtesy of Home Planners, Inc.)*

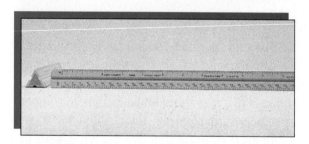

FIGURE 3–2 Triangular architect's scale *(Courtesy of Koh-I-Noor Rapidograph, Inc.)*

of the scales. The eleven scales found on the architect's triangular scale are:

Full scale

1/8″	= 1′-0″		1/4″	= 1′-0″
3/8″	= 1′-0″		3/4″	= 1′-0″
1/2″	= 1′-0″		1″	= 1′-0″
1 1/2″	= 1′-0″		3″	= 1′-0″
3/32″	= 1′-0″		3/16″	= 1′-0″

Two scales are combined on each face, except for the full-size scale which is fully divided into sixteenths. The combined scales are compatible because one is twice as large as the other and their zero points and extra-divided units are on opposite ends of the scale.

The fractional number near the zero at each end of the scale indicates the unit length in inches that is used on the drawing to represent one foot of the actual building. The extra unit near the zero end of the scale is subdivided into twelfths of a foot, or inches, as well as fractions of inches on the larger scales.

To read the triangular architect's scale, turn it to the 1/4-inch scale. The scale is divided on the left from the zero towards the 1/4 mark so that each line represents one inch. Counting the marks from the zero toward the 1/4 mark, there are twelve lines marked on the scale. Each one of these lines is one inch on the 1/4″ = 1′-0″ scale.

The fraction 1/8 is on the opposite end of the same scale. This is the 1/8-inch scale and is read from the right to the left. Notice that the divided unit is only as large as the one on the 1/4-inch end of the scale. Counting the lines from the zero toward the 1/8 mark, there are only six lines. This means that each line represents two inches at the 1/8-inch scale.

Metrics in Construction

As the nation changes to the use of metric units of measure instead of inches and feet, it will also become necessary for construction work-

FIGURE 3–3   A pictorial drawing of an office building *(Courtesy of NYNEX Properties Company; Einhorn Yaffee Prescott, Architect & Engineering, P.C.)*

ers to change. Most sectors of industry are undergoing this change at the present time. However, because of the standardization of material sizes and the permanence of much older construction, the construction industry relies mainly on inches and feet for units of measure. When metric materials and designs are used, metric scales are used on drawings.

To provide all of the information necessary to construct what the architects and engineers have designed, several kinds of drawings are required. These drawings include plans, elevations, sections, and details. There are two methods used to draw three-dimensional objects: pictorial drawings and multi-view drawings.

A **pictorial drawing** represents how an object looks to the eye. It is shown in only one position. A pictorial drawing can be used to show a client what the proposed structure will look like, **Figure 3–3**.

A **multi-view drawing** shows two or more views of an object. A typical three-view drawing includes the front, top, and one side of the object, **Figure 3–4**. Multi-view projection is used for all architectural working drawings.

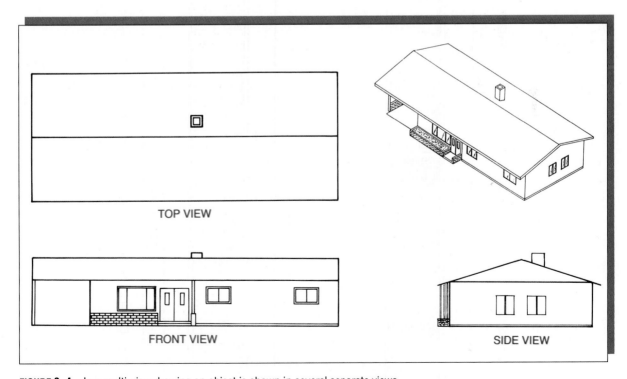

TOP VIEW

FRONT VIEW

SIDE VIEW

FIGURE 3–4   In a multi-view drawing an object is shown in several separate views.

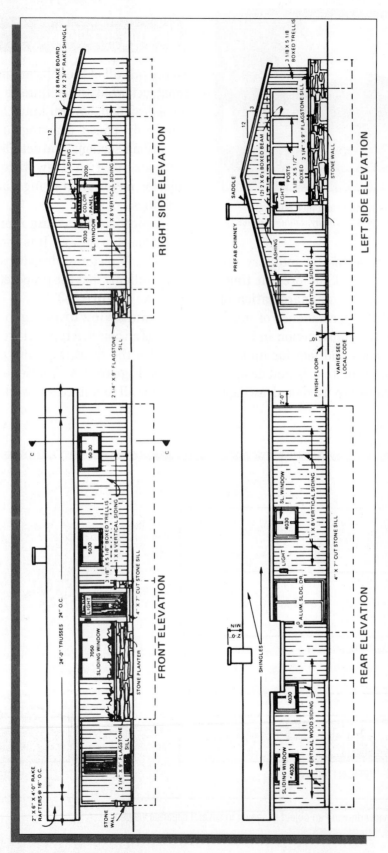

FIGURE 3–5 Elevations from a set of working drawings. *(Courtesy of Home Planners, Inc.)*

Elevations

Elevations are drawings of a building as it is seen from different sides. Different elevations are indicated by different names. The front of a building is drawn as the front elevation, the right side is drawn as the right-side elevation, the side on the left is drawn as the left-side elevation, and the back of the building is drawn as the rear elevation, **Figure 3–5**.

Plan Views

Assume that a horizontal cut is made through the building, about three feet from the floor, and the top section is removed. Looking straight down from the top at the remaining part of the building is what is seen on the *floor plan*, **Figure 3–6**. The cut through the building is made at the proper height so that it passes through doors, windows, and other wall openings. The floor plan shows the location of the walls, all door and window openings, and their sizes. A separate plan is drawn for the basement and each floor.

Section Views

Think of the original building as it was before the horizontal line was cut through it to make the floor plans. Now assume that a vertical cut is made through the building and one part is removed. Looking into the building, one sees a **section view**, **Figure 3–7**.

Where sufficient detail cannot be shown on regular plans, elevations, and section views, **detail drawings** are made. These are drawn to a larger scale and include complete

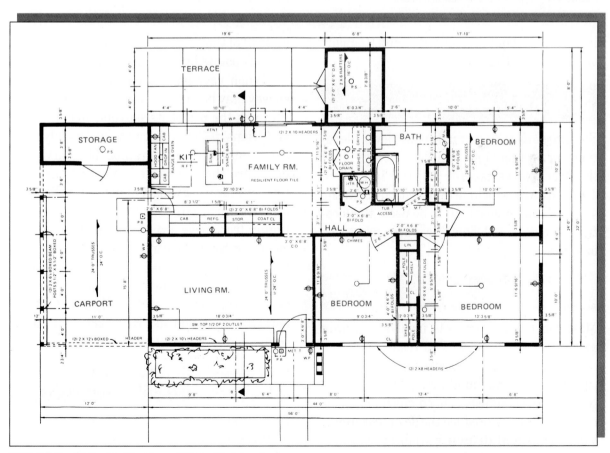

FIGURE 3–6 Floor plan *(Courtesy of Home Planners, Inc.)*

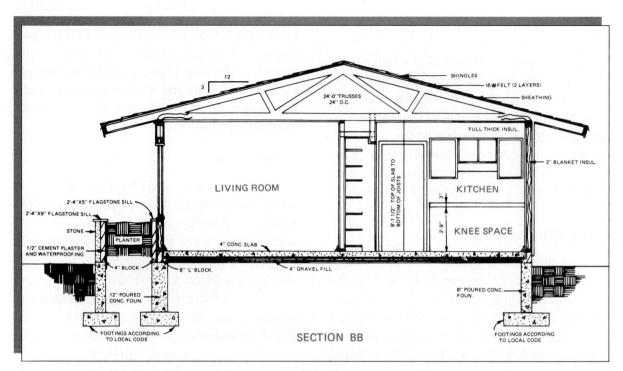

FIGURE 3–7 Typical section view of a house *(Courtesy of Home Planners, Inc.)*

information about the construction of that part of the project, **Figure 3–8**.

Structural Drawings

Structural drawings can be plan views, elevations, or section views. These drawings show the structural parts of the project, **Figure 3–9**. These drawings include information about the size and kind of material to be used, the location of parts, and how the parts are to be fastened.

Civil Drawings

Most construction projects involve a certain amount of earth work. For residential construction this may be excavating (digging) for the basement and foundation. Civil constructions, such as highways, need much more cutting and filling to produce the desired contour. **Topographical drawings** indicate the contour and layout of the land. The elevation (rise

and fall of the land) and layout is measured at various points by surveyors, **Figure 3–10**.

When very large areas are involved, the survey may be made from aerial photographs. Through the use of stereoscopic instruments, variations in elevation can be accurately measured on aerial photographs. The surveyor's notes are then converted into drawings. On topographical drawings, sometimes called contour drawings, **contour lines** show the elevation in feet above sea level. The difference in elevation from one contour line to the next is called the *vertical contour interval*. It is usually given on the drawing. For example, if the vertical contour interval in **Figure 3–11** is 2 feet, the elevation at "B" is 286 feet (three contour lines or 6 feet above "A").

Symbols and Abbreviations

It is not practical to show every detail of some features of a construction project by drawing them. Architectural symbols have been de-

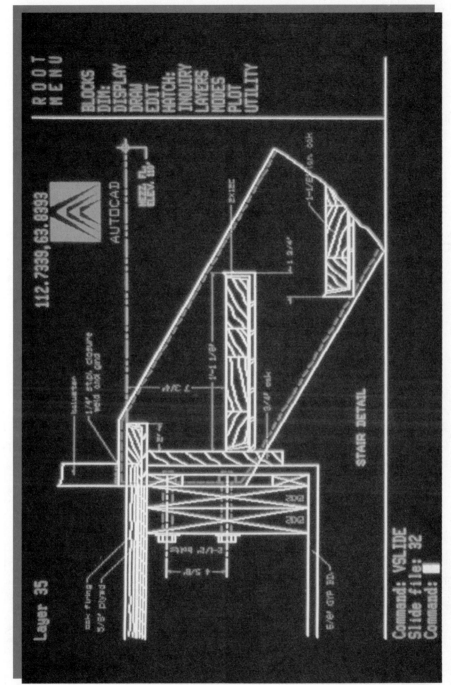

FIGURE 3–8 A construction detail shows complete information about a small piece of the construction *(Courtesy of Autodesk, Inc.)*

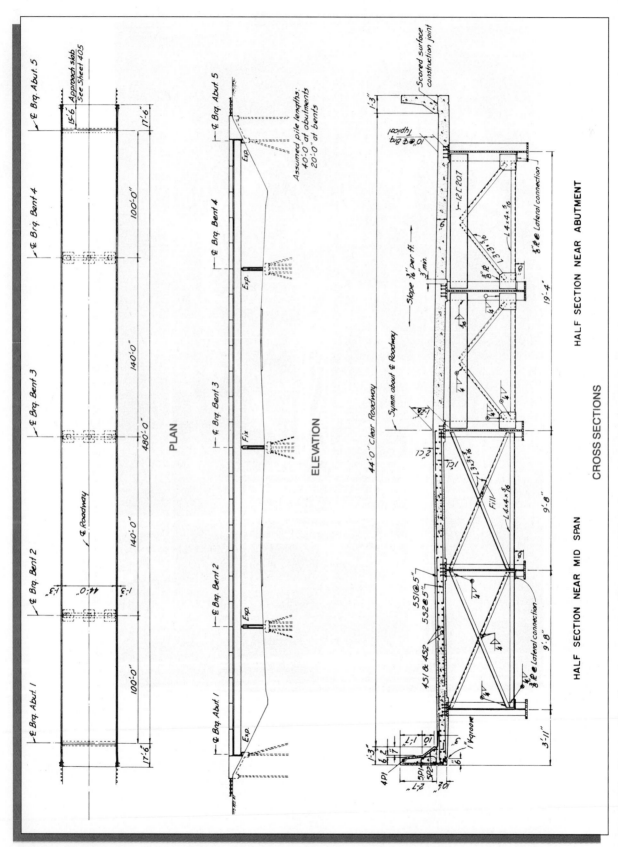

FIGURE 3-9 Structural drawings show weld symbols, size and kind of material, and placement of structural parts.

Patricia Peirsol

Occupation:
Architect

How long:
25 Years

Typical day on the job:

Patricia has her own private architectural office. She specializes in the design of any type of building. Every day is different, but the work includes: writing proposals, meeting with clients to get information about their needs, creating schematic drawings on a CAD system, developing designs, (adding detailed information to the schematics), preparing contract documents, advising clients as they get bids from contractors, visiting construction sites to monitor the work, and running the office (paying bills, ordering supplies, writing invoices, promoting the business, etc.).

Sometimes it is necessary to turn away a prospective client. A major construction project is a complex undertaking, and if there is not a basic trust between the owner and the architect, it is not worth trying to work together.

Patricia makes the point that the architect is the owner's representative, not the contractor's.

Education or training:

A bachelor's degree is a minimum requirement to be licensed as an architect. Today, most architectural programs require five years to earn a bachelor of architecture degree. Not all architects are alike. You can specialize in the artistic aspects of architecture, the engineering and structural aspects, management, or construction.

Previous jobs in construction:

The best route for an architect is to work for as many other architectural firms as possible before starting a private practice. Most architectural firms like to hire architects that have worked for a variety of firms and have broad experience.

Future opportunities:

Patricia looks forward to networking with other architects. This might be just a way to learn what others are doing and to share design ideas or it might result in actually collaborating on projects.

Working conditions:

Most of her work is done in an architectural office, with typical office furnishings, but there is some site visitation involved.

Best aspects of the job:

The design process has become a way of life for her. You get a really good feeling of accomplishment when you "solve the puzzle" for a client.

Disadvantages or drawbacks of the occupation:

There is a lot of tedious work to be done in an architectural office. In a big office, the junior architects get most of the tedious work. As you become more senior, there is usually it, but when you are the only person in the business, you have to do the tedious part yourself.

FIGURE 3–10   Land surveyors measure the size, shape, and location of parcels of land. *(Courtesy of the Lietz Company, copyright XX 1986)*

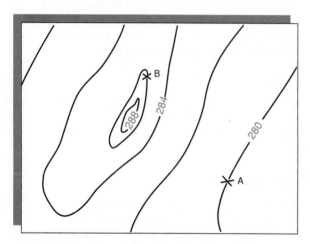

FIGURE 3–11   Topographical drawing with vertical contour interval of two feet.

veloped to show many parts, kinds of construction, and materials, **Figure 3–12**. These symbols are placed on the drawings to provide more complete information. They are standardized, as are many other drawing practices, by the American National Standards Institute (ANSI). A **standard** is a specific agreed-upon way of doing something. This standardization is necessary so that each symbol means the same thing to everyone.

Computer-Aided Design and Drafting

CAD has made a major change in the way drawings are prepared. Using CAD, the designer enters information into the computer one time, then recalls and revises that information to produce the final design. The computer is a powerful problem-solving tool. Now, much of the computation that used to take most of the designer's time can be done very quickly. A CAD workstation consists of a computer, computer peripherals (attachments), and software, **Figure 3–13**.

Computers

A digital computer is an electronic device that performs only a few types of operations. It categorizes information into one of two categories. These categories are generally represented by 1s and 0s. All information, called **data**, processed by the computer is represented by 1s and 0s as they are entered into the system. Then, on command from the operator, the computer calls this data up from memory to be processed, combined, or separated according to instructions from the operator. All of this handling of data is done very fast.

The actual part of the computer that receives, stores, and processes data is very small, **Figure 3–14**. Without attachments, the chip would be worthless. When an architect buys a CAD system, the computer includes a cabinet of some kind, a keyboard for communicating with the chip, and receptacles through which other components can be connected. Many computers also come with a **CRT** (cathode ray tube) display screen. This is the part that looks like a television screen. The CRT is a peripheral: an output device. It is one of the devices through which the data processed or stored in the computer can be communicated to the operator.

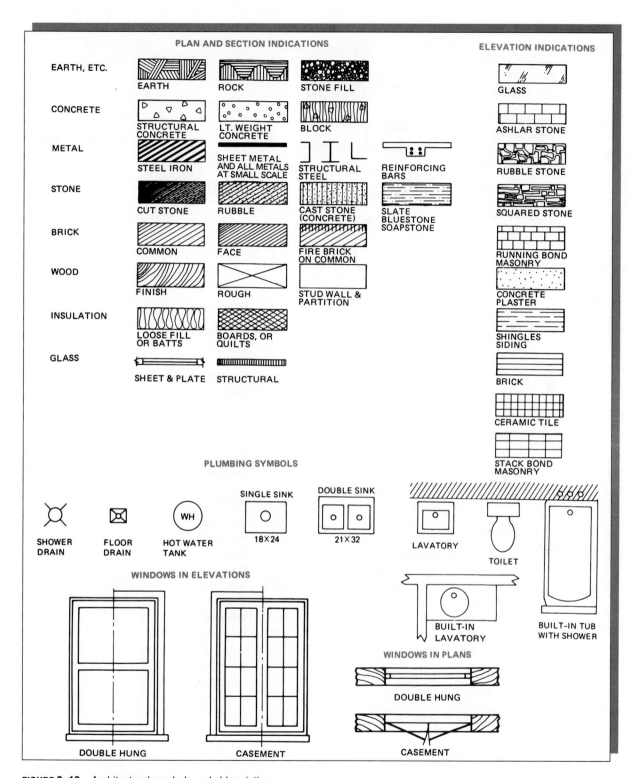

FIGURE 3–12 Architectural symbols and abbreviations

ELECTRICAL SYMBOLS

SWITCH OUTLETS

S	-	SINGLE POLE SWITCH	S_4 -	FOUR WAY SWITCH
S_2	-	DOUBLE POLE SWITCH	S_D -	AUTOMATIC DOOR SWITCH
S_3	-	THREE WAY SWITCH	Scb-	CIRCUIT BREAKER

CONVENIENCE OUTLETS

DUPLEX OUTLET

WEATHERPROOF
WP

RANGE OUTLET

SPECIAL PURPOSE

LIGHTING PANEL

POWER PANEL

POWER TRANSFORMER

PUSH BUTTON

TELEPHONE

GENERAL OUTLETS

CEILING	WALL	
		OUTLET DROP CORD
(D)		
(S)	(S)	PULL SWITCH
(J)	(J)	JUNCTION BOX

ABBREVIATIONS USED ON WORKING DRAWINGS

AWG	American Wire Gauge		GL	Glass
B	Bathroom		HB	Hose Bibb
BR	Bedroom		C	Hundred
BD	Board		INS	Insulation
BM	Board Measure		INT	Interior
BTU	British Thermal Unit		KD	Kiln Dried
BLDG	Building		K	Kitchen
CLG	Ceiling		LAV	Lavatory
C to C	Center to Center		LR	Living Room
CL or C̶	Centerline		MLDG	Molding
CLO	Closet		OC	On Center
COL	Column		REF	Refrigerator
CONC	Concrete		R	Riser
CFM	Cubic feet per minute		RM	Room
CU YD	Cubic Yard		SPEC	Specification
DR	Dining Room		STD	Standard
ENT	Entrance		M	Thousand
EXT	Exterior		T & G	Tongue and Groove
FIN	Finish		UNFIN	Unfinished
FL	Floor		WC	Water Closet
FTG	Footing		WH	Water Heater
FDN	Foundation		WP	Waterproof
GA	Gauge		WD	Wood

FIGURE 3–12 (Continued)

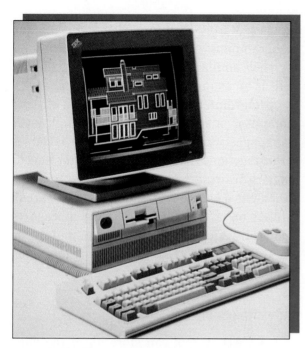

FIGURE 3–13 Microcomputer CAD system *(Courtesy of Autodesk, Inc.)*

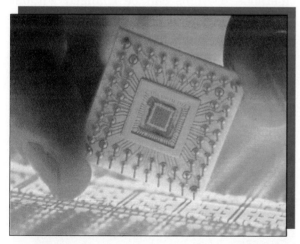

FIGURE 3–14 This electronic chip contains all of the circuits for the computing section of a computer. *(Courtesy of the Sperry Corp.)*

Input Devices

Any peripheral used by the CAD operator to give instructions to the computer is an **input device**. There are three commonly used kinds of input devices. The keyboard, although not a peripheral, is an input device.

FIGURE 3–15 A mouse is a commonly used computer input device. *(Courtesy of Mouse Systems)*

FIGURE 3–16 Digitizing tablet

A **mouse**, **Figure 3–15**, is a hand-held device which, when moved around on a flat surface, causes a marker on the CRT to move. By depressing one of the buttons on the mouse when the marker (called a cursor) is on a particular command, the computer is instructed to follow that command. The mouse can also be used to indicate positions on the CRT where lines or objects are to be located.

A **digitizer**, **Figure 3–16**, is used like a mouse, but on a digitizing tablet. Through electronic circuits, the computer senses the position of the puck (the hand-held part of the

FIGURE 3–17 The plotter draws the data stored in the computer. *(Courtesy of Hewlett-Packard Company)*

digitizer) on the tablet. By attaching a drawing to the tablet, the digitizer can be used to input all of the points and lines of the drawing into the computer.

Output Devices

The most used **output device** is the CRT. However, to create working drawings and other design documents, the computer output must be drawn or printed on paper. Most CAD output for finished drawings is through a **plotter**, **Figure 3–17**. A plotter holds a sheet of paper or drafting film and several pens. The computer either instructs the paper to move under the pens, or the pens to move over the paper. The computer also tells the plotter which pen to use. The various pens can either be of different widths, to create thick and thin lines, or different colors.

Software

When computers were first invented, computer operators actually had to communicate with their computers in **binary code** (1s and 0s). This meant that only very highly trained people could use computers. Today, computers come with instructions already built into them, so that they can operate with high-level lan-

guages. A *high-level language* is a computer programming code that includes words and symbols similar to those used in English. BASIC is an example of a high-level language.

Programmers, who are skilled in writing computer instructions in these high-level languages, write software. **Software** is the set of instructions that tells the computer how to make a drawing, how to store an inventory list, or how to do word processing. The software is stored on a disk or tape. Tape storage is generally found only on minicomputers or larger systems. Disks are widely used for microcomputers.

Advantages of CAD

A CAD system can store all of the data necessary to design a construction project. It thus offers some big advantages over manual methods. Designs can be changed without having to be completely redrawn. Parts of a design done for one project can be saved, then called up for use in a later project. Drawings can be done in layers, so that a basic design outline can be used for several purposes, **Figure 3–18** and **3–19**. By using a CAD function called zoom, a design can be shown as close up or as far away as desired, **Figure 3–20** and **Figure 3–21**.

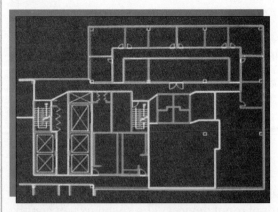

FIGURE 3–18 A basic floor plan *(Courtesy of Intergraph Corporation)*

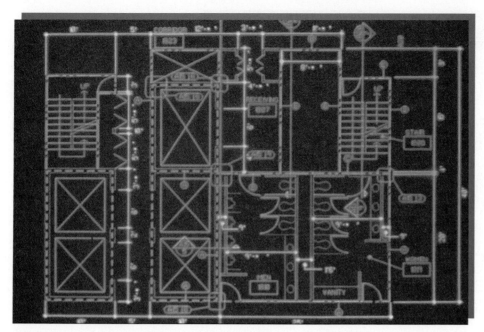

FIGURE 3–19 Dimensions and plumbing symbols are added to the floor plan in another layer. *(Courtesy of Intergraph Corporation)*

FIGURE 3–20 Interior view of an office created on CAD *(Courtesy of Intergraph Corporation)*

Sketching

Often the ability to sketch an object or an idea is as valuable as the ability to use spoken language. In construction much of the information necessary to design or build a structure can most easily be communicated by a drawing or sketch. If you have taken or are taking a course in drafting, you will use much of what you have learned in that course in your study of construction technology. However, it is frequently more practical to use a freehand sketch rather than a precisely drafted drawing. With a few simple hints, you will quickly learn to communicate a lot of information with sketches, **Figure 3–22**.

Sketching Straight Lines

Most of us can draw a short line freehand and it will appear to be straight. Long lines, however, present a difficulty. No matter how we try, a long line drawn with a single stroke will appear wavy and bent. Repeated practice will improve the quality of line work. There are

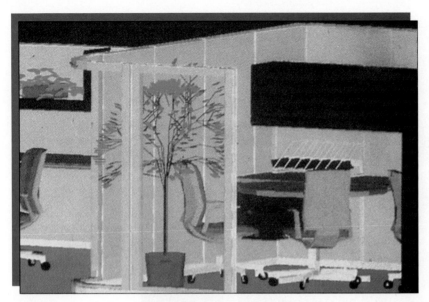

FIGURE 3–21 Zooming magnifies the office so that all detail can be seen. *(Courtesy of Intergraph Corporation)*

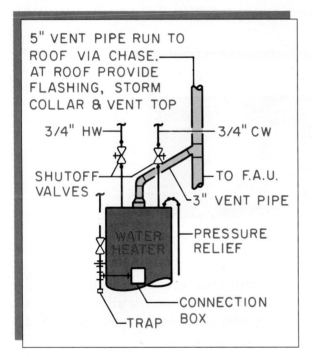

5" VENT PIPE RUN TO ROOF VIA CHASE. AT ROOF PROVIDE FLASHING, STORM COLLAR & VENT TOP

3/4" HW

3/4" CW

SHUTOFF VALVES

TO F.A.U.

3" VENT PIPE

WATER HEATER

PRESSURE RELIEF

CONNECTION BOX

TRAP

FIGURE 3–22 Freehand drawings are sometimes used by professionals to communicate technical information. *(From Huth, Understanding Construction Drawings, copyright 1983, Delmar Publishers Inc.)*

several tricks that the beginner can use to help in developing proficiency in drawing long lines. One trick is to use a series of dots along the path of the line to be drawn. To illustrate, draw a line freehand about 6 inches long. If it is drawn in one stroke, the line will almost certainly be wavy.

Instead of sketching the entire line, lay out a row of dots about 1 inch apart, **Figure 3–23**. Now hold the paper up and sight along this row of dots. Any dot that is out of line shows up immediately and can easily be shifted so that all the dots are in line. To draw a light line connecting each of the dots is a relatively simple job. As a final step, go over the entire line with a single stroke, making it the desired weight. With additional practice, the beginner will be able to widen the distance between dots and, eventually, eliminate them entirely.

Horizontal lines are best drawn from left to right if you are right-handed, **Figure 3–24**. Left-handed sketchers should draw from right to left. These lines should be drawn with a forearm rather than a wrist movement. While pivoting from the wrist may be a somewhat easier movement for the beginner, it will cause a curve when drawing longer lines. When a series of lines are to be drawn, it is a good practice to first sketch all of the lines lightly. When you are sure that your sketch is correct, the lines can all be darkened at the same time.

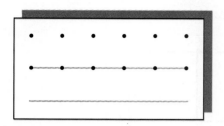

FIGURE 3–23 Sketching straight lines

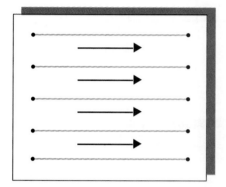

FIGURE 3–24 Horizontal lines

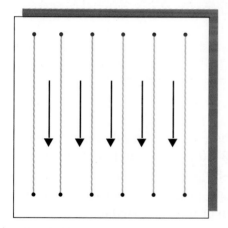

FIGURE 3–25 Vertical lines

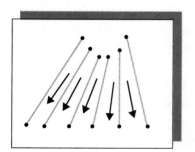

FIGURE 3–26 Sloping lines

This eliminates much erasing and keeps the sketch clean.

Vertical lines are best drawn from the top down, **Figure 3–25**. As in drawing horizontal lines, the forearm should be used as a pivot point. Long, vertical lines can be drawn by using the row of dots as an aid. To draw lines parallel, as they are shown in **Figure 3–25**, is a difficult job for the beginner. Many people use a piece of paper as a rule to mark off equal distances at each end of a line. A line sketched through these two points will be parallel to the first line. Again, repeated practice will increase the sketcher's proficiency to the point at which the use of aids will not be necessary.

Sloping lines can be drawn in the same manner as horizontal or vertical lines, **Figure 3–26**. An advantage of freehand sketching is that the paper is not fastened down in a stationary position, and the sketcher is free to turn the paper in any direction. This makes it simpler to draw lines that might otherwise be in an awkward position.

Circles and Arcs

Circles and parts of circles (arcs) are more difficult to sketch than straight lines. To simplify this operation, there are several techniques which can be used.

One of the main causes of poorly sketched circles is the fact that beginners try to draw the entire circle by sight and in one stroke. Like all other drawing details, a circle or an arc should be carefully laid out before it is drawn.

Notice that the circle in **Figure 3–27** bears some geometric resemblance to a square. Its length and height are the same size as are those of the square. If the square were circumscribed about the circle as shown, it can be seen that the two figures are tangent to each other at four points. These points are the intersections of the centerlines of the circle and the midpoints of the sides of the square.

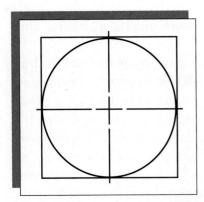

FIGURE 3–27 A square circumscribed about a circle

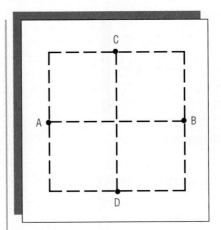

FIGURE 3–28

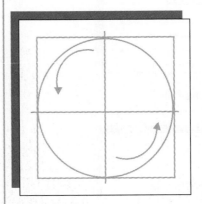

FIGURE 3–29

Since the square is relatively easy to sketch, it can be used as the beginning point for sketching the circle. The procedure for sketching a circle is as follows:

Step 1 Sketch a square of the same size as the diameter of the required circle.

Step 2 Mark off the midpoint of each of the four sides (A, B, C and D).

Step 3 Sketch lines A-B and C-D. These lines represent the vertical and horizontal centerlines of the circle. They divide the square into four smaller squares, **Figure 3–28**.

Step 4 In each of the four squares, sketch an arc that is tangent to two sides of the square. Sketch these arcs with light, feathered strokes as shown in **Figure 3–29**.

Step 5 When the four separate arcs have been sketched roughly, make any necessary sketching strokes to blend them completely into one smooth curve.

Step 6 Go over the outline of the circle with a single stroke so that the circle is drawn with the same line weight as the rest of the sketch.

Sketching an Irregular Shape

Sketching an irregularly shaped object requires the use of many reference points. For such sketches, the use of graph paper, as shown in **Figure 3–30**, greatly simplifies the drawing. **Figure 3–30** shows the steps to follow in sketching an object or a workpiece with a nonuniform or irregular shape.

Step 1 Lay out a series of equally spaced vertical and horizontal lines across the view to be sketched. The space between the lines will depend on the complexity of the shape. For the most accurate reproduction, the lines should be fairly close together. This condition will the greatest number of reference points.

Step 2 On a blank sheet of graph paper that has the same size squares as the

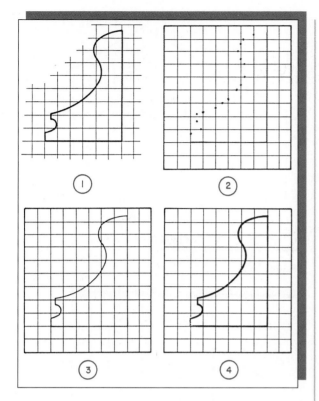

FIGURE 3–30   Sketching an irregular shape

original, plot each of the reference points as it appears on the original.

Step 3   Sketch in lightly the irregular shape so that it passes through each of the reference points.

Step 4   Darken the outline so that it blends in with the other lines of the sketch.

This method of sketching can also be used to make a sketch to a reduced or an enlarged scale. If the sketch is made on graph paper that has smaller or larger squares than the original layout, the finished sketch will be in perfect proportion to the original but will be either smaller or larger as desired.

Sketching Isometric Shapes

Isometric drawings combine three views of an object into one picture. The front, one side, and top are usually shown so that it looks like the object is viewed from an angle where all three of these surfaces can be seen from one position.

Start an isometric sketch from the three axes: the vertical axes, which is common to the front and the end elevations (Line 1), and the two horizontal lines which represent the bottom surface of both of these views (Lines 2 and 3 of **Figures 3–31** and **3–32**). To make an isometric sketch of the notched block shown in the illustration, follow steps outlined below.

Step 1   Draw a light, long, vertical line to represent Line 1. Position it on the paper so that there is room for the rest of the sketch. Draw long, light guidelines to represent Lines 2 and 3. Draw these lines, representing the horizontal axes, at an angle of 30 degrees from the horizontal direction. To help estimate a 30-degree angle, remember that it is one-third of a right angle, **Figure 3–33**.

Step 2   Mark off the overall height of the notched block on Line 1, the overall length on Line 2, and the overall width on Line 3 as indicated. Draw light lines through these points to complete the solid, rectangular outline which represents the basic shape of the notched block, **Figure 3–34**.

Step 3   Mark off the length and height of the notch on the lines of the front elevation. Through these points sketch in the vertical and horizontal lines of the notch. Project the ends of these lines as shown so that the surfaces of the notch appear in all views of the isometric sketch, **Figure 3–35**.

Step 4   Remove all unnecessary guidelines from the sketch and go over the outline with solid, single-stroke object lines, **Figure 3–36**.

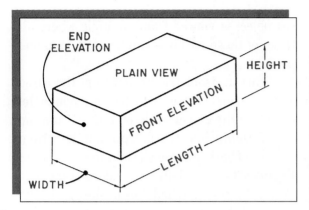

FIGURE 3–31 Isometric drawing

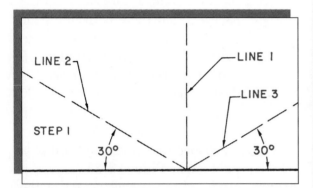

FIGURE 3–32 Isometric axes

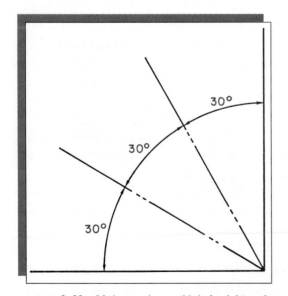

FIGURE 3–33 30 degrees is one-third of a right angle

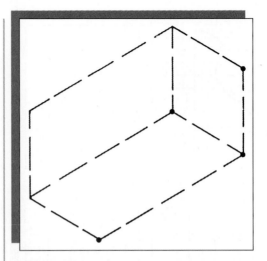

FIGURE 3–34

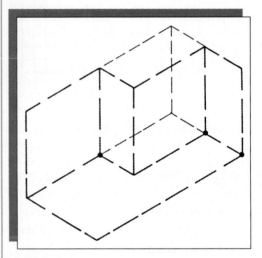

FIGURE 3–35

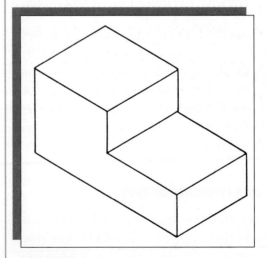

FIGURE 3–36

Models

Models are sometimes built of large construction projects. These models are built to scale and may be complete in every detail. They can be made of paper, cardboard, wood, clay, and plastic. Models effectively show the client what the architect has designed, **Figure 3–37**. When it is necessary to test the solutions to engineering problems, engineering models can be used. Engineering models, made with the actual construction materials and design, can be used to test the wind resistance of tall structures, the operation of mechanical devices, and the strength of bridges. Computer models are often used for preliminary testing, **Figure 3–38**.

FIGURE 3–37   Architectural models show a client what a structure will look like. *(Courtesy of Einhorn, Yaffee, Prescot)*

FIGURE 3–38   Engineers sometimes build engineering models to test structures and mechanical design. *(Courtesy of Einhorn, Yaffee, Prescot)*

ACTIVITIES

Designing a Small Store

After considering the client's needs and other factors, an architect makes rough sketches and notes to describe the proposed structure. From this information, an architectural drafter makes a finished set of plans for the project.

In this activity, assume you are an architect for a client who wants a small auto parts store designed. Visualize the project, making rough sketches and notes. Use these notes as an architectural drafter would to draw a floor plan, front elevation, and side elevation.

Equipment and Materials

17″ × 22″ paper
Architect's scale
Drafting triangle
6H and 2H pencils
Eraser

Procedure

Draw a floor plan, front elevation, and side elevation. Use a scale of 1/4″ = 1′-0″. The total area of the store is between 1,400 and 1,800 square feet. It must include the following:

1. Front and rear entrances
2. Customer/display area
3. Counter
4. Storage (stock) area of at least 1,000 square feet
5. Office
6. Furnace room of atleast 60 square feet
7. Lavatory

Drafting Suggestions

1. Architects and drafters use T square and triangles to draw straight lines and square corners. Use the plastic triangle as a straightedge and as a guide for making square corners if a T-square is not available.
2. It is considered bad practice to use an architect's scale as a straightedge.

3. Lay out all lines with the 6H pencil, then darken with the 2H pencil.
4. Refer often to illustrations in this unit for symbols and drawing style.

Model Making

After designing a project, an architect or engineer sometimes constructs a model of the proposed structure. This shows the client exactly what it will look like.

Equipment and Materials

Cardboard
Tape
Glue
Knife
Poster paints
Balsa wood or bass wood

Procedure

Construct an architectural model of the building described in the first activity. This should appear as much like the exterior of the building as possible.

Applying Construction Across the Curriculum

Communications

Obtain a set of construction drawings from a construction company or from your teacher and write a list of specific information that can be obtained from those drawings. For each bit of information list one occupation that would use that information and how the information would be used. Include as many occupations as possible.

Mathematics

List ten dimensions which you have measured from a model. If an architectural model is not available, dimensions from any model, such as a model car or airplane, will do. Find out what the scale is for the model you measured. For each dimension measured from the model, calculate what the full-size dimension would be on the real object.

REVIEW

A. Questions. Give a brief answer for each question. These questions are based on the drawings in **Figures 3–5** and **3–6.**

1. How far beyond the end walls of the house does the roof overhang?
2. What material is used to cover the exterior of the house?
3. How high does the chimney project above the top of the roof?
4. What are the overall dimensions of the living room?
5. What size is the bathroom door?
6. How many electrical outlets are located in the carport?
7. How many windows does the house have?
8. How far is the center of the main entrance from the outside wall of the carport?
9. What is the overall length and width of the house?
10. Where is the water heater located?

B. Matching. Match the device or function in Column II with the correct statement in Column I.

<table>
<tr><th>Column I</th><th>Column II</th></tr>
<tr><td>1. The information that is processed by a computer</td><td>a. disk</td></tr>
<tr><td>2. A hand-held device for moving a cursor locating a point on a CRT</td><td>b. plotter
c. BASIC</td></tr>
<tr><td>3. A computer peripheral that maneuvers pens and paper to produce drawings</td><td>d. data
e. digitizer, mouse, and keyboard</td></tr>
<tr><td>4. A high-level language</td><td>f. digitizer</td></tr>
<tr><td>5. Input devices</td><td>g. zoom</td></tr>
<tr><td>6. Output devices</td><td>h. plotter and CRT</td></tr>
<tr><td>7. Device used to input an existing drawing into a CAD system</td><td>j. mouse</td></tr>
<tr><td>8. CAD function that allows the operator to create a close-up of part of a drawing</td><td>i. binary code</td></tr>
<tr><td>9. Computer language consisting of 1s and 0s</td><td></td></tr>
<tr><td>10. The storage device most commonly used for microcomputer programs</td><td></td></tr>
</table>

CHAPTER 4

Specifications and Contracts

OBJECTIVES

After completing this chapter, you should be able to:

▼ describe the purpose and organization of construction specifications;

▼ explain the importance of construction contracts to owners, designers, and contractors; and

▼ define fixed-sum contracts and cost-plus contracts.

KEY TERMS

specifications

CSI format

word processing

data base

bid

estimator

direct cost

overhead cost

square-foot estimate

real estate

principal

mortgage

contract

fixed-sum contract

cost-plus contract

Specifications

Working drawings contain as much information as possible about the materials to be used, the location of parts, and the size of parts for a construction project. However, it is impossible to include all of the information that is necessary. For example, it is impractical to show on the drawings the grade of lumber to be used for the roof framing of a house. **Construction specifications**, commonly called *specs*, are written documents that give detailed information not shown on the working drawings, **Figure 4–1**. Specs describe the

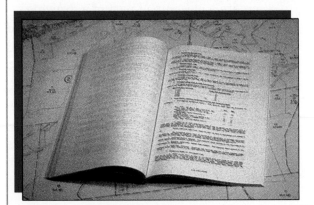

FIGURE 4–1 Construction specifications are written instructions.

quality and type of material to be used, and the method for putting the materials together.

Specification writing requires a great deal of knowledge about construction practices and construction materials. Thorough knowledge of building codes and regulations governing construction is also needed. On smaller projects, the specification is brief and may be written by the architect who designed the structure. The Federal Housing Administration's (FHA) *Description of Materials*, **Figure 4–2**, is an example of a brief form of specifications.

U. S. DEPARTMENT OF HOUSING AND URBAN DEVELOPMENT
FEDERAL HOUSING ADMINISTRATION

FHA Form 2005
VA Form 26-1852
Rev. 2/75

For accurate register of carbon copies, form may be separated along above fold. Staple completed sheets together in original order.

Form Approved
OMB No. 63–R0055

DESCRIPTION OF MATERIALS

No. _____
(To be inserted by FHA or VA)

☐ Proposed Construction

☐ Under Construction

Property address _____ City _____ State _____

Mortgagor or Sponsor _____ _____
                        (Name)   (Address)

Contractor or Builder _____ _____
                       (Name)   (Address)

INSTRUCTIONS

1. For additional information on how this form is to be submitted, number of copies, etc., see the instructions applicable to the FHA Application for Mortgage Insurance or VA Request for Determination of Reasonable Value, as the case may be.
2. Describe all materials and equipment to be used, whether or not shown on the drawings, by marking an X in each appropriate check-box and entering the information called for in each space. If space is inadequate, enter "See misc." and describe under item 27 or on an attached sheet. THE USE OF PAINT CONTAINING MORE THAN ONE HALF OF ONE PERCENT LEAD BY WEIGHT IS PROHIBITED.
3. Work not specifically described or shown will not be considered unless

required, then the minimum acceptable will be assumed. Work exceeding minimum requirements cannot be considered unless specifically described.
4. Include no alternates, "or equal" phrases, or contradictory items. (Consideration of a request for acceptance of substitute materials or equipment is not thereby precluded.)
5. Include signatures required at the end of this form.
6. The construction shall be completed in compliance with the related drawings and specifications, as amended during processing. The specifications include this Description of Materials and the applicable Minimum Property Standards.

1. EXCAVATION:

Bearing soil, type _____

2. FOUNDATIONS:

Footings: concrete mix _____ ; strength psi _____ Reinforcing _____

Foundation wall: material _____ Reinforcing _____

Interior foundation wall: material _____ Party foundation wall _____

Columns: material and sizes _____ Piers: material and reinforcing _____

Girders: material and sizes _____ Sills: material _____

Basement entrance areaway _____ Window areaways _____

Waterproofing _____ Footing drains _____

Termite protection _____

Basementless space: ground cover _____ ; insulation _____ ; foundation vents _____

Special foundations _____

Additional information: _____

3. CHIMNEYS:

Material _____ Prefabricated *(make and size)* _____

Flue lining: material _____ Heater flue size _____ Fireplace flue size _____

Vents *(material and size)*: gas or oil heater _____ ; water heater _____

Additional information: _____

4. FIREPLACES:

Type: ☐ solid fuel; ☐ gas-burning; ☐ circulator *(make and size)* _____ Ash dump and clean-out _____

Fireplace: facing _____ ; lining _____ ; hearth _____ ; mantel _____

Additional information: _____

5. EXTERIOR WALLS:

Wood frame: wood grade, and species _____ ☐ Corner bracing. Building paper or felt _____

Sheathing _____ ; thickness _____ ; width _____ ; ☐ solid; ☐ spaced _____" o. c.; ☐ diagonal; _____

Siding _____ ; grade _____ ; type _____ ; size _____ ; exposure _____"; fastening _____

Shingles _____ ; grade _____ ; type _____ ; size _____ ; exposure _____"; fastening _____

Stucco _____ ; thickness _____"; Lath _____ ; weight _____ lb.

Masonry veneer _____ Sills _____ Lintels _____ Base flashing _____

FIGURE 4–2   Brief specification form

Masonry: ☐ solid ☐ faced ☐ stuccoed; total wall thickness _____"; facing thickness _____"; facing material _____

Backup material _____; thickness _____"; bonding _____

Door sills _____ Window sills _____ Lintels _____ Base flashing _____

Interior surfaces: dampproofing, _____ coats of _____; furring _____

Additional information: _____

Exterior painting: material _____; number of coats _____

Gable wall construction: ☐ same as main walls; ☐ other construction _____

6. FLOOR FRAMING:

Joists: wood, grade, and species _____; other _____; bridging _____; anchors _____

Concrete slab: ☐ basement floor; ☐ first floor; ☐ ground supported; ☐ self-supporting; mix _____; thickness _____";

reinforcing _____; insulation _____; membrane _____

Fill under slab: material _____; thickness _____". Additional information: _____

7. SUBFLOORING: *(Describe underflooring for special floors under item 21.)*

Material: grade and species _____; size _____; type _____

Laid: ☐ first floor; ☐ second floor; ☐ attic _____ sq. ft.; ☐ diagonal; ☐ right angles. Additional information: _____

8. FINISH FLOORING: *(Wood only. Describe other finish flooring under item 21.)*

LOCATION	ROOMS	GRADE	SPECIES	THICKNESS	WIDTH	BLDG. PAPER	FINISH
First floor _____							
Second floor _____							
Attic floor _____ sq. ft.							

Additional information: _____

9. PARTITION FRAMING:

Studs: wood, grade, and species _____ size and spacing _____ Other _____

Additional information: _____

10. CEILING FRAMING:

Joists: wood, grade, and species _____ Other _____ Bridging _____

Additional information: _____

11. ROOF FRAMING:

Rafters: wood, grade, and species _____ Roof trusses (see detail): grade and species _____

Additional information: _____

12. ROOFING:

Sheathing: wood, grade, and species _____; ☐ solid; ☐ spaced _____" o.c.

Roofing _____; grade _____; size _____; type _____

Underlay _____; weight or thickness _____; size _____; fastening _____

Built-up roofing _____; number of plies _____; surfacing material _____

Flashing: material _____; gage or weight _____; ☐ gravel stops; ☐ snow guards

Additional information: _____

13. GUTTERS AND DOWNSPOUTS:

Gutters: material _____; gage or weight _____; size _____; shape _____

Downspouts: material _____; gage or weight _____; size _____; shape _____; number _____

Downspouts connected to: ☐ Storm sewer; ☐ sanitary sewer; ☐ dry-well. ☐ Splash blocks: material and size _____

Additional information: _____

14. LATH AND PLASTER

Lath ☐ walls, ☐ ceilings: material _____; weight or thickness _____ Plaster: coats _____; finish _____

Dry-wall ☐ walls, ☐ ceilings: material _____; thickness _____; finish _____;

Joint treatment _____

15. DECORATING: *(Paint, wallpaper, etc.)*

ROOMS	WALL FINISH MATERIAL AND APPLICATION	CEILING FINISH MATERIAL AND APPLICATION
Kitchen _____		
Bath _____		
Other _____		

Additional information: _____

16. INTERIOR DOORS AND TRIM:

Doors: type _____; material _____; thickness _____

Door trim: type _____; material _____ Base: type _____; material _____; size _____

Finish: doors _____; trim _____

Other trim *(item, type and location)* _____

Additional information: _____

FIGURE 4-2   (Continued)

17. WINDOWS:

Windows: type _____; make _____; material _____; sash thickness _____

Glass: grade _____; ☐ sash weights; ☐ balances, type _____; head flashing _____

Trim: type _____; material _____ Paint _____; number coats _____

Weatherstripping: type _____; material _____ Storm sash, number _____

Screens: ☐ full; ☐ half; type _____; number _____; screen cloth material _____

Basement windows: type _____; material _____; screens, number _____; Storm sash, number _____

Special windows _____

Additional information: _____

18. ENTRANCES AND EXTERIOR DETAIL:

Main entrance door: material _____; width _____; thickness ___". Frame: material _____, thickness ___"

Other entrance doors: material _____; width _____; thickness ___". Frame: material _____; thickness ___"

Head flashing _____ Weatherstripping: type _____; saddles _____

Screen doors: thickness ___"; number _____; screen cloth material _____ Storm doors: thickness ___"; number _____

Combination storm and screen doors: thickness ___"; number _____; screen cloth material _____

Shutters: ☐ hinged; ☐ fixed. Railings _____, Attic louvers _____

Exterior millwork: grade and species _____ Paint _____; number coats _____

Additional information: _____

19. CABINETS AND INTERIOR DETAIL:

Kitchen cabinets, wall units: material _____; lineal feet of shelves _____; shelf width _____

  Base units: material _____; counter top _____; edging _____

  Back and end splash _____ Finish of cabinets _____; number coats _____

Medicine cabinets: make _____; model _____

Other cabinets and built-in furniture _____

Additional information: _____

20. STAIRS:

STAIR	TREADS		RISERS		STRINGS		HANDRAIL		BALUSTERS	
	Material	Thickness	Material	Thickness	Material	Size	Material	Size	Material	Size
Basement										
Main										
Attic										

Disappearing: make and model number _____

Additional information: _____

21. SPECIAL FLOORS AND WAINSCOT:

	LOCATION	MATERIAL, COLOR, BORDER, SIZES, GAGE, ETC.	THRESHOLD MATERIAL	WALL BASE MATERIAL	UNDERFLOOR MATERIAL
FLOORS	Kitchen				
	Bath				

	LOCATION	MATERIAL, COLOR, BORDER, CAP. SIZES, GAGE, ETC.	HEIGHT	HEIGHT OVER TUB	HEIGHT IN SHOWERS (FROM FLOOR)
WAINSCOT	Bath				

Bathroom accessories: ☐ Recessed; material _____; number _____; ☐ Attached; material _____; number _____

Additional information: _____

22. PLUMBING:

FIXTURE	NUMBER	LOCATION	MAKE	MFR'S FIXTURE IDENTIFICATION NO.	SIZE	COLOR
Sink						
Lavatory						
Water closet						
Bathtub						
Shower over tub △						
Stall shower △						
Laundry trays						

△☐ Curtain rod   △☐ Door   ☐ Shower pan: material _____

FIGURE 4–2   (Continued)

Water supply: ☐ public; ☐ community system; ☐ individual (private) system.★
Sewage disposal: ☐ public; ☐ community system; ☐ individual (private) system.★
★Show and describe individual system in complete detail in separate drawings and specifications according to requirements.
House drain (inside): ☐ cast iron; ☐ tile; ☐ other _____ House sewer (outside): ☐ cast iron; ☐ tile; ☐ other _____
Water piping: ☐ galvanized steel; ☐ copper tubing; ☐ other _____ Sill cocks, number _____
Domestic water heater: type _____; make and model _____; heating capacity _____
_____ gph. 100° rise. Storage tank: material _____; capacity _____ gallons.
Gas service: ☐ utility company; ☐ liq. pet. gas; ☐ other _____ Gas piping: ☐ cooking; ☐ house heating.
Footing drains connected to: ☐ storm sewer; ☐ sanitary sewer; ☐ dry well. Sump pump; make and model _____
_____; capacity _____; discharges into _____

23. HEATING:
☐ Hot water. ☐ Steam. ☐ Vapor. ☐ One-pipe system. ☐ Two-pipe system.
   ☐ Radiators. ☐ Convectors. ☐ Baseboard radiation. Make and model _____
   Radiant panel: ☐ floor; ☐ wall; ☐ ceiling. Panel coil: material _____
   ☐ Circulator. ☐ Return pump. Make and model _____; capacity _____ gpm.
   Boiler: make and model _____ Output _____ Btuh.; net rating _____ Btuh.
Additional information: _____
Warm air: ☐ Gravity. ☐ Forced. Type of system _____
   Duct material: supply _____; return _____ Insulation _____, thickness _____ ☐ Outside air intake.
   Furnace: make and model _____ Input _____ Btuh.; output _____ Btuh.
   Additional information: _____
☐ Space heater; ☐ floor furnace; ☐ wall heater. Input _____ Btuh.; output _____ Btuh.; number units _____
   Make, model _____ Additional information: _____
Controls: make and types _____
Additional information: _____
Fuel: ☐ Coal; ☐ oil; ☐ gas; ☐ liq. pet. gas; ☐ electric; ☐ other _____; storage capacity _____
   Additional information: _____
Firing equipment furnished separately: ☐ Gas burner, conversion type. ☐ Stoker: hopper feed ☐; bin feed ☐
   Oil burner: ☐ pressure atomizing; ☐ vaporizing _____
   Make and model _____ Control _____
   Additional information: _____
Electric heating system: type _____ Input _____ watts; @ _____ volts; output _____ Btuh.
   Additional information: _____
Ventilating equipment: attic fan, make and model _____; capacity _____ cfm.
              kitchen exhaust fan, make and model _____
Other heating, ventilating, or cooling equipment _____

24. ELECTRIC WIRING:
Service: ☐ overhead; ☐ underground. Panel: ☐ fuse box; ☐ circuit-breaker; make _____ AMP's _____ No. circuits _____
Wiring: ☐ conduit; ☐ armored cable; ☐ nonmetallic cable; ☐ knob and tube; ☐ other _____
Special outlets: ☐ range; ☐ water heater; ☐ other _____
☐ Doorbell. ☐ Chimes. Push-button locations _____ Additional information: _____

25. LIGHTING FIXTURES:
Total number of fixtures _____ Total allowance for fixtures, typical installation, $ _____
Nontypical installation _____
Additional information: _____

26. INSULATION:

Location	Thickness	Material, Type, and Method of Installation	Vapor Barrier
Roof			
Ceiling			
Wall			
Floor			

27. MISCELLANEOUS: (Describe any main dwelling materials, equipment, or construction items not shown elsewhere; or use to provide additional information where the space provided was inadequate. Always reference by item number to correspond to numbering used on this form.) _____
_____
_____
_____
_____
_____
_____
_____

FIGURE 4–2 (Continued)

HARDWARE: *(make, material, and finish.)* _____
_____
_____

SPECIAL EQUIPMENT: *(State material or make, model and quantity.   Include only equipment and appliances which are acceptable by local law, custom and applicable FHA standards.   Do not include items which, by established custom, are supplied by occupant and removed when he vacates premises or chattles prohibited by law from becoming realty.)*_____
_____
_____
_____
_____
_____
_____
_____
_____
_____

PORCHES:
_____
_____

TERRACES:
_____
_____
_____

GARAGES:
_____
_____
_____

WALKS AND DRIVEWAYS:

Driveway: width _____ ; base material _____ ; thickness _____ "; surfacing material _____ ; thickness _____ "
Front walk: width _____ ; material _____ ; thickness _____ ". Service walk: width _____ ; material _____ ; thickness _____ "
Steps: material _____ ; treads _____ "; risers _____ ". Cheek walls _____

OTHER ONSITE IMPROVEMENTS:
(Specify all exterior onsite improvements not described elsewhere, including items such as unusual grading, drainage structures, retaining walls, fence, railings, and accessory structures.)
_____
_____
_____

LANDSCAPING, PLANTING, AND FINISH GRADING:

Topsoil _____ " thick: ☐ front yard; ☐ side yards; ☐ rear yard to _____ feet behind main building.
Lawns *(seeded, sodded, or sprigged)*: ☐ front yard _____ ; ☐ side yards _____ ; ☐ rear yard_____
Planting: ☐ as specified and shown on drawings; ☐ as follows:

_____ Shade trees, deciduous, _____ " caliper.	_____ Evergreen trees. _____ ' to _____ ', B & B.
_____ Low flowering trees, deciduous, _____ ' to _____ '	_____ Evergreen shrubs. _____ ' to _____ ', B & B.
_____ High-growing shrubs, deciduous, _____ ' to _____ '	_____ Vines, 2-year _____
_____ Medium-growing shrubs, deciduous, _____ ' to _____ '	_____
_____ Low-growing shrubs, deciduous, _____ ' to _____ '	_____

IDENTIFICATION.—This exhibit shall be identified by the signature of the builder, or sponsor, and/or the proposed mortgagor if the latter is known at the time of application.

Date_____ Signature _____

Signature _____

FHA Form 2005
VA Form 26–1852

FIGURE 4–2   (Continued)

However, on larger projects and in large architectural firms, *specification writers* with a background in architectural engineering and experience in the construction industry perform this task.

The amount of detail described by the specifications can vary depending on the size of the construction project. On small jobs, and where the reputation of the contractor is known, the specifications may be brief. In 1966, the construction industry adopted the Construction Specification Institute's **(CSI) Format** *for Construction Specification*. This format uses a division-section organization. All specifications are divided into sixteen divisions, **Figure 4–3**.

Divisions 2 through 16 deal with the actual construction of the project. They are arranged as nearly as possible in the order of work. This makes it easy for estimators and other construction personnel to follow the specifications in the order that the work is to be done. "Division 1—General Requirements" is an overall division describing such things as contracts; relationships between the owner, architect, and contractor; scheduling of work; temporary utilities; and project closeout.

Within these divisions, individual units of work are treated as sections. For example, one section of "Division 8—Doors and Windows" is Metal Windows. This section contains all of the detailed specifications for the metal windows to be installed in the project. **Figure 4–4** is an example of one section of specifications following this format.

Word Processing Specifications

The introduction of word processing into the architectural and engineering professions has been especially useful in specification writing. **Word processing** is the use of a computer to produce, add to, remove from, and reorganize written text. CAD makes it easy to store a drawing part for later use or to modify a drawing. Word processing does the same thing for the written word. Nearly all medium to large sized construction companies and architectural offices now use this computer application, **Figure 4–5**.

The specification writer can use a word processor to save everything he or she creates in a computer data base. A computer **data base** is like a library that catalogs and stores large amounts of written material. When any part of that material is needed, it can be quickly located and used. For example, once the specifications for finish carpentry in **Figure 4–5** are stored in the computer's data base, they can be called up for use on any project. If the kind of wood for exterior trim needs to be changed to Douglas fir, that can be done without retyping any other part of Section 06200.

With each project the construction office specifies, the data base gets larger. This means that greater variety of specifications can be assembled without the need for extensive new work.

Bidding

Once the owner and architect have finalized plans, working drawings, and specifications for a project, they must select a contractor to construct it. A contractor coordinates the building of a project according to a written agreement for a specified sum of money. On small jobs, selecting a contractor may simply mean discussing the job with a contractor known to either the owner or architect. On large construction projects, contractors normally bid on the job.

A **bid** is an offer to complete a construction project for a specified amount of money and within a certain period of time. Making a bid requires an estimate of the time and cost needed to complete the project according to the specifications. The bidding process requires a large

DIVISION 1—GENERAL REQUIREMENTS

01010 SUMMARY OF WORK
01100 ALTERNATIVES
01150 MEASUREMENT & PAYMENT
01200 PROJECT MEETINGS
01300 SUBMITTALS
01400 QUALITY CONTROL
01500 TEMPORARY FACILITIES
      & CONTROLS
01600 MATERIAL & EQUIPMENT
01700 PROJECT CLOSEOUT

DIVISION 2—SITE WORK

02010 SUBSURFACE EXPLORATION
02100 CLEARING
02110 DEMOLITION
02200 EARTHWORK
02250 SOIL TREATMENT
02300 PILE FOUNDATIONS
02350 CAISSONS
02400 SHORING
02500 SITE DRAINAGE
02550 SITE UTILITIES
02600 PAVING & SURFACING
02700 SITE IMPROVEMENTS
02800 LANDSCAPING
02850 RAILROAD WORK
02900 MARINE WORK
02950 TUNNELING

DIVISION 3—CONCRETE

03100 CONCRETE FORMWORK
03150 FORMS
03200 CONCRETE REINFORCEMENT
03250 CONCRETE ACCESSORIES
03300 CAST-IN-PLACE CONCRETE
03350 SPECIALLY FINISHED
      (ARCHITECTURAL) CONCRETE
03360 SPECIALLY PLACED CONCRETE
03400 PRECAST CONCRETE
03500 CEMENTITIOUS DECKS
03800 GROUT

DIVISION 4—MASONRY

04100 MORTAR
04150 MASONRY ACCESSORIES
04200 UNIT MASONRY
04400 STONE
04500 MASONRY RESTORATION &
      CLEANING
04550 REFRACTORIES

DIVISION 5—METALS

05100 STRUCTURAL METAL FRAMING
05200 METAL JOISTS
05300 METAL DECKING
05400 LIGHTGAGE METAL FRAMING
05500 METAL FABRICATIONS
05700 ORNAMENTAL METAL
05800 EXPANSION CONTROL

DIVISION 6—WOOD & PLASTICS

06100 ROUGH CARPENTRY
06130 HEAVY TIMBER CONSTRUCTION
06150 TRESTLES
06170 PREFABRICATED STRUCTURAL
      WOOD
06200 FINISH CARPENTRY
06300 WOOD TREATMENT
06400 ARCHITECTURAL WOODWORK
06500 PREFABRICATED STRUCTURAL
      PLASTICS
06600 PLASTIC FABRICATIONS

DIVISION 7—THERMAL & MOISTURE PROTECTION

07100 WATERPROOFING

07150 DAMPPROOFING
07200 INSULATION
07300 SHINGLES & ROOFING TILES
07400 PREFORMED ROOFING & SIDING
07500 MEMBRANE ROOFING
07570 TRAFFIC TOPPING
07600 FLASHING & SHEET METAL
07800 ROOF ACCESSORIES
07900 SEALANTS

DIVISION 8—DOORS & WINDOWS

08100 METAL DOORS & FRAMES
08200 WOOD & PLASTIC DOORS
08300 SPECIAL DOORS
08400 ENTRANCES & STOREFRONTS
08500 METAL WINDOWS
08600 WOOD & PLASTIC WINDOWS
08650 SPECIAL WINDOWS
08700 HARDWARE & SPECIALTIES
08800 GLAZING
08900 WINDOW WALLS/CURTAIN
      WALLS

DIVISION 9—FINISHES

09100 LATH & PLASTER
09250 GYPSUM WALLBOARD
09300 TILE
09400 TERRAZZO
09500 ACOUSTICAL TREATMENT
09540 CEILING SUSPENSION SYSTEMS
09550 WOOD FLOORING
09650 RESILIENT FLOORING
09680 CARPETING
09700 SPECIAL FLOORING
09760 FLOOR TREATMENT
09800 SPECIAL COATINGS
09900 PAINTING
09950 WALL COVERING

DIVISION 10—SPECIALTIES

10100 CHALKBOARDS & TACKBOARDS
10150 COMPARTMENTS & CUBICLES
10200 LOUVERS & VENTS
10240 GRILLES & SCREENS
10260 WALL & CORNER GUARDS
10270 ACCESS FLOORING
10280 SPECIALTY MODULES
10290 PEST CONTROL
10300 FIREPLACES
10350 FLAGPOLES
10400 IDENTIFYING DEVICES
10450 PEDESTRIAN CONTROL DEVICES
10500 LOCKERS
10530 PROTECTIVE COVERS
10550 POSTAL SPECIALTIES
10600 PARTITIONS
10650 SCALES
10670 STORAGE SHELVING
10700 SUN CONTROL DEVICES
      (EXTERIOR)
10750 TELEPHONE ENCLOSURES
10800 TOILET & BATH ACCESSORIES
10900 WARDROBE SPECIALTIES

DIVISION 11—EQUIPMENT

11050 BUILT-IN MAINTENANCE
      EQUIPMENT
11100 BANK & VAULT EQUIPMENT
11150 COMMERCIAL EQUIPMENT
11170 CHECKROOM EQUIPMENT
11180 DARKROOM EQUIPMENT
11200 ECCLESIASTICAL EQUIPMENT
11300 EDUCATIONAL EQUIPMENT
11400 FOOD SERVICE EQUIPMENT
11480 VENDING EQUIPMENT
11500 ATHLETIC EQUIPMENT
11550 INDUSTRIAL EQUIPMENT
11600 LABORATORY EQUIPMENT
11630 LAUNDRY EQUIPMENT
11650 LIBRARY EQUIPMENT

11700 MEDICAL EQUIPMENT
11800 MORTUARY EQUIPMENT
11830 MUSICAL EQUIPMENT
11850 PARKING EQUIPMENT
11880 WASTE HANDLING EQUIPMENT
11870 LOADING DOCK EQUIPMENT
11880 DETENTION EQUIPMENT
11900 RESIDENTIAL EQUIPMENT
11970 THEATER & STAGE EQUIPMENT
11990 REGISTRATION EQUIPMENT

DIVISION 12—FURNISHINGS

12100 ARTWORK
12300 CABINETS & STORAGE
12500 WINDOW TREATMENT
12550 FABRICS
12600 FURNITURE
12670 RUGS & MATS
12700 SEATING
12800 FURNISHING ACCESSORIES

DIVISION 13—SPECIAL CONSTRUCTION

13010 AIR SUPPORTED STRUCTURES
13050 INTEGRATED ASSEMBLIES
13100 AUDIOMETRIC ROOM
13250 CLEAN ROOM
13350 HYPERBARIC ROOM
13400 INCINERATORS
13440 INSTRUMENTATION
13450 INSULATED ROOM
13500 INTEGRATED CEILING
13540 NUCLEAR REACTORS
13550 OBSERVATORY
13600 PREFABRICATED STRUCTURES
13700 SPECIAL PURPOSE ROOMS &
      BUILDINGS
13750 RADIATION PROTECTION
13770 SOUND & VIBRATION CONTROL
13800 VAULTS
13850 SWIMMING POOLS

DIVISION 14—CONVEYING SYSTEMS

14100 DUMBWAITERS
14200 ELEVATORS
14300 HOISTS & CRANES
14400 LIFTS
14500 MATERIAL HANDLING SYSTEMS
14570 TURNTABLES
14600 MOVING STAIRS & WALKS
14700 TUBE SYSTEMS
14800 POWERED SCAFFOLDING

DIVISION 15—MECHANICAL

15010 GENERAL PROVISIONS
15050 BASIC MATERIALS & METHODS
15180 INSULATION
15200 WATER SUPPLY & TREATMENT
15300 WASTE WATER DISPOSAL &
      TREATMENT
15400 PLUMBING
15500 FIRE PROTECTION
15600 POWER OR HEAT GENERATION
15650 REFRIGERATION
15700 LIQUID HEAT TRANSFER
15800 AIR DISTRIBUTION
15900 CONTROLS & INSTRUMENTATION

DIVISION 16—ELECTRICAL

16010 GENERAL PROVISIONS
16100 BASIC MATERIALS & METHODS
16200 POWER GENERATION
16300 POWER TRANSMISSION
16400 SERVICE & DISTRIBUTION
16500 LIGHTING
16600 SPECIAL SYSTEMS
16700 COMMUNICATIONS
16850 HEATING & COOLING
16900 CONTROLS & INSTRUMENTATION

FIGURE 4–3 CSI format for specifications. *(Courtesy of Construction Specifications Institute)*

DIVISION 6—WOOD AND PLASTICS

Section 06200—FINISH CARPENTRY

General: This section covers all finish woodwork and related items not covered elsewhere in these specifications. The contractor shall furnish all materials, labor, and equipment necessary to complete the work, including rough hardware, finish hardware, and specialty items.

Protection of Materials: All millwork (finish woodwork*) and trim is to be delivered in a clean and dry condition and shall be stored to insure proper ventilation and protection from dampness. Do not install finish woodwork until concrete, masonry, plaster, and related work is dry.

Materials: All materials are to be the best of their respective kind. Lumber shall bear the mark and grade of the association under whose rules it is produced. All millwork shall be kiln dried to a maximum moisture content of 12%.

1. Exterior trim shall be select grade white pine, S4S.
2. Interior trim and millwork shall be select grade white pine, thoroughly sanded at the time of installation.

Installation: All millwork and trim shall be installed with tight fitting joints and formed to conceal future shrinkage due to drying. Interior woodwork shall be mitered or coped at corners (cut in a special way to form neat joints*). All nails are to be set below the surface of the wood and concealed with an approved putty or filler.

*(explanations in parentheses have been added to aid the student.)

FIGURE 4–4 Specifications following the CSI format

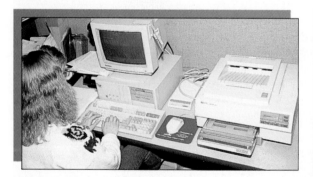

FIGURE 4–5 Word processors are used for writing specifications.

investment of time and money itself. The planned project is listed in newspapers or construction periodicals, **Figure 4–6**. Contractors normally subscribe to these publications so they are aware of what projects are available for bid.

A contractor who intends to bid on a project receives working drawings and specifications from the architect or owner. It is important for the contractor to consider the size of the project. Large firms that specialize in heavy construction do not normally bid on small, single build-

LEGAL NOTICE

MBE/WBE subcontractors and suppliers are invited to submit bids on the Plant Science Building, SUCF Project No. 16058; Space Science Building; Savage Hall, Cornell University, Mechanical Work. Proposals To the Prime Contractor, Tougher Industries, Inc., 175 Broadway, P.O. Box 4067, Albany, N.Y. 12204 - Phone (518) 465-3426. Plan and specifications may be examined at the office of Tougher Industries, Inc. in Albany. Bids are to be submitted as soon as possible. Contact Anthony Celeste.
Tougher Industries, Inc. is an Affirmative Action/Equal Employment Opportunity Employer.
TU-Dec. 11, 12, 13, 14, 15, 16, 17, 18, 19, 20, 21, 22, 23, 24, 25—15t (0610)

FIGURE 4–6 An invitation to bid on a project. *(Courtesy of Tougher Industries, Inc.)*

ings. Likewise, a small construction company would not bid on a very large project.

The next step is for the contractor's **estimator** to determine how much labor and material will be needed to complete the project. The money paid for this labor and material is called **direct cost**.

In preparing the estimate, a percentage of the direct cost is added for operating expenses that are not linked directly to that project. These costs, called **overhead costs**, include office costs, insurance, and maintenance of equipment. In addition to direct costs and overhead, a percentage is also added for profit, **Figure 4–7**.

Sometimes, in order to make rough estimates in a minimum of time, square-foot and cubic-foot estimates are used. For a **square-foot estimate**, the square-foot area of all floor space is multiplied by an average price for that type of construction.

When the contractor has determined the price for the project, it is submitted as a bid to the architect or owner. All bids remain selaed until a predetermined date. On that day, all bids are opened. The contractor submitting the lowest bid for completing the project, according to the plans and specifications, is awarded the job.

Financing

Often, it is not possible or not practical to pay cash for **real estate** (land and buildings). Even though a large company may have the cash needed to buy the real estate, the company may choose to borrow the money. Borrowing money for a real estate investment leaves more money available for other uses. Individuals rarely have enough cash available to buy a home.

Real estate loans are usually made for part of the total cost of the real estate. For example, the borrower may be required to have 20 percent of the total. The lender loans the remaining 80 percent to the borrower. The borrower agrees to make payments (usually monthly) to the lender. These payments are made up of a percentage of the unpaid balance of the loan (**principal**) and interest. The first payments are often mostly interest, so the principal is paid very slowly at first. As the loan nears maturity (the final payment), the portion of the payments applied to the principal increase.

EXAMPLE:
$125,000   property value
$\times 80\%$ maximum loan amount
$100,000   amount of actual loan

$100,000 loan with 9% interest to be repaid in 180 equal payments (monthly for 15 years) of $1,014.27
—payment #1: $750.00 interest + $264.27 principal
—payment #180: $7.55 interest + $1,006.72 principal

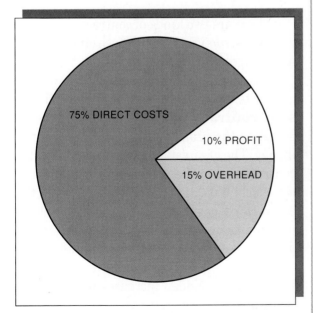

FIGURE 4–7 On one construction project 75% of the cost was for direct expenses to the contractor; 15% was for overhead, such as office expenses and maintenance of equipment; and 10% was profit.

Lenders require some form of guarantee that the loan payments will be made. The most conventional guarantee for real estate loans is called a **mortgage**. The mortgager (lender) is allowed to assume ownership of the real estate if the payments are not made. The mortgager can then sell the real estate to repay the loan. Because real estate loans are usually made for less than the full value of the property, the mortgager is sure to be able to repay the loan from the sale of the real estate.

Contracts

When the job is awarded, a legal agreement is signed which states that the contractor will perform the work and the owner will pay for the work. This agreement is called a **contract**. A verbal agreement between two people is a legal contract, but construction contracts are almost always in the form of written documents signed by all parties involved, **Figure 4–8**, these contracts include provisions for:

▼ *Completion schedule.* It is customary to specify the date by which the project is to be completed. On some large projects, dates are specified for completion of the various stages of construction.

▼ *Schedule of payments.* Sometimes the contractor receives a percentage of the contract price after the completion of each stage of construction. A typical schedule of payments for a house is 20 percent when the foundation is completed; 30 percent when the structure is enclosed; 30 percent when the rough wiring, rough plumbing, and rough heating and air conditioning are installed; and the remaining 20 percent when the house is completed. Another method is for the contractor to receive partial payment for the work done each month.

▼ *Responsibilities of all parties involved.* The owner is usually responsible for having the property surveyed. The architect may be responsible for administering the contract. The contractor is reponsible for the construction and security of the site during the construction period.

▼ *Insurance.* Certain kinds of insurance are required during the construction. The contractor is required to have liability insurance. This protects the contractor against being sued for accidents occurring on the site. The owner is required to have property insurance.

▼ *Worker's compensation.* This is another form of insurance that provides income for the contractor's employees if they are injured at work.

▼ *Termination of the contract.* The contract describes conditions under which the contract can be ended. Contracts can be terminated if one party fails to comply with the contract, when one of the parties is disabled or dies, and for several other reasons.

There are two kinds of contracts in use for most construction. Each of these offers certain advantages and disadvantages.

Fixed-sum (sometimes called lump-sum) contracts are used most often. With a **fixed-sum contract**, the contractor agrees to complete the project for a certain amount of money. The greatest advantage of this kind of contract is that the owner knows in advance the exact final cost of the project. However, the contractor does not know what hidden problems may be encountered. Therefore the contractor's price must be high enough to cover unforeseen circumstances, such as excessive rock in the excavation or sudden increases in the cost of materials.

A **cost-plus contract** is one in which the contractor agrees to complete the work for the actual cost, plus a percentage for overhead and

Building Construction Agreement

This Agreement made and entered into this _____ day of _____, 19 _____.

between _____

_____

called "OWNER" whose address is _____

and _____

Called "CONTRACTOR" whose address is _____

_____

It is hereby agreed:

1. DESCRIPTION OF WORK: Contractor shall furnish all labor, materials, and services to construct and complete in good, expeditious,

workmanlike and substantial manner a _____

hereafter called "project" upon the following described real property: _____

_____

_____

OWNER shall locate and point out the property lines to CONTRACTOR and shall provide boundary stakes by a licensed land surveyor or registered civil engineer if in doubt as to boundaries.

2. PLANS, SPECIFICATIONS AND PERMITS: The project shall be constructed according to the project plans and specifications which have been examined and accepted by the OWNER and which are hereby incorporated by reference and made a part of this Agreement.

Unless otherwise specifically provided in the plans or specifications, CONTRACTOR shall obtain and pay for all required building permits, and OWNER shall pay assessments and charges required by public bodies and utilities for financing or repaying the cost of sewers, storm drains, water service, and other utilities, including sewer and storm drain reimbursement charges, revolving fund charges, hook-up charges and the like.

3. PAYMENT: OWNER shall pay CONTRACTOR the agreed price of $ _____

_____

together with any additional sums as may be provided for herein, in installments as follows: _____

_____

_____

Final payment shall be made thirty (30) days after Notice of Completion has been recorded or upon issuance of a lien free endorsement by a title company to lender.

Upon execution of this Agreement the final payment shall be placed in escrow for CONTRACTOR's benefit. This Agreement constitutes authority for the escrow to pay CONTRACTOR thirty (30) days after recordation of Notice of Completion.

If payments are to be made through a Construction Lender, OWNER represents that the construction loan fund is sufficient to pay the contract price. OWNER will do every thing possible to expedite all payments. OWNER hereby irrevocably authorizes Construction Lender to make all payments directly to CONTRACTOR when due. The name and address of the Construction Lender is _____

_____ of

If corrective or repair work of a minor nature remains to be accomplished by CONTRACTOR after the project is ready for occupancy CONTRACTOR shall perform such work expeditiously and OWNER shall not withhold any payment pending the completion of such work.

If major items of corrective or repair work remain to be accomplished after the project is ready for occupancy the cost of which aggregates more than one (1%) percent of the gross agreed price as determined by the CONTRACTOR, then OWNER pending completion of such work, may withhold in escrow a sufficient amount to pay for completion of such work, but no more.

4. COMPLETION: CONTRACTOR shall begin work within _____ days after the site is ready for CONTRACTOR and the required

building permit has been issued and the construction loan, if any, has been recorded and shall complete the project within _____ working days, subject to permissable delays as described in Paragraph 5.

5. DELAY: CONTRACTOR shall be excused for any delay in completion of the project caused by acts of God, acts of OWNER or OWNER's agent, employee or independent contractor, stormy weather, labor trouble, acts of public utilities, public bodies or inspectors, extra work, failure of the Owner to make progress and extra work payments promptly, or other contingencies unforeseen by CONTRACTOR and beyond the reasonable control of CONTRACTOR.

6. LABOR AND MATERIALS: CONTRACTOR shall pay all valid charges for labor and material incurred by CONTRACTOR and used in the construction of the project but is excused from this obligation in any period during which OWNER is in arrears in making progress and extra work payments to CONTRACTOR.

7. AGREEMENT, PLANS, AND SPECIFICATIONS: The Agreement, plans and specifications are intended to supplement each other. In case of conflict, however, the specifications shall control the plans, and the provisions of this Agreement shall control both.

8. EXTRA WORK, CHANGES AND INTEREST: (a) Changes shall be made only when a change order is signed by CONTRACTOR and OWNER.

(b) Should OWNER, Construction Lender, or any proper governmental body or building inspector direct any modification or addition to the work covered by this Agreement, the cost shall be added to the agreed price.

(c) Expense incurred because of unusual or unanticipated ground conditions (such as fill, rock, or ground water) shall be paid for as extra work.

(d) Any extra cost incurred by CONTRACTOR at the instance of OWNER shall be paid for as extra work.

(e) Payment for extras and changes shall include the actual cost of labor and materials plus CONTRACTOR's fee and when feasible shall be paid for in advance, otherwise on receipt of statement and there shall be no retention of funds related to extras or changes.

(f) All moneys due CONTRACTOR from OWNER whether for the agreed price or for changes or extras shall bear interest at the rate of ten (10%) percent per annum.

FIGURE 4–8 Construction contract. *(Courtesy of the Building Industry Association of Superior California, Inc.)*

9. ALLOWANCES: If the Agreed price includes "allowances," and the cost of performing the work covered by the allowance is either greater or less than the allowance, then the agreed price shall be increased or decreased accordingly. Unless otherwise requested by OWNER in writing, CONTRACTOR shall use his own judgment in accomplishing work covered by an allowance. If OWNER requests that work covered by an allowance be accomplished in such a way that the cost will exceed the allowance, CONTRACTOR shall comply with OWNER's request provided that OWNER pays the additional cost on receipt of statement.

10. NOTICE OF COMPLETION AND OCCUPANCY: OWNER agrees to sign and record a Notice of Completion within five (5) days after the project is complete and ready for occupancy.

11. AGENCY: If OWNER fails to so record Notice of Completion, then OWNER hereby appoints CONTRACTOR as OWNER's agent to sign and record a Notice of Completion on behalf of OWNER. This agency is irrevocable and is an agency coupled with an interest.

12. OCCUPANCY: CONTRACTOR may use such reasonable force or legal means as is necessary to deny occupancy of the project by OWNER or anyone else until CONTRACTOR has received all payments due under this Agreement and until Notice of Completion has been recorded. In the event OWNER occupies the building or any part thereof before CONTRACTOR has received all payment due under this Agreement, such occupancy shall constitute full and unqualified acceptance of all CONTRACTOR's work by OWNER and OWNER agrees that such occupancy shall be a waiver of any and all claims against CONTRACTOR.

13. INSURANCE AND DEPOSITS: OWNER shall procure at his own expense and before the commencement of any work hereunder, fire insurance with coverage for construction, vandalism and malicious mischief; such insurance to be in a sum at least equal to the agreed construction price with loss, if any, payable to any beneficiary under any deed of trust covering the project. Such insurance shall name the CONTRACTOR, all sub-contractors and all suppliers as an additional insured for the protection of OWNER, CONTRACTOR, sub-contractors, and construction lender as their interests may appear.

Should OWNER fail to do so, CONTRACTOR may, but is not required to, procure such insurance as an extra and as agent for and at the expense of OWNER.

If the project is destroyed or damaged by any accident, disaster, or calamity, such as fire, storm, flood, landslide or subsidence, earthquake or by theft or vandalism, any work done by CONTRACTOR in rebuilding or restoring the project shall be paid for by OWNER as extra work.

CONTRACTOR shall carry Workmen's Compensation Insurance for the protection of CONTRACTOR's employees during the progress of the work. OWNER shall obtain and pay for insurance against injury to his own employees and persons under OWNER's direction and persons on the job site at OWNER's invitation.

14. RIGHT TO STOP WORK: If OWNER fails to make any payment whether for the agreed price, changes or extras, CONTRACTOR may stop work until the payments are made and shall not be deemed to have breached this Agreement by reason of such stoppage. OWNER's obligation to make payments is an express condition precedent to CONTRACTOR's obligation to proceed with the work. CONTRACTOR may at his option elect to terminate the Agreement and treat the failure to pay as a breach. Time is of the essence of this Agreement.

15. ARBITRATION: Any controversy or claim arising out of or relating to this Agreement, or the breach thereof, shall be settled by arbitration in accordance with the Construction Industry Arbitration Rules of the American Arbitration Association, and judgment upon the award may be entered in any Court having jurisdiction.

16. WARRANTIES: CONTRACTOR warrants against any defects in workmanship or materials for a period of one (1) year from date of completion. There are no other warranties express or implied.

17. LIMITATIONS: No action of any character arising out of or relating to this Agreement or the performance thereof, shall be commenced by either party against the other more than two (2) years after the completion or cessation of work under this Agreement.

18. ATTORNEYS FEES: In the event of any arbitration or litigation between the parties concerning the work hereunder or any event relating thereto, the party prevailing in such dispute shall be entitled to reasonable attorneys fees.

19. CLEAN-UP: Upon completion of the work, CONTRACTOR will remove debris and surplus material created by his operation from OWNER's property and leave it in a neat and clean condition.

20. ASSIGNMENT: Neither party may assign this Agreement without the written consent of the other party, except for the right of CONTRACTOR to engage such sub-contractors as he deems necessary.

21. NOTICES: Any notice required or permitted under this Agreement may be given by ordinary mail at the addresses contained herein, but such addresses may be changed by written notice by one party to the other from time to time. Notice shall be deemed received one day after deposited in the mail, postage prepaid.

22. ACKNOWLEDGEMENT BY OWNER. OWNER acknowledges he is aware that:

A. NOTICE — "Under the Mechanics' Lien Law, any contractor, subcontractor, laborer, supplier or other person who helps to improve your property but is not paid for his work or supplies, has a right to enforce a claim against your property. This means that, after a court hearing, your property could be sold by a court officer and the proceeds of the sale used to satisfy the indebtedness. This can happen even if you have paid your own contractor in full, if the subcontractor, laborer, or supplier remains unpaid."

B. NOTICE TO OWNER — "Under the Mechanics' Lien Law, any contractor, subcontractor, laborer, materialman or other person who helps to improve your property and is not paid for his labor, services or material, has a right to enforce his claim against your property.
"Under the law, you may protect yourself against such claims by filing, before commencing such work or improvement, an original contract for the work of improvement or a modification thereof, in the office of the county recorder of the county where the property is situated and requiring that a contractor's payment bond be recorded in such office. Said bond shall be in an amount not less than fifty percent (50%) of the contract price and shall, in addition to any conditions for the performance of the contract, be conditioned for the payment in full of the claims of all persons furnishing labor, services, equipment or materials for the work described in said contract."

23. _____

The parties have executed this Agreement the day and year first mentioned herein.

CONTRACTORS are required by law to be licensed and regulated by the Contractors' State License Board. Any questions concerning a Contractor may be referred to the registrar of the board whose address is: Contractors' State License Board, 1020 N Street, Sacramento, California, 95814.

_____	_____
OWNER	CONTRACTOR
_____	_____
OWNER	OFFICER
AGREED TO:	LICENSE NO. _____
_____	
Construction Lender	

FIGURE 4–8 (Continued)

Richard Theriault

Occupation:

Building Inspector, Town of Guilderland Building Department

How long:

6 years as assistant building inspector and building inspector

Typical day on the job:

Rick works from 9:00 AM until 10:30 AM in the Building Department office reviewing building plans for code compliance and providing technical assistence to property owners and builders. Technical assistance is limited to answering questions about building codes and how to comply with them. This does not include suggestions for how to design or build something a better way. Rick then conducts ten to fifteen field inspections from 10:30 AM until 2:00 PM. Builders make an appointment for an inspection at least 24 hours ahead of time. Rick usually returns to the office about 2:00 PM to record the results of his field inspections, issue certificates of occupancy, write letters to inform builders, owners, and architects of discrepancies, and answer questions.

Education or training:

Rick took a construction technology course in high school. He then earned an associate degree in construction engineering technology from a community college. He has also attended seminars and workshops for municipal building inspectors.

Previous jobs in construction:

Rick worked for a general contractor as punch-out foreman, job foreman, and job superintendent for two and a half years after attending community college.

Future opportunities:

Rick would like to advance to the job of Chief Building Inspector or Zoning Administrator.

Working conditions:

He spends half of his time in a clean, comfortable, well-lighted office. He spends the other half of his time in the field on construction sites, ranging from sunny spring and summer conditions to spring mud and subzero winter weather.

Best aspects of the job:

His job allows Rick to develop a lot of contacts with interesting people and gives him a chance to see a variety of construction techniques. He enjoys contact with the public.

Disadvantages or drawbacks of the occupation:

People aren't always happy when you tell them that their work is in violation of the code. Property owners don't always understand that the building inspector and the building code are there to protect them from possible hazards.

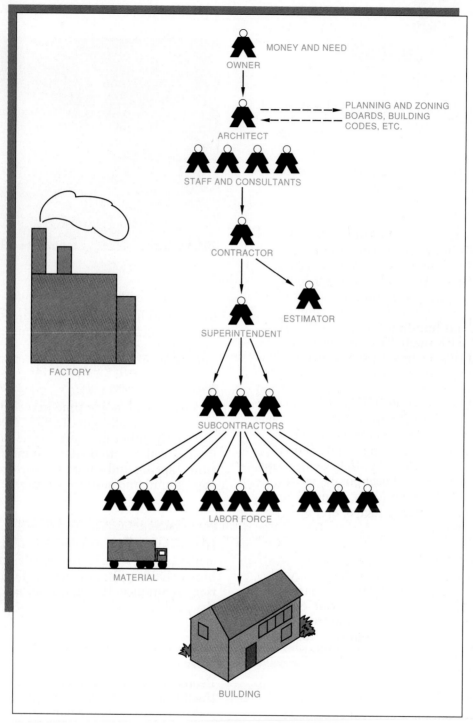

FIGURE 4–9 A construction project requires the services of many trades and professions.

profit. The advantage of this type of contract is that the contractor does not have to allow for unforeseen problems, so the final price is apt to be less. A cost-plus contract is also useful when changes are likely to be made during the course of construction. The main disadvantage of this kind of contract is that the owner does not know exactly what the cost will be until the project is completed.

Contractors

According to the signed agreement, a general contractor is responsible for directing the entire construction operation from start to finish. The contractor supplies the workers, equipment, and materials for the project and guarantees to complete the work in the time specified in the contract, **Figure 4–9**.

It is not always practical for a contractor's construction company to employ all of the specialists needed for every project. Instead, the contractor may hire *subcontractors* to perform certain parts of the work. Subcontractors specialize in electrical work, plumbing, heating and air-conditioning, or other trades, **Figure 4–10**. The bidding process is used to select a subcontractor. The cost for each subcontractor is included in the contractor's bid to the owner.

Contractors must have a thorough knowledge of the kind of work they are doing. Gen-

FIGURE 4–10   The painting on a project is often done by a subcontractor. *(Courtesy of Richard T. Kreh, Sr.)*

eral construction contractors must be familiar with all areas of construction as well as understand good business practices. Subcontractors must also be able to manage their businesses, but their skills are geared for their particular trade.

Most contractors begin their careers in construction by working for other contractors. The knowledge necessary to work in construction can be acquired in vocational schools, technical schools, apprentice programs, or through on-the-job training.

ACTIVITIES

Flow Charts

Draw a flow chart showing the steps of a construction project from the time the architect completes the design until the general contractor signs the contract. Wherever possible, name the construction career involved. A suggested design for this chart is shown in **Figure 4–11** Charts similar to this, called critical-path networks, are often used by contractors to schedule work on a construction project.

Contracts

Clearly worded contracts that accurately describe each party's responsibilities are necessary so that future questions can be answered. The owner, the architect, the contractor, and each party's attorney need to know exactly what the others expect. It is never safe to assume that the other party knows they are expected to do something that is not clearly spelled out in the contract. In this activity you are to write a contract between an owner and a general contractor for the construction of a tool shed or a project your class is constructing. Do not attempt to use flowery language. Write the contract in

your own words, being sure to cover the following items:

▼ Name of the contractor and the owner.
▼ A brief description of the work to be done. You might want to refer to the drawings, if any exist for the project.
▼ The scope of the work. For example, is the contractor responsible for clearing the site where the shed will be placed, painting the interior, etc.?
▼ Who will obtain a building permit from the town or city?
▼ Date the work is to be completed.
▼ How much will be paid? When?
▼ What is the procedure to be followed if the owner is not satisfied when the work is done?
▼ Who is responsible for delivery to the final site?

Compare your contract with others in your class. Is there anything you would like to change or add to your contract? If you write the contract with a word processing program on a computer, it will be easy to make changes and improve your contract.

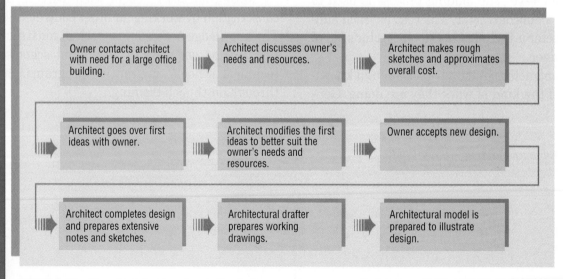

FIGURE 4–11   Flow chart showing steps in the design process

Applying Construction Across the Curriculum

Communications

Obtain a set of specifications for a construction project and read one section carefully. In your own words describe the specifications for one small part of the project. Choose a very small segment of the project so that you will be able to look up the information that you do not understand. The objective here is to understand and communicate a small bit of the specifications, not to display your knowledge of construction.

Communications and Social Studies

Find a newspaper article that reports on some problem involving a construction project. Explain what the construction contract might have to do with that problem. Do you think there might have been a way to include something in the contract that would have prevented this problem?

Mathematics

A property owner wishes to obtain a mortgage loan to construct a house. The total cost of construction will be $120,000. The property owner will make a 10% down payment. The interest rate of the loan will be 9%. What amount must the owner borrow? If the 9% interest is divided equally among the 12 monthly payments to be made each year, the monthly interest rate is 0.75%. How much interest will be paid in the first monthly payment?

REVIEW

Multiple Choice.   Select the best answer for each of the following questions.

1. On which construction document is a summary of the work to be performed likely to be found?
   - a. Specifications
   - b. Working drawings
   - c. Contract
   - d. Estimate

2. Which of the following personnel has the primary responsibility for predicting the amount of time required to do the finished carpentry work on a construction project?
   - a. Carpenter
   - b. Specification writer
   - c. Architect
   - d. Estimator

3. According to the CSI format, which is the number of the division for sound control if there is no equipment or conveying system to be included?
   - a. 11
   - b. 12
   - c. 13
   - d. 14

4. Which of the following is considered an advantage to the owner in a cost-plus contract?
   - a. The exact price of the contract is known before construction begins.
   - b. If unforeseen problems arise, the contractor receives extra pay.
   - c. The contractor does not need to allow extra to cover unforeseen problems.
   - d. The owner is assured of better quality work.

5. Which of the following is considered an advantage to the owner in a fixed-sum contract?
   - a. The contractor allows enough extra to cover any unforeseen problems.
   - b. It is possible to make changes on the working drawings after work has begun.
   - c. The owner is assured of better quality work.
   - d. The exact price of the contract is known before construction begins.

6. If the average cost per square foot for a certain type of construction is $57.00, what is the cost for a building measuring 140 feet by 270 feet?
   - a. $37,000.00
   - b. $2,154,600.00
   - c. $113,400.00
   - d. $11,986,000.00

7. Which of the following is an example of overhead cost?
   - a. Heating the contractor's office
   - b. Cost of materials for a construction project
   - c. Profit
   - d. Subcontractors' fees

8. What is a schedule of payments?
   - a. A method for determining how much the general contractor is to receive
   - b. A method for determining how much each subcontractor is to receive
   - c. A breakdown of when the contractor is to receive payments
   - d. A method of keeping payroll records

9. How does the general contractor usually estimate the cost of the subcontractor's work?

   a. The subcontractor's bids determine these costs.

   b. The general contractor figures these costs.

   c. The general contractor's estimator figures these costs.

   d. The architect figures these costs into the design.

10. What construction document describes the quality of materials to be used?

   a. Contract

   b. Working drawings

   c. Bid

   d. Specifications

CHAPTER 5

Business of Construction

OBJECTIVES

After completing this chapter, you should be able to:

▼ describe sole proprietorships, partnerships, and corporations;

▼ describe three kinds of estimates;

▼ construct a simple PERT network diagram;

▼ define collective bargaining; and

▼ list the steps in a typical grievance procedure.

KEY TERMS

sole proprietorship

general partnership

limited partnership

corporation

board of directors

stockholder

estimate

quantity take-off

PERT network

merit worker

union

collective bargaining

grievance

Forms of Ownership

Construction companies vary in size from small, single-person companies to very large international organizations. However, the size of the company does not necessarily indicate the form of ownership.

Sole Proprietorship

The **sole proprietorship** is the easiest form of ownership to understand, **Figure 5–1**. The two words in the name of this form clearly describe it. "Sole" means only one, or single. The "proprietor" of a business is the owner and operator. So, a sole proprietorship is a business whose owner and operator are the same person. Sole proprietorship construction companies are usually small companies in which the owner is one of the main workers.

Each form of business ownership has advantages and disadvantages. The advantages of the sole proprietorship are that the owner has complete control over the business and there is a minimum of government regulation. If the company is successful, the owner receives high profits. However, in the sole proprietorship, the owner is liable for any debts of the

FIGURE 5–1 Building trades workers sometimes start their own sole proprietorships.

company. The owner can be sued for the company and the owner suffers all of the losses of the company.

Partnership

A partnership is similar to a sole proprietorship, but with two or more owners. In a **general partnership** each partner shares the profits and losses of the business in proportion to the partner's share of investment. General partnerships are common among engineering and architectural companies where each partner is an expert in a different field.

In a general partnership, each partner can be held liable for all of the debts of the company. The advantage of this form of ownership is that the partners share the expense of starting the business. Also, partnerships, like sole proprietorships, are not controlled by extensive government regulations.

A variation of the general partnership is the **limited partnership**. A limited partner is one who invests in the business, receives a proportional share of the profit or loss, but has limited liability. In other words, a limited partner can only lose his or her investment.

Every limited partnership must have one or more general partners who run the business. The general partners in a limited partnership have unlimited liability. They can be personally sued for any debts of the company.

Corporation

In a **corporation**, **Figure 5–2**, a group of people own the company. Another, usually smaller, group of people manage the business. The owners buy shares of stock, **Figure 5–3**. The value of each share of stock increases or decreases according to the success of the business. The stock of many large corporations is bought and sold in public stock exchanges, **Figure 5–4**. Most smaller corporations and many large corporations are privately held. A

FIGURE 5–2 This is the world headquarters for the Bechtel Group, Inc., one of the largest construction corporations in the world. *(Courtesy of the Bechtel Group, Inc.)*

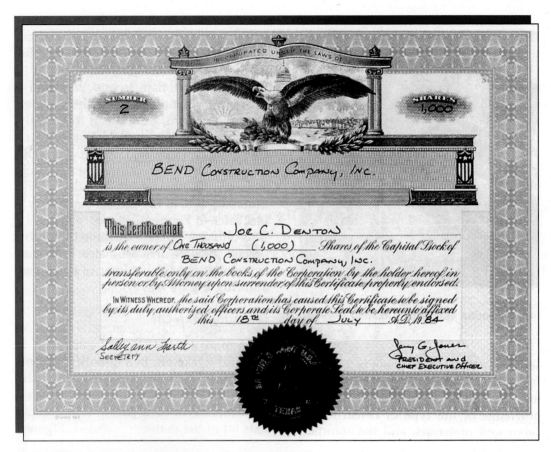

FIGURE 5–3 Owners of corporations have shares of stock in the corporation.

FIGURE 5–4 New York Stock Exchange *(Courtesy of the New York Stock Exchange)*

privately held corporation is one in which stock is owned only by a select group of investors. Privately held stock cannot be bought and sold through the major stock exchanges.

A corporation is managed by its **board of directors**. The board of directors is appointed by the **stockholders** (the owners) at an annual meeting. In some smaller corporations, all of the owners are on the board of directors. The directors meet regularly to form the policies and major operating procedures for the company. The day-to-day operation of the company is the responsibility of the president, who is named by the directors.

In a corporation, no person has unlimited liability. The owners can only lose the money they invested in shares of stock. The owners of a corporation are not responsible for the debts of the corporation. The corporation itself is the legal body and is responsible for the debts of the company. This protection against personal liability is one of the greatest advantages of a corporation.

Larry Tune

Occupation:

Owner; Wheels, Places, 'N Things, Inc.

How long:

20 Years

Typical day on the job:

Wheels, Places, 'N Things is a diversified construction contracting company. It contracts for trucking and earth work on highway construction and new construction and renovation of medium size commercial construction. Larry's day starts very early.

He and his job superintendents review the plans for the day's work before the work crews arrive on the job. If any special equipment, such as compressors or rock drills will be needed, he arranges for that equipment early in the day. He points out that scheduling is a major part of his job. Larry works with his estimator to prepare bids for new contracts. The estimator does quantity take offs and prices materials, but Larry has the job experience to know how long a job should take and how it should be done.

A financial guarantee, made by an independent bonding company, that acceptable work will be completed on schedule is called a *bond*. Bonding is very important to a medium size contractor, because it is the best assurance an owner or developer can have that the contractor's work will be acceptable. Small contractors do not have the background to allow them to get bonding, so bonding separates the small contractors from the bigger businesses. Bonding is required for most state and federal jobs.

Education or training:

Larry learned all aspects of the construction business on the job. He recently hired a couple of smaller contractors so his business would benefit from their experience and knowledge.

Previous jobs in construction:

He started in 1973, with one truck, doing small construction trucking jobs.

Future opportunities:

He has recently broadened his business from construction trucking to general contracting. This has provided many growth opportunities and he looks forward to continued steady growth.

Working conditions:

Larry does the planning and scheduling early in the morning so that he can be on the job sites during working hours. Then, after the work crews have left for the day, he returns to the office to write bids, answer mail, and do the many administrative tasks associated with running a business.

Best aspects of the job:

He gets a good feeling when he proves to himself and to others that he can do the job well.

Disadvantages or drawbacks of the occupation:

Making the business grow requires a lot of time. Sometimes Larry wonders if the time he invests in building up his business will actually pay off.

Because there is no individual who can be held accountable for the actions of the company, the government has stricter regulations for corporations. Also, corporations are much more expensive to form than are partnerships and sole proprietorships.

Estimating

Before a construction company even begins to discuss the contract for a job, some form of estimate must be done. A construction estimate is a prediction of how much time, materials, and money will be needed to complete the job. All estimates are based on knowledge of the type of construction, the cost of labor and materials, and the standard operating procedures of the company. Construction estimators are usually experienced technicians, **Figure 5–5**.

Area and Volume Estimates

A quick, but sometimes inaccurate guess at the cost of a project can be made on the basis of the square-foot area or cubic-foot volume of the project. Area and volume approximations can be used to get an initial idea of the cost of a project, but these methods are not adequate for contract pricing.

An area estimate is one in which a cost per square foot is multiplied by the number of

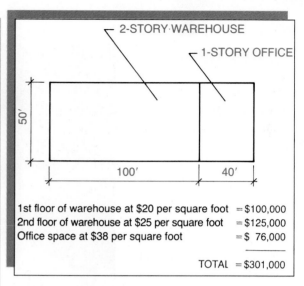

FIGURE 5–6   Area estimate

square feet involved. For example, if it typically costs $50 per square foot to build a warehouse, a 12,000 square-foot warehouse will cost $600,000. The area estimate can be a little more precise by considering the type of construction and estimating different parts of the project separately. For an example, refer to **Figure 5–6**.

Volume estimates are sometimes slightly more precise than are area estimates. A volume estimate is based on the total volume of space enclosed by a building, **Figure 5–7**.

FIGURE 5–5   Construction estimator *(Courtesy of Timberline Software Corp.)*

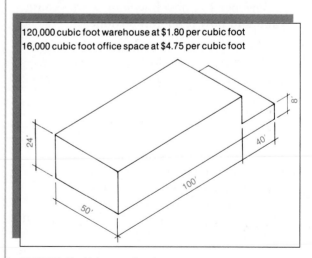

FIGURE 5–7   Volume estimate

Quantity Take-Off

The first step in accurate estimating is a quantity take-off. In a **quantity take-off**, the estimator goes through the entire construction process mentally. As the estimator considers each stage of construction in each part of the project, he or she lists all of the materials needed, **Figure 5–8**. With a complete quantity take-off of materials, it is a simple matter to determine the total material costs for the project.

The next step in a quantity take-off estimate is to determine what the labor cost will be. The cost of labor is figured according to the amount of material used. For example, assume 100 feet of 2×4 wall framing can normally be installed with .3 hour of unskilled labor. Then 6 hours of unskilled labor will be needed to build walls requiring 2,000 feet of 2×4s.

Computerized Estimating

The computer can make the estimating process quicker and can yield more information than a manual estimate. A computerized estimate is also much easier to revise if the need arises—as it almost always does. The estimator keeps records of each project the company has completed. This record of historical costs is called a data base. Some companies exist just to publish up-to-date data bases of construction data. Construction companies purchase these data bases to check the accuracy of their own records or for use when the company's own data base does not have all of the necessary information. Some computerized data bases of construction cost data are sold in special formats to be read by estimating software. The estimator is still required to have good knowledge of construction methods, costs, and scheduling. Estimating programs produce detailed quantity take offs and summary sheets that organize the final estimate in a useful form, **Figure 5–9**.

Scheduling

Accurate scheduling is very important to construction companies. Construction contracts require that the project be completed by a certain deadline. There is often a penalty (a fine) for each day beyond the deadline. Penalties for construction delays can be as high as several thousand dollars per day on large projects. In order to meet the deadlines, the company must know how many hours are required for each item of work. It is also very important that materials are ordered early enough so they will arrive at the site when the workers need them.

Ordering materials too early, however, is expensive. Some materials cannot be stored easily at a construction site. Consider, for example, the problems of storing carpeting before a building is weathertight. Also, the company will want to earn interest on its money until the money must be used to pay for the materials.

PERT (Program Evaluation Review Technique)

PERT is a method for showing all of the steps in construction on a graph. This graph is called a **PERT network**. Six definitions are needed for a discussion of PERT. These definitions may not be obvious at first. They will become clear if you refer back to them as you study this section.

▼ *Activity*—a single piece of work requiring one crew, which will not change from start to finish, and one supply of materials. A single activity can always be finished without interruption for any other activity once it is started.

▼ *Event*—A precise point in time. An event can be the beginning of an activity or the end of an activity.

Sheet 7 of 15

estimating data sheet

JOB TITLE 1460 East Ave. NAME A. Hahn

Grade

	INTERIOR TRIM			
	Bedroom # 2	–	All Select Pine Trim	
2	Window Stools		3/4 × 2 1/2 × 3'-4"	
	Total Stools		6'-8"	Lin. Ft.
2	Window Aprons		3/4 × 2 1/4 × 3'-4"	
	Total Aprons		6'-8"	Lin. Ft.
2	Casings		11/16 × 2 1/4 × 3'-4"	
4	Casings		11/16 × 2 1/4 × 4'-0"	
	Total Window Csg.		22'-8"	Lin. Ft.
1	Door Jamb		4 5/8 × 2'-6" × 6'-8"	
1	Door Jamb		4 5/8 × 6'-0" × 6'-8"	
2	Casings		11/16 × 2 1/4 × 2'-6"	
2	Casings		11/16 × 2 1/4 × 6'-6"	
8	Casings		11/16 × 2 1/4 × 7'-0"	
	Total Door Csg.		74'-0"	Lin. Ft.
	Baseboard		11/16 × 3 1/4 × 58'-0"	
	Closet Rod - chrome		8'-0"	
	Closet Shelf #2 Com. Pine		3/4 × 12 × 8'-0"	

FIGURE 5–8 Material takeoff

```
R. L. McCoy Construction        Estimating Bill of Materials          3-15-98
                                   Timberline Headquarters              4:27 pm

    DESCRIPTION                  ORDER QTY             UNIT PRICE           AMOUNT

Conc-Concrete Prod
-------------------
   3500 psi concrete             457.00  cuyd            45.00          20,565.00
   3000 psi concrete              81.00  cuyd            42.50           3,442.50
                                                                       ---------------
                                                                        24,007.50


Conc-Form Material
-------------------
   Footing Forms                1,092.00 sqft              .65             709.80
   Wall Forms                    8,736.00 sqft              .30           2,620.80
   Ledge Blockouts                531.00  sqft              .50             265.50
                                                                       ---------------
                                                                         3,596.10


Conc-Misc Products
-------------------
   Vapor Barrier 6 Mil Poly        13.00  roll            38.50             500.50
   Keyway                       1,040.00  lnft              .30             312.00
   Protect & Cure                 240.00  sq              2.00             480.00
   Floor Hardener                  69.00  gal            10.50             724.50
   Strip/Oil Forms              8,320.00  sqft              .05             416.00
   Perimeter Notch/Siding       1,040.00  lnft              .20             208.00
                                                                       ---------------
                                                                         2,641.00
```

FIGURE 5-9 This bill of materials was generated by a computerized estimating program. *(Courtesy of Timberline Software Corp.)*

▼ *Optimistic time*—the shortest time in which it is reasonable to expect an activity to be performed. The optimistic time is usually indicated by t_a.

▼ *Probable time*—the amount of time it is most apt to take for completion of an activity. Usually indicated by t_b.

▼ *Pessimistic time*—the longest time that might be required for an activity. Usually indicated by t_c.

▼ *Expected time*—the time that can be expected for completion of an activity. The expected time, indicated by t_e, is found by a mathematical formula. The formula for t_e need not be used to understand what PERT is, but it must be used to actually work with a PERT network.

$$t_e = t_a + 4t_b + t_c / 6$$

EXAMPLE:

5_a = 15 days
5_b = 18 days
t_c = 20 days
$t_e = 15 + (4 \times 18) + 20 / 6$
$t_e = 15 + 72 + 20 / 6$
$t_e = 107 / 6$
t_e = 17.8 days

To construct a PERT network, begin by listing all of the activities in the project. List the last activities first. Then list all of the activities that must be completed in order to perform the last one. Work backward through the entire project until all activities have been listed. These can be shown in a PERT diagram.

A sample PERT network diagram is shown in **Figure 5–10**. (Note: The actual PERT network for a construction project would be much larger.) In a PERT diagram, circles are used to represent events. Arrows represent the activities required to get from one event to another. Along each activity arrow are three numbers. The first number is the optimistic time (t_a). The second number is the probable time (5_b). The third number is the pessimistic time (t_c).

The procedure for constructing a PERT network follows, **Figure 5–11**.

1. In the first column of a table, start with the ending event and list each event, working downward to the first event. (There must be one ending event and one starting event.)
2. In the second column list each event that must be completed before the event in the first column can be started.
3. In the next three columns list the t_a, t_b, and t_c for each activity. (Remember an activity is the line between two events.)
4. Calculate the t_e for each activity.

The PERT times listed by this method can be used to determine how much time will be needed for each part of the project, when materials should be scheduled to arrive, and where personnel can be reassigned for an earlier completion date.

Event	Previous Event	t_a	t_b	t_c	t_e
8	7	6	7	8	7
	6	8	10	13	10.2
	5	18	20	23	20.2
7	2	19	24	30	24.1
6	4	3	4	6	4.2
	3	10	12	15	12.2
5	4	3	5	6	4.8
4	1	19	22	24	21.8
3	1	9	12	16	12.2
2	1	11	14	18	14.2

FIGURE 5–11   PERT table

Personnel

Construction is sometimes called a *labor-intensive industry*. This means that a large amount of human labor is required for a construction project. Because construction is labor intensive, it is especially important for construction managers to be good personnel managers.

Sources of Personnel

Over half of the construction workers in America are non-union. Non-union personnel apply for their jobs on their own. In any bargaining with their employers, each of these non-union workers represents himself or herself. There are two forces that regulate the treatment of these **merit workers**.

Federal, state, and local governments regulate the pay and working conditions of many American workers. The federal government requires that all workers be paid a minimum wage, which is currently $4.25 per hour. The Occupational Safety and Health Administration (OSHA) requires that certain safety standards be met, **Figure 5–12**.

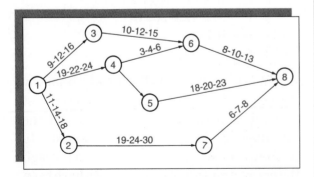

FIGURE 5–10   PERT network diagram

Gerry Cross

Occupation:

Construction Cost Estimator
General Electric Power Generation
Service Division

How long:

11 Years

Typical day on the job:

Gerry spends most days in the office, with regular office hours, but he often has to work until as late as 8:00 PM. He uses a computer data base to determine the number of hours of labor, the amount of materials, and the cost of supplies needed to install gas- and steam-turbine power plants. General Electric's data base is very large and has information that would be useful to their competitors, so that information is closely guarded. A finished estimate might be as short as 20 pages or as long as 120. Estimates are used to write proposals for the installation of new turbines.

Education or training:

Gerry went through a construction drafting apprenticeship at General Electric. He also studied several engineering and management courses at General Electric's headquarters and at outside schools and colleges.

Previous jobs in construction:

Gerry worked as an electrical draftsman for six years, a design draftsman for four years, and an electrical designer for four years.

Future opportunities:

Gerry hopes to become a project estimator, in charge of all aspects of estimating a particular project. Other possibilities for him are proposal manager or marketing services manager.

Working conditions:

Gerry works in an office, with occasional ional visits to construction sites throughout the United States.

Best aspects of the job:

The estimating phase of a job is done within a couple of weeks, so there is constant turnover of projects and variety in the work. It is exciting to be involved in the early stages of a project, then wait to see the project finished. Since Gerry works in a large corporation there are plenty of opportunities to take courses and get additional training.

Disadvantages or drawbacks of the occupation:

Whenever a job loses money or does not turn out as planned, the estimator is the person who gets blamed. Often, the estimator is blamed for errors by designers or poor work in the installation.

FIGURE 5–12 OSHA requires all workers on a construction site to wear proper hardhats.

Although the government regulates many aspects of construction work, another force is even more powerful. Competition is the basis of our economy. Competition in hiring prevents an employer from paying too little. Competition for the price of the finished product also prevents a company from overpaying its employees. If the company tries to pay too little or treat its employees poorly, it will have a hard time hiring good workers. If a company tries to pay too much or offer expensive fringe benefits, it will not be able to build projects within budget. Two hundred years of competition have created an industry that controls itself, **Figure 5–13**.

When workers feel the need to work together to obtain better working conditions,

FIGURE 5–13 This company is proud of its safety program.

they can organize into a union. A **union** is an organization of workers for the sake of more strength in negotiating for pay, working conditions, and settling disputes.

Collective Bargaining

Collective bargaining is the process in which representatives from the union and the management of the company agree on the terms of employment for all members of the union. The end result of collective bargaining is a contract. The contract describes all of the benefits, working conditions, pay rates, and how disputes will be handled, **Figure 5–14**. The union contract is for a specific length of

▼ Regular wages
▼ Overtime
▼ Vacation
▼ Sick leave
▼ Health insurance
▼ Life insurance
▼ Grievance procedure
▼ Pension program
▼ Duration of contract
▼ Procedure for new contract
▼ Guaranteed help (laborers)
▼ Use of apprentices
▼ Tools and equipment to be provided
▼ Safe working conditions
▼ Use of non-union labor

FIGURE 5–14 Some of the topics usually covered by a union contract.

Bruce Patterson

Occupation:

Controller, Davis-Giovinazzo Construction Company

How long:

4 Years

Typical day on the job:

The controller is the person who is responsible for controlling the flow of cash and managing the finances of the business on a day-to-day basis. Construction companies are involved in projects that require very large amounts of money and it is possible to make or lose tremendous sums in a single day. Therefore, Bruce monitors the company's cash flow on a daily basis. He must verify cash balances daily so that decisions can be made about future purchases and other cash requirements.

Bruce manages an accounting department of three people.

Education or training:

Bruce's background is in accounting. He earned a BA degree in accounting and has taken additional courses in construction management. He thinks that a 2-year program in construction and a 2-year program in accounting would be good preparation for a construction company controller.

Previous jobs in construction:

Bruce does not have prior experience in construction. He has worked as a controller for a maintenance company and as the assistant financial director for a non-profit organization. His first job after college was as an accountant for a hotel.

Future opportunities:

Bruce looks forward to becoming more established in the construction industry—maybe someday owning his own construction company.

Working conditions:

Bruce works in a comfortable office.

Best aspects of the job:

He likes being exposed to all aspects of the construction business. This gives him a lot of valuable experience as he strives to become better established in the industry.

Disadvantages or drawbacks of the occupation:

Bruce does not see a lot of drawbacks, but says that there is always the fear that his job could be eliminated if business is bad.

Boldt Builds...

In 1889, Martin Boldt used a horse and wagon to haul wood shaped by hand for his construction operation. Three generations later, the Oscar J. Boldt Construction Company has grown into a national construction giant.

With offices located in Appleton, Wausau and Milwaukee, Wisconsin; as well as Cloquet, Minnesota; and Oklahoma City, Oklahoma; the company is the largest general contractor and construction manager in Wisconsin. It is among the top 75 construction managers in the country and has ranked among the top 25% of contractors in the United States for the past 15 years. Utilizing over $20 million of construction equipment, the company is currently working on projects across the country valued at more than $200 million. One of the keys to the company's success has been diversification. The company has enjoyed considerable success by providing its services to a variety of industries.

Located in the heart of paper country in Wisconsin's Fox River Valley, Boldt has built an impressive record of service to the paper industry throughout the country. Boldt Construction has worked with most of the nation's major paper companies. It has performed construction, machinery installation or consulting assignments in California, Ohio, New Jersey, Alabama, Texas, Virginia, Tennessee, Minnesota, Arkansas, South Carolina, Michigan, New York, Georgia, Kentucky, Oklahoma, Connecticut and Canada.

Boldt also operates as a commercial industrial builder. Boldt has worked on projects, both large and small, ranging from construction of the headquarters for the Aid Association for Lutherans, the largest fraternal benefit society in the world, to rebuilding a men's clothing store destroyed by fire in just three months and in time for the holiday season. Currently, Boldt is building the 90,000 square foot headquarters of The Miller Group Limited. Miller Electric Manufacturing, one of the Miller Group companies, is a major manufacturer of welding machines and portable electrical generators.

Machinery installation services are a natural extension of the company's general contracting capabilities. Boldt has become a specialist in moving and erecting numerous types of manufacturing equipment for a variety of industries.

To help cities with power generation and waste treatment problems, Boldt Construction has worked on congeneration, hydroelectric, coal, gas, oil-fired and nuclear power plants, and has built industrial and municipal waste treatment facilities.

The company is also involved in several other important areas of the construction business. With an unusual understanding of the health care industry, Boldt is one of the nation's largest health care builders. The company is also proud of its civil accomplishments and has been involved in the construction of many schools, churches, and community organization projects. Boldt has also worked with numerous commercial and industrial real estate development firms who have used the company in the design, financing, and construction stages of various projects.

One of the most unusual Boldt enterprises, The Berg Corporation, is under contract to the Chicago Northwestern Transportation Company and stands by 24 hours a day to respond to train derailments in South Dakota, Minnesota, Nebraska, Iowa and Wisconsin. Berg has also been called on to move large structures such as a 450 ton tugboat and a 980 ton airport control tower. ■

(Courtesy of Miller Electrical Mfg. Co., Appleton, WI)

time, such as one year or three years. At the end of the contract period a new contract is negotiated. If the union members do not feel that they can get a satisfactory contract from the company management, they can strike. Actually, very few contract negotiations involve strikes.

Grievances

One of the topics covered by the labor contract is the grievance procedure. A **grievance** is a complaint that the contract has been violated. The contractual grievance procedure describes the exact steps to be taken in settling a dispute. The typical steps are:

1. The employee discusses the problem with his or her supervisor.
2. If the grievance cannot be settled by step 1, the employee gives the supervisor a written notice of the grievance.
3. If step 2 does not settle the dispute, the employee presents the grievance to the union representative. The union representative goes to the next higher level of management.
4. If the grievance still remains, the union takes it to the company's top management.
5. The last step is to present the grievance to an outside person, called an *arbitrator*. The arbitrator is an expert at settling grievances and is chosen by the union and management together. The decision of the arbitrator is final.

A time limit is specified for each step in the grievance procedure. For example, the employee may have ten days in which to start the procedure. If no settlement is reached in a week, the employee may have ten more days in which to go to the next step. As the grievance goes on to higher steps, the time allowed usually gets longer.

ACTIVITIES

Forms of Ownership

Using the yellow pages in your phone book, the newspaper, and your local chamber of commerce, find one example of each form of ownership: sole proprietorship, partnership, and corporation. With your class organized into three groups, each group should obtain information about one type of company. List the following information about the company your group is studying.

1. Sole Proprietorship
   a. Company name
   b. Owner's name
   c. Type of business or construction
   d. Number of full-time employees
2. Partnership
   a. Company name
   b. Names of general partners
   c. Number of limited partners, if any
   d. Type of business or construction
   e. Number of full-time employees
3. Corporation
   a. Company name
   b. Public or private corporation
   c. Chairperson's name
   d. Number of members on the board of directors
   e. President's name
   f. Type of business or construction
   g. Number of full-time employees

Area Estimate

In this activity you will estimate the cost of building a new school to replace your school. If a floor plan is available, your teacher can help you find the overall dimensions. (You may want to review Chapter 3.) If the floor plan is not available, you can measure the outside of the building to determine its area.

Equipment and Materials

50-foot or 100-foot tape measure
Large paper (11″ x 17″)
Pencil
Calculator (optional)

Procedure

1. Begin by making a sketch to show the outline or general shape of the building. Even if you are using a working drawing, make a sketch with only the needed information.
2. Measure each outside wall of the building and show these dimensions on your sketch. In any area with more than one floor level, write the number of floors on your sketch.

CAUTION: Do not attempt to climb fire escape ladders or gain access to roof areas. Do not violate any school rules to complete this assignment.

3. If the building is not a rectangle, draw lines on your sketch so it appears as two or more rectangles, **Figure 5–15.**

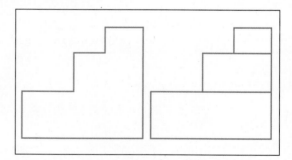

FIGURE 5–15

4. Calculate the area of the building (length times width of each rectangle). List the area of second floors separately. If your school has more than three floors, you may use only the first three.
5. Estimate the replacement cost of your school at the rate of $80.00 per square foot.

PERT Network

If you are working on a group project in class, construct a PERT network for part of it. Choose a stage with six to ten activities. Remember, a PERT network starts with a single event and ends with a single event.

If you are not working on a group project, construct a PERT network for the following activities in the construction of a doghouse. This will be a group project so assume there are three people working at the same time on different parts of the project. The numbers in parentheses are hours.

1. Make working drawings. (1.5–2–3)
2. Get materials from storeroom or building materials store. (.5–1–2)
3. Cut out all plywood pieces. (1.5–2.5–4)
4. Cut framing pieces for walls. (.5–1.5–2.5)
5. Assemble walls. (2–3–4)
6. Cut rafters for roof. (2–3–4)
7. Assemble roof. (1–2–3)
8. Apply shingles to roof. (2–2.5–3)
9. Paint outside. (1–1.5–2)

Applying Construction Across the Curriculum

Social Studies

Find one example of each of the major forms of business ownership in your area: sole proprietorship, partnership, and corporation. Why do you think that form of ownership was chosen for each of the businesses?

Social Studies and Communications

Research the procedure for forming a limited partnership or a corporation in your state. Write a report outlining the steps and include copies of any forms or legal documents you found.

Mathematics

Measure your classroom or laboratory and calculate the area of the floor in square feet. Contact a contractor or building materials supplier in your community to find out what a reasonable cost per square foot would be for the type of floor you have (concrete, hardwood, etc.). Find out the cost of materials and the cost of labor, assuming one journeyman and one laborer or helper could do the work. How much money would you estimate for that floor? (Do not include form work for concrete, framing, etc., only the cost of the surface material on the floor.)

REVIEW

Multiple Choice.   Select the best answer for each of the following questions.

1. In which form of ownership is the owner liable for all of the debts of the company?
   a. Corporation
   b. Limited partnership
   c. Sole proprietorship
   d. None of these
2. What is the greatest advantage of a general partnership?
   a. Partners share the expense of operating the business.
   b. Partners are only responsible for their share of the company's debts.
   c. A general partnership does not pay federal income taxes.
   d. None of these

3. Which form of ownership is most carefully controlled by the government?
   - a. Sole proprietorship
   - b. Corporation
   - c. General partnership
   - d. Limited partnership

4. Which type of estimate can generally be done most quickly?
   - a. Area
   - b. Volume
   - c. Quantity take-off
   - d. Contractor's guide

5. Which type of estimate gives the most accurate results?
   - a. Area
   - b. Volume
   - c. Quantity take-off
   - d. Contractor's guide

6. Which type of estimate would probably be used to given an owner a rough idea about the cost of a new building proposal?
   - a. Area
   - b. Trade
   - c. Quantity take-off
   - d. Contractor's guide

7. Why do construction managers use PERT or another similar techniques?
   - a. To determine how long a project will take
   - b. To ensure that materials arrive when they are needed
   - c. To determine where personnel can best be used
   - d. All of these

8. Which of the following would be an event in PERT?
   - a. Placing concrete for a foundation
   - b. Completion of concrete placement
   - c. Allowing the concrete to cure for seven days
   - d. All of these

9. Which of the following helps to determine the working conditions of construction workers?
   - a. OSHA
   - b. Unions
   - c. Competition
   - d. All of these

10. Which of the following best applies to collective bargaining?
    - a. All employees meet with the employer to discuss wages.
    - b. A court-appointed arbitrator decides the wages.
    - c. One contract covers working conditions for all employees.
    - d. A union representative submits a formal complaint to top management.

Section Two

Material Resources

Materials are one of the major kinds of resources for the construction industry. To understand the processes of construction technology or the impact of those processes, it is necessary to know the properties of the materials and how to work with them.

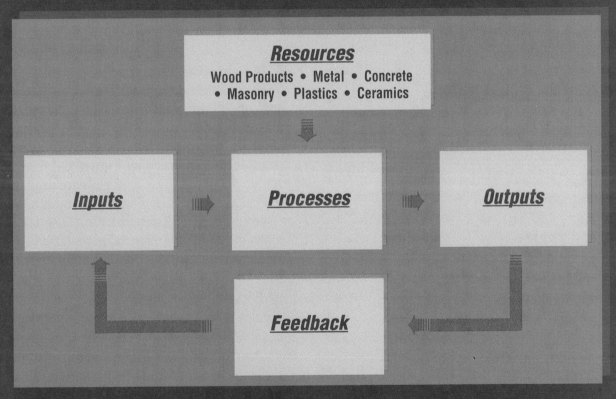

Chapter 6
Concrete and Masonry
covers concrete and masonry, including how the materials are produced, their most important properties, and basic processes for working with materials.

Chapter 7
Metal and Wood Products
outlines the production of common steel materials and their properties, the making of aluminum and its properties, the harvesting and primary processing of forest products, and their important properties.

Chapter 8
Gypsum, Glass, and Plastics
covers most of the widely used construction materials not included in Chapters 6 and 7. These are gypsum products, glass, polymers (plastics), sealants, and adhesives.

CHAPTER 6

Concrete and Masonry

OBJECTIVES

After completing this chapter, you should be able to:

▼ outline the process for converting natural resources into concrete; and

▼ describe the most important physical properties of concrete and masonry materials.

KEY TERMS

aggregate

portland cement

air entraining

void

slump

compressive strenth

tensile strength

rebar

mortar

common brick

face brick

During the planning stages of any construction project, designers must determine what material to use for every part of a structure. Almost every material available to industry can be used to some advantage in construction, **Figure 6–1**. However, only a few of these materials are used in the structural parts (giving basic strength) of most construction. Concrete, for instance, is an important structural material in construction. Architects, engineers, and specification writers

FIGURE 6–1 Hundreds of products are used in the construction of a building. *(Courtesy of Niagara Mohawk Power Corporation)*

must rely on their knowledge of these materials to select those with the most desirable properties for each use.

Concrete

Concrete has been used for centuries as a basic building material. Concrete consists of hard particles, called **aggregates**. These are held together by cement. Modern concrete can be designed to give great strength and an attractive appearance, **Figure 6–2**.

Portland Cement

Most modern concrete is made with **portland cement**, so named because it is like a rock found on the Isle of Portland. A special type of portland cement, called **air-entraining** portland cement, is used where freezing and thawing is a problem. Small amounts of special materials are added to the cement resulting in millions of tiny air bubbles in the cured concrete. This reduces the effects of freezing and thawing.

Aggregates

All concrete contains fine and coarse aggregates. The fine aggregate is normally

FIGURE 6–2 Concrete provides strength and can be cast into any shape. *(Courtesy of the New York State Commerce Department)*

sand. Coarse aggregates are crushed stone and gravel more than 1/4 inch in diameter. For special purposes, lightweight aggregate can be used to develop concrete weighing less than one-fourth the weight of normal concrete. Some common lightweight aggregates are expanded *shale*, *vermiculite*, and *perlite*. These are made from natural rocks that expand, like popcorn, when they are heated. They are not normally used in concrete where weight is not an important factor.

Mixing Concrete

Several variables affect the proportions of materials needed for a good concrete mixture. The maximum size of the coarse aggregate is determined by the size of the concrete member (part). It is generally best to use the largest coarse aggregates that result in a finished product with no **voids** (air pockets). The amount of fine aggregate needed varies with the size of coarse aggregate used, **Figure 6–3**. The aggregate should be carefully measured and thoroughly mixed with the cement before the water is added.

The amount of water used in concrete is very important. If too little water is used, the concrete does not develop to its full strength. If too much water is used, the fine and coarse aggregates separate. Weak concrete is the result. The exact amount of water needed varies depending on the water contained in the aggregate. Only water that is clean enough to drink should be used. Dirt or oil interferes with the reaction between the cement and water.

Slump tests measure the stiffness of concrete mixtures. Freshly mixed concrete is placed in a slump cone. The cone is then carefully emptied onto a flat surface. The number of inches that the concrete settles is the **slump** of that mix, **Figure 6–4**. Water should not be added to increase the slump. A cement mason, construction crew supervisor,

Proportions for Various Mixes of Concrete	Cement (cu. ft.)	Coarse Aggregate (cu. ft.)	Fine Aggregate (cu. ft.)	Gal. Water	
				Wet Sand	Dry Sand
3/4″ max. aggregate	1	2 1/4	2	5	6
1″ max. aggregate	1	3	2 1/4	5	6
1 1/2″ max. aggregate	1	3 1/2	2 1/2	5	6

FIGURE 6–3 Typical concrete mixes

FIGURE 6–4 The number of inches the concrete sags after being released from the cone. *(Courtesy of Portland Cement Association)*

or construction inspector usually conducts a slump test.

Reinforced Concrete

Concrete is a valuable building material because of its permanence and high **compressive strength** (resistance to crushing). However, concrete has low **tensile strength** (resistance to pulling apart). To increase the tensile strength of concrete members, they are reinforced with steel rods or mesh, **Figures 6–5** and **6–6**.

Bars used for concrete reinforcement are usually deformed. This means that they have raised projections on the surface so they will not move in the cured concrete. These bars are commonly referred to as **rebar**. Rebar is available in sizes from #2 to #18. The number size indicates the diameter in eighths of an inch. Thus, a #5 rebar has a diameter of 5/8 inch. Wire mesh is specified by the gauge

FIGURE 6–5 Steel rods are used to reinforce concrete. *(Courtesy of Niagara Mohawk Power Corporation)*

FIGURE 6–6 Wire mesh is used to reinforce light-duty concrete slabs.

diameter of the wire and the spacing of the wires. Mesh made with 10-gauge wire in both directions forming 6-inch squares is designated as 6 × 6-10/10 mesh.

In applications where bending loads are placed on concrete members, part of the member is subjected to tensile forces, **Figure 6–7**. It is important that the steel reinforcement be positioned correctly to withstand that force. Usually the size and placement of reinforcement is specified by an engineer. Ironworkers, or rodsetters, place the reinforcing steel in position in the concrete form.

Masonry

Masonry construction involves joining units of stone, concrete, clay, and similar materials with mortar. Several kinds of stone, including limestone, marble, and granite, are used in masonry construction. All masonry units, except stone, are made at another site and transported to the construction area.

Mortar

Mortar for masonry construction is made up of portland cement, hydrated lime, and sand mixed with water. The proportions of the mortar mix are stated in that order and by volume. For example, 1:1-1/2:4 mortar is made up of 1 part portland cement, 1-1/2 parts lime, and 4 parts sand. On some jobs masonry cement is used instead of portland cement and lime. Masonry cement is a pre-blended mix of portland cement and hydrated lime.

The amount of water varies depending on the condition of the materials being used. Mortar mixed with dry sand requires more water than mortar mixed with wet sand. Also, some masonry units absorb water from the mortar faster than others. The best way to

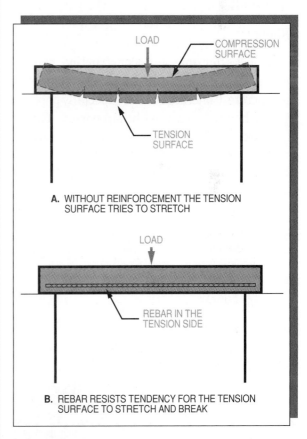

FIGURE 6–7 As a load is applied to the beam, the bottom surface attempts to stretch and the top surface tries to become shorter.

FIGURE 6–8 Concrete blocks are frequently used for building foundations because of their high compressive strength. *(Courtesy of Richard T. Kreh, Sr.)*

determine the proper amount of water is through trial and error. Just enough water should be used so the mortar sticks to the trowel after a vertical snap of the wrist.

Block

Blocks for masonry construction can be made of several materials, but the most common is concrete. Concrete block is especially popular where heavy loads must be supported. This is because of the high compressive strength of

concrete and the ease of handling individual blocks, **Figure 6–8**.

Concrete blocks come in assorted sizes and shapes, **Figure 6–9**. The size of a concrete block is specified by nominal dimensions. Nominal dimensions are not the actual size. They are either smaller or larger depending upon the material. An $8'' \times 8'' \times 16''$ nominal size block, for instance, is actually 3/8 inch smaller in all dimensions. This allows for a 3/8-inch mortar joint.

Different types of block are available to give a pleasing appearance or for special use, **Figure 6–10**. Some blocks are made with lightweight aggregates. These lightweight blocks are used for walls where there is inadequate support from beneath for heavier materials. Scored blocks are also available to give a special appearance.

Brick

Bricks are small rectangular units made of fired (baked in a kiln) inorganic materials such as clay or shale.

There are several kinds of bricks, but most can be classified as either common brick, face

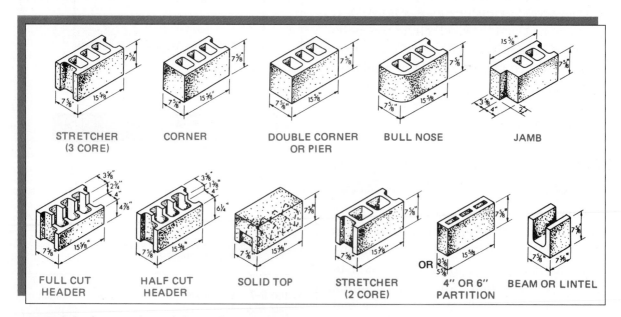

STRETCHER (3 CORE) CORNER DOUBLE CORNER OR PIER BULL NOSE JAMB

FULL CUT HEADER HALF CUT HEADER SOLID TOP STRETCHER (2 CORE) 4" OR 6" PARTITION BEAM OR LINTEL

FIGURE 6–9 Common sizes and shapes of concrete blocks

Donald Chrysler

Occupation:

Mason

How long:

8 years

Typical day on the job:

Donald works from 7:00 AM to 3:30 PM. The contractor plans the use of materials and personnel on the job. The mason's job involves either placing and finishing concrete or laying bricks and concrete blocks. Donald has to do plenty of heavy lifting and physical exertion in his work.

Education or training:

Donald started as a laborer for a swimming pool company. They gave him the chance to become a plasterer and later, a mason. Although most union masons enter the trade through an apprenticeship, Donald learned masonry on the job without a formal apprenticeship.

Previous jobs in construction:

He worked as a laborer, then as a plasterer for twelve years.

Future opportunities:

Donald hopes to become a foreman, supervising three or four other masons and laborers.

Working conditions:

Most work is done outdoors, but when the weather is bad, there is plenty of indoor work available, mostly doing maintenance and remodelling for General Electric.

Best aspects of the job:

There is something different to work with every few days: brick, block, concrete slabs, walks, walls, grouting base plates, etc.

Disadvantages or drawbacks of the occupation:

It's very exhausting to pour concrete in the sun on hot summer days.

brick, fire brick, or special brick. Common bricks are the frequently seen **red bricks** used where special shapes and colors are not required, **Figure 6–11**. Face bricks are similar to **common bricks,** but have a finer surface finish. Fire brick is similar in size and shape to common brick but is used in areas that come in contact with fire or extreme heat. Special bricks include such types as paving bricks and odd shapes. Several brick sizes are shown in **Figure 6–12.**

Masons

Masons are skilled in building with masonry units and mortar and finishing con-

FIGURE 6–11 Bricks are durable and create a pleasing appearance.

crete surfaces. Stone masons build the stone exterior of structures. They work with two types of stone—natural cut (such as marble, granite, and limestone) and artificial stone made from cement, marble chips, or other masonry materials.

Bricklayers build walls, partitions, fireplaces, and other structures with brick, concrete blocks, and other masonry units. They also install fire-brick linings in industrial furnaces.

Cement masons finish concrete surfaces, such as patios and floors. They know which chemical additives speed or slow the setting time. They understand the effects of heat, cold, and wind on the curing process. Cement masons also recognize by sight and touch what is happening in the cement mixture so they can prevent structural defects.

Masons and bricklayers are assisted by tenders or helpers. To name just a few of their duties, these laborers supply the mason with masonry units and other materials, mix mortar, and set up and move scaffolding.

FIGURE 6–10 Several styles and uses of decorative concrete blocks *(Courtesy of National Concrete Masonry Association)*

	SIZES OF MODULAR BRICK							
Unit Designation	Nominal Dimensions, in.			Joint Thickness, in.	Manufactured Dimensions, in.			Modular Coursing, in.
	T	H	L		T	H	L	
Standard Modular	4	2 2/3	8	3/8	3 5/8	2 1/4	7 5/8	3C = 8
				1/2	3 1/2	2 1/4	7 1/2	
Engineer	4	3 1/5	8	3/8	3 5/8	2 13/16	7 5/8	5C = 16
				1/2	3 1/2	2 11/16	7 1/2	
Economy 8 or Jumbo Closure	4	4	8	3/8	3 5/8	3 5/8	7 5/8	1C = 4
				1/2	3 1/2	3 1/2	7 1/2	
Double	4	5 1/3	8	3/8	3 5/8	4 15/16	7 5/8	3C = 16
				1/2	3 1/2	4 13/16	7 1/2	
Roman	4	2	12	3/8	3 5/8	1 5/8	11 5/8	2C = 4
				1/2	3 1/2	2 1/4	11 1/2	
Norman	4	2 2/3	12	3/8	3 5/8	2 1/4	11 5/8	3C = 8
				1/2	3 1/2	2 1/4	11 1/2	
Norwegian	4	3 1/5	12	3/8	3 5/8	2 13/16	11 5/8	5C = 16
				1/2	3 1/2	2 11/16	11 1/2	
Economy 12 or Jumbo Utility	4	4	12	3/8	3 5/8	3 5/8	11 5/8	1C = 4
				1/2	3 1/2	3 1/2	11 1/2	
Triple	4	5 1/3	12	3/8	3 5/8	4 15/16	11 5/8	3C = 16
				1/2	3 1/2	4 13/16	11 1/2	
SCR brick	6	2 2/3	12	3/8	5 5/8	2 1/4	11 5/8	3C = 8
				1/2	5 1/2	2 1/4	11 1/2	
6-in. Norwegian	6	3 1/5	12	3/8	5 5/8	2 13/16	11 5/8	5C = 16
				1/2	5 1/2	2 11/16	11 1/2	
6-in. Jumbo	6	4	12	3/8	5 5/8	3 5/8	11 5/8	1C = 4
				1/2	5 1/2	3 1/2	11 1/2	
8-in. Jumbo	8	4	12	3/8	7 5/8	3 5/8	11 5/8	1C = 4
				1/2	7 1/2	3 1/2	11 1/2	

FIGURE 6–12 Actual and nominal sizes of several types of modular bricks.

ACTIVITIES

Slump Test

A cement mason on a construction site often performs a slump test on a concrete mixture to test its stiffness. Prepare a concrete mixture from the proportions given. Then perform a slump test to determine the mixture's stiffness. If the slump is between one and four inches, use the batch to complete the next activity.

Equipment and Materials

1/4-cubic foot portland cement
1/2-cubic foot clean sand
3/4-cubic foot of 3/4-inch aggregate
Slump cone
Folding rule or tape measure

Procedure

CAUTION: Safety glasses should be worn whenever you work with mortar or concrete. This is to prevent it from splashing in your eyes. Gloves should also be worn to protect against chemical burns which can result from prolonged contact with portland cement products.

1. Thoroughly mix the sand and the coarse aggregate with the portland cement.
2. Mix in five quarts of clean water. If the sand is very dry, it may be necessary to mix in an additional 1/2 quart of water.
3. Fill the slump cone with freshly mixed concrete and *consolidate* (work out the air bubbles) with a rod.
4. Carefully empty the cone onto a flat surface.
5. Measure the vertical distance from a rod laid across the top of the empty cone to the top of the concrete sample, **Figure 6–13**. This is the slump of the batch. If the slump is between 1 and 4 inches, use this batch for the next activity.

FIGURE 6–13 Measure the distance from the straightedge laying on top of the slump cone to the top of the concrete. *(Courtesy of Portland Cement Association)*

Reinforcing Concrete

(This activity requires seven days for the concrete to cure.)

Assume you are an ironworker. Following the procedure outlined, add reinforcement to the forms provided. This reinforced steel will increase the tensile strength of the concrete beam.

Equipment and Materials

2 forms, approximately
   1-1/2″ W × 3″ D × 23″ L, **Figure 6–14**.
Waxed paper or plastic wrap
2 pieces of #2 rebar, 24 inches long
2 chairs to support 2 bars at a height of 3/4 inch
Concrete to fill forms
4-pound hammer

CAUTION: Safety glasses must be worn for this activity.

1. Line forms with waxed paper or plastic wrap to prevent cured concrete from sticking to forms.
2. Place two pieces of rebar in one of the forms. The rebar should be positioned 3/4 inch from the bottom of the form

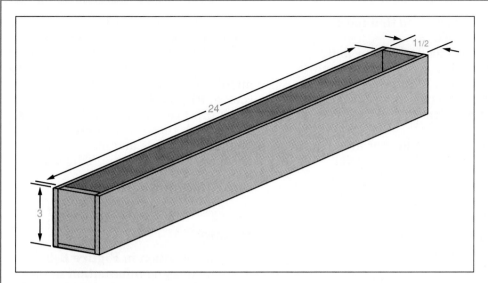

FIGURE 6–14 Form for experimental beam

on chairs made of stiff wire, **Figure 6–15.**

3. Fill both forms with concrete and consolidate. Be careful not to displace the rebar.

4. Write your name in the top of the fresh concrete to help locate the top surface after the concrete is removed from the form.

5. Allow seven days for the concrete to cure.

6. After seven days, position each concrete beam between two supports. The reinforced beam should be positioned with the rebar on the bottom.

7. Strike each beam at the midpoint with a 4-pound hammer. Notice the difference between the beams as they are broken.

Masonry

Bricklayers use mortar and trowel to lay bricks. They must work carefully to ensure all joints are a uniform thickness and each *course* (row) is straight. In this activity you will mix mortar and lay three courses of bricks.

Equipment and Materials

9 common bricks

1/4 cubic foot hydrated lime
1 cubic foot clean sand
2″ × 4″ × 4′ piece of lumber
Brick mason's trowel

CAUTION: Wear safety glasses and gloves when working with mortar. Lime is an irritant.

1. Mix lime and sand thoroughly. When dry ingredients are well blended, mix

FIGURE 6–15 Cross-sectional view of form for experimental beam, with reinforcement in place. Chairs (supports) may be made of stiff wire.

in only enough water so that mortar sticks to the brick.

2. Pick up a trowel load of mortar and snap your wrist to set the mortar and remove the excess, **Figure 6–16**. Spread a 1/2-inch bed of mortar on a 36-inch length of the $2'' \times 4'' \times 4'$ piece of lumber.

3. Place one brick in the mortar at one end. Tap the brick with the trowel handle to obtain a 3/8-inch mortar joint.

4. Butter (spread) mortar onto one end of another brick and place this end against the end of the first brick. Tap this brick down to obtain a 3/8-inch joint.

5. Continue this process until four bricks are laid out in a straight line on the $2'' \times 4'' \times 4'$ lumber with 3/8-inch joints.

6. After each brick is laid, use a slicing motion of the trowel to clean off the excess mortar. This mortar can be reused on the next brick.

7. Repeat the process for a second course. The ends of the bricks in the second course should line up with the middle of the bricks in the first course, **Figure 6–17**.

8. Lay a third course with the end joints in line with those of the first course. Masons use levels and mason's lines to ensure a straight line. This will be covered in a later chapter.

Applying Construction Across the Curriculum

Mathematics

Using the information in **Figure 6–3**, complete the missing information in the following table:

Coarse Aggregate Max Size	3/4″
Cement (cu. ft.)	3
Coarse Aggregate (cu. ft.)	3/4″
Fine Aggregate (cu. ft.)	1″
Gal. Water (Dry Sand)	12

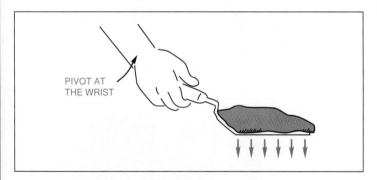

FIGURE 6–16   Mortar on a trowel is set by a vertical snap of the wrist.

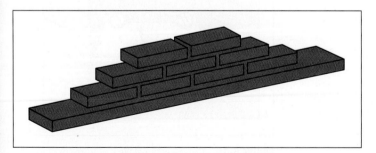

FIGURE 6–17   The joints in each course of bricks should line up with the centers of the bricks in the preceding course.

Mathematics

How many concrete 8″ × 8″ × 16″ concrete blocks would be required to build a masonry wall 8 feet high and 52 feet long? How many blocks would be needed if the wall was 9′-6″ high? How many jumbo bricks measuring 4″ × 4″ × 8″ would be required for the 8 foot high wall?

Science

Obtain samples of as many kinds of concrete aggregate as possible and mount them on a display board. Describe the important properties of each aggregate and where that type of aggregate can be found.

Science

Explain the physical change that takes during freezing and why entraining air in concrete helps prevent damage from freezing.

REVIEW

Multiple Choice.   Select the best answer for each of the following questions.

1. Which of the following is an ingredient in portland cement?
    a. Coarse aggregate
    b. Hydrated lime
    c. Water
    d. Gypsum
2. Which of the following is included in a typical concrete mix?
    a. Alumina
    b. Hydrated lime
    c. Gypsum
    d. Fine aggregate
3. What is the diameter of #4 rebar?
    a. 1/4 inch
    b. 1/2 inch
    c. 1 inch
    d. 1-1/4 inches
4. Which type of portland cement would probably be used where it will be exposed to frequent freezing and thawing?
    a. Air-entrained
    b. High-density
    c. Low-density
    d. Sulphate
5. Where should rebar be placed in a beam?
    a. Near the top
    b. Near the bottom
    c. In the center
    d. Rebar is not used in beams
6. Which of the following terms is associated with concrete?
    a. Clay
    b. Mortar
    c. Course
    d. Aggregate

7. Which of the following is the most important reason for wearing gloves when working with masonry products?

   a. For protection against skin irritation

   b. To keep your hands clean

   c. To make handling rough masonry units more comfortable

   d. Because perspiration slows the curing of mortar

8. Which of the following is a characteristic of concrete and masonry material?

   a. High tensile strenth

   b. High compressive strength

   c. High bending strength

   d. Changes dimensions on contact with moisture

CHAPTER 7

Metal and Wood Products

OBJECTIVES

After completing this chapter, you should be able to:

▼ recognize ferrous and nonferrous metals, and describe several uses of each;

▼ outline the process of converting standing timber into lumber and wood products; and

▼ describe the most important physical properties of steel, aluminum, wood, and wood products.

KEY TERMS

ferrous metal	warp
nonferrous metal	knot
alloy	dimensional stability
ingot	veneer
blooming mill	plywood
ductile	hardboard
galvanized	particleboard
softwood	waferboard
hardwood	
plain sawed	
quarter sawed	
kiln	

Metal and wood products are two important groups of structural materials used in the construction industry. These materials can be manufactured into a variety of shapes and forms with almost limitless properties.

Steel

Metal is either ferrous or nonferrous. Metal containing a large percentage of iron is called **ferrous metal**. Metal with little or no iron is **nonferrous metal**. Ferrous metal is used as a structural building material because of its high strength, **Figure 7–1**. Steel is an **alloy** (mixture) of iron and a small amount (not over 2 percent) of carbon and other materials.

Shaping Steel

When steel is made, it is formed into huge slabs called **ingots**. Ingots of steel are further processed into a variety of shapes, **Figure 7–2**. The shaping process used and the ingredients of the steel determine the properties of the steel.

FIGURE 7–1 Structural steel is frequently used for building frames because of its high strength and comparatively light weight.

FIGURE 7–2 A variety of structural steel shapes will be used to construct this building. *(Courtesy of Nissan Motor Manufacturing Corp., USA)*

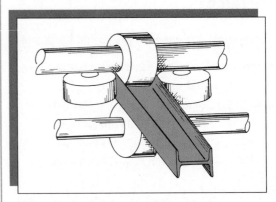

FIGURE 7–3 Structural steel shapes are formed by hot rolling. The space between the rollers is reduced on each successive pass.

Blooming Mill. The steel ingots are heated and rolled in a blooming mill. A **blooming mill** is a set of rollers that rolls the ingots into a large slab called a bloom. Rolling in the blooming mill also tends to compress the coarse grain structure of the cast ingot. This makes the steel tougher.

Hot Rolling. The bloom may be reheated or it may go directly from the blooming mill to the rolling mill. In the rolling mill it is passed through a series of rollers to produce the desired shape, **Figure 7–3**. The red-hot steel is passed through the rollers several times. The size is reduced on each pass. When the desired size and shape is reached, the piece is cooled and cut to length. Because the red-hot steel contracts as it cools, there may be small variations in the size of hot-rolled parts.

Cold Rolling. To produce steel with closely controlled dimensions and a smooth surface, steel may be cold rolled. The steel is cleaned in a chemical solution, then rolled without heat. As it is cold rolled, it increases in strength but becomes less **ductile**. Ductility refers to the ability of any material to withstand bending and forming.

Cold Drawing. Wire is formed by pulling cold steel through openings called die, **Figure 7–4**. The wire is reduced in size each time it passes through one of a series of dies. This produces high-strength wire of an accurate size.

Pipe. Pipe is classified as seamless or welded. Seamless pipe is formed by forcing a steel rod over a pointed device called a mandrel. A set of rollers around the outside controls the diameter of the pipe. Welded pipe is formed by rolling flat strips into a round shape, then welding the seam.

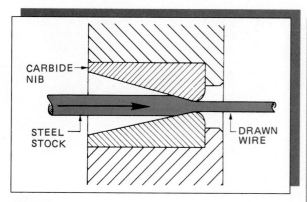

FIGURE 7–4 Wire is formed by pulling steel through a drawing die.

Types of Steel

The carbon content of steel determines its hardness and strength. Steel with high carbon content is generally stronger and more brittle. The classifications of steel according to carbon content are shown in **Figure 7–5**.

By alloying other elements with steel, different properties can be developed. Manganese, nickel, copper, chromium, and vanadium are among the most common alloying elements. For example, by alloying chromium with the steel, stainless steel is produced. Stainless steel is very strong. Because of the oxide film that forms on its surface, it does not stain or corrode. Mettalurgists coordinate and test the quality of steel during its manufacture.

Light-gauge steel (steel that has been rolled into thin sheets) is easy to form and has reasonably good strength characteristics. This makes it an excellent alternative to wood framing, where only moderate strength is required and factory-cut sizes are desirable. To prevent rusting, which would result from normal exposure to atmospheric moisture, light-gauge metal framing is **galvanized** (coated with an alloy of tin and zinc). Light-gauge metal framing is especially common in small commercial buildings and for non load bearing sections in larger buildings, **Figure 7–6**.

Aluminum

Aluminum is the most widely used nonferrous metal. It is usually quite easy to tell the difference between aluminum and ferrous metals. Aluminum is very light in color, almost white. It is much lighter weight than ferrous metals. With the exception of stainless steel, which is more silver in color, ferrous metals rust. Aluminum does not rust. Under certain conditions, it does develop a white, powdery substance on its surface.

The final step in the aluminum-making process is to shape the aluminum into a useful product. Aluminum can be rolled into structural shapes or sheets like steel, **Figure 7–7**. It can be drawn into wire or formed into pipe and tubing. Many aluminum products are extruded.

TYPE	CHARACTERISTICS
Very Mild	0.05 to 0.15 percent carbon. Soft, tough steel used for sheets, wire, and rivets
Mild Structural	0.15 to 0.25 percent carbon. Ductile, machinable steel used for buildings and bridges
Medium	0.25 to 0.35 percent carbon. Stronger and harder than mild structural, used for machinery and construction.
Medium-hard	0.35 to 0.65 percent carbon steel used where it is subject to wear and abrasion.
Spring	0.85 to 1.05 percent carbon steel used in springs
Tool	1.05 to 1.20 percent carbon. Very hard and strong steel used for making cutting tools.

FIGURE 7–5 The classifications of carbon steel

Gateway Arch...

Jefferson National Expansion Memorial, on the old St. Louis riverfront, is a national monument. It celebrates the westward expansion of the United States. The most famous part of the memorial is the Gateway Arch.

The stainless-steel faced arch spans 630 feet between the outer faces of its triangular legs at ground level. Its top soars 630 feet into the sky. It takes the shape of an inverted catenary curve—a shape such as would be formed by a heavy chain hanging freely between two supports.

Each leg is an equilateral triangle. The sides are each 54 feet long at ground level, tapering to 17 feet at the top. The legs have double walls of steel 3 feet apart at ground level and 7-3/4 inches apart above the 400-foot level. Up to the 300-foot mark, the space between the walls is filled with reinforced concrete. Beyond that point, steel stiffeners are used.

The double-walled, triangular sections were placed one on top of another. They were then welded inside and out to build up the legs of the arch. Sections ranged in depth from 12 feet at the base to 8 feet for the two keystone sections. The complex engineering design and construction is completely hidden from view. All that can be seen is its sparkling stainless steel outside skin and inner skin of carbon steel. They combine to carry the gravity and wind loads to the ground. The arch has no real structural skeleton. Its inner and outer steel skins give it its strength and performance.

Reinforced concrete foundations, sunk 60 feet into the ground, extend 30 feet into bedrock. These contribute greatly to the structural strength of the arch. The structural engineers of the arch report that under a wind load of 55 pounds per square foot (equivalent to a 150-mph wind), the arch will deflect at the top only 18 inches from side to side.

Entrance to the arch is from an underground museum. The museum is rectangular in shape, and is located directly beneath the arch. Visitors are carried from the museum level below to the observation platform at the top of the arch by a 40-passenger train made up of eight five-passenger capsules in each leg. Operating at the rate of 340 feet per minute, the ride takes 7 minutes for the round trip. The observation platform is 65 feet by 7 feet. Plate-glass windows provide views in the east and west directions. There is also a conventional passenger elevator in each leg as far as the 372-foot level. Stairways with 1,076 steps in each leg rise from the base to the top of the arch. The elevators and stairways are for maintenance and emergency use only.

To prepare the site for the arch foundations and the museum, the general contractor for the project had to excavate 300,000 cubic feet of earth and rock. Alloy-steel tensioning bars or tendons (252 for each leg) extend 34 feet below the top of the foundations. These tendons anchor the structure securely to its base. At ground level, only the two outside corners of each triangular base are pre-stressed—by two groups of 63 steel bars in each corner.

In cross section, each arch leg is a double-walled equilateral triangle. Each leg has a hollow core 48 feet wide at the base, tapering to 15-1/2 feet at the top. The inner skin is of carbon steel, 3/8 inch thick except at the corners where it is 1-3/4 inch thick to provide greater stiffness. The outside surface was fabricated from 900 tons of polished stainless steel in panels 1/4 inch thick. The panels vary in size from 6 × 18 feet to 6 × 5-1/2 feet. The outer and inner walls were fabricated in sections and bolted together at the Pittsburgh and Warren, Pennsylvania, plants of the Pittsburgh–Des

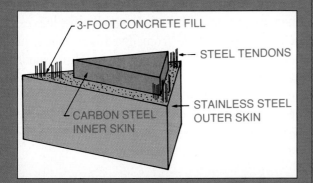

Moines Steel Company, steel fabricators and erectors for the arch.

The completed arch sections were shipped to St. Louis on railroad cars. Two "sandwich" wall sections rode side by side. Their stainless sides faced each other but were held apart by steel uprights covered with wood and neoprene. Steel rods welded to the carbon-steel plates and to the steel sides of the cars secured each section.

A crane lifted a completed section from the welding pad onto a specially designed railroad car. The car had a 42 × 52-foot deck made up of 24 steel beams. The car was equipped with an outrigger that rode on rubber-tired wheels and supported one corner of the triangular arch section. A tractor pushed the railroad car along special tracks to a position near the arch leg.

Specially designed creeper cranes climbing the arch legs lifted a completed section into position 4 inches above the previously placed section. The cranes then set the section on three 35-ton screw jacks. The jacks positioned the section accurately, leaving a gap between the sections just wide enough for the welds. The first section on each leg was attached to the foundation by anchor bolts 5/8 inch and 1 inch in diameter. ■

Courtesy of the Jefferson National Expansion Memorial NHS/National Park Service)

FIGURE 7–6 Light-gauge metal framing is popular because it is inexpensive and can be erected quickly.

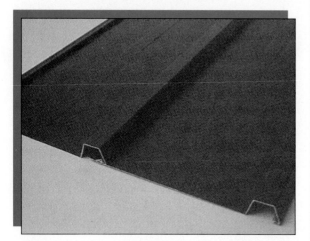

FIGURE 7–7 Aluminum roofing panels and siding are formed from sheets. *(Courtesy of Reynolds Metals Company)*

In extrusion, the hot, soft aluminum is forced through a die to form a long continuous shape, **Figure 7–8**. Extruding is somewhat like squeezing toothpaste from a tube.

Wood Products

Wood has long been a major building material. Cave dwellers probably used wood in their earliest construction efforts. Sawmills have been in operation for nearly four hundred years. Forest products are especially important to modern industry. The earth's forests are our only renewable source of construction material.

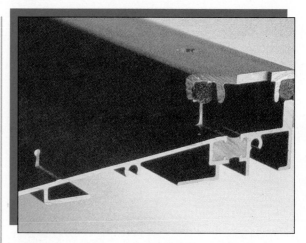

FIGURE 7–8 This aluminum threshold is extruded. *(Courtesy of Morgan Products Ltd.)*

Modern industry uses dozens of products from trees, but only major structural materials will be included here.

Classifications of Wood

Wood is classified as either hardwood or softwood. However, these classifications do not indicate the actual hardness of the wood. Many softwoods are as hard as some hardwoods. **Softwoods** come from coniferous (cone-bearing) trees, or those with needle-like leaves. **Hardwoods** come from deciduous trees, or those trees that lose their leaves. Hardwoods are used mainly for furniture, cabinets, paneling, interior trim, and specialty items, **Figure 7–9**. Softwoods are used for general construction and structural purposes, **Figure 7–10**.

Lumbering

Wood goes through many steps involving a great number of occupations before it is ready for use at the construction site, **Figure 7–11**. Trees are selected for cutting by foresters. Marked trees are then felled, cut up by loggers, and dragged out of the forest. The logs are stacked at a central point. They are then loaded into trucks and delivered to the mill. At the mill, the logs are cut into *cants* (large square

FIGURE 7–9 Hardwood is used for (A) floors, (B) stair railings, and (C) cabinets. *((A) Photo by Randall Perry, (C) Courtesy of Wood-Mode, Inc.)*

(A)

(B)

(C)

SPECIES	PRINCIPAL USES
Douglas Fir	Timbers, pilings, and plywood
Engelman Spruce	Wide range of uses from rough construction to interior finish
Incense Cedar	Closet lining, poles, and siding
Lodgepole Pine	Timbers, poles, and railroad ties. Some general-purpose lumber
Ponderosa Pine	General lumber for residential construction, trim, and furniture
Redwood	Posts, fences, siding, and bridge timbers
Western Hemlock	General construction lumber
Western Larch	Heavy construction lumber
White Pine	Trim, millwork, paneling, and furniture

FIGURE 7–10   Common softwoods in construction

timbers). The cants are sawed into lumber by the head sawyer or operator of the saw mill. The amount of lumber obtained from the logs depends on the sawyer's skills and knowledge.

Most softwoods for general construction are **plain-sawed** or flat-sawed, meaning the saw cut is parellel to one side of the log, **Figure 7–12**. Other pieces are **quarter sawed**, so the saw cut is perpendicular to the growth rings of the tree. Plain-sawed lumber generally displays a more attractive grain and is less expensive to produce than quarter sawed, **Figure 7–13**. However, it tends to warp and show other defects. Quarter-sawed lumber is less apt to warp.

Moisture Content of Wood

Wood is made up of long tubular cells that contain a large amount of water in a living tree. Much drying is required to produce usable lumber. The moisture content of wood is specified as a percentage of its oven-dry weight. For example, wood at 15 percent moisture content weighs 15 percent more than wood that has been thoroughly dried in an oven.

Lumber can be air dried by stacking it in covered piles for several months. Kiln drying is a more common method for modern lumber production. With this method, lumber is stacked in large oven-like rooms, called **kilns**. In the kilns, the lumber is exposed to warm air and controlled humidity, **Figure 7–14**. Kiln drying requires only a few days and results in better quality lumber.

Grading

Most lumber contains some defects. As lumber dries it has a tendency to **warp**, that is, become distorted in shape, **Figure 7–15**. **Knots** occur where branches grow from the trunk of a tree. Knots are classified according to their size and whether they are firm or loose in the lumber. Stains may be caused by problems in drying. Decay may be caused by fungi or insects. All of these defects affect the quality of the lumber.

Softwood lumber, for most construction purposes, is graded according to its strength. In other words, the highest grade lumber should be the strongest. The appearance of framing lumber is not a consideration in grading. Stronger boards are, however, usually more free of defects than lower grade boards. The strength of lumber is affected by defects such

A. Forest timber is considered by many to be our only renewable resource.

B. Timber fellers use chainsaws to cut down (fell) trees.

C. The limbs are removed and the felled timber is moved into staging areas.

D. Logs are transported to the mill on trucks.

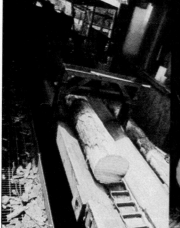

E. At the mill much of the work is taken over by machines.

F. However, skilled workers, such as this sawyer, are still required to control the machines.

G. Grading lumber is another step that requires a great amount of skill and knowledge.

FIGURE 7–11　From forest to finished lumber *(Courtesy of Weyerhaeuser Co.)*

Jeff Shockley

Occupation:
Building Materials Salesperson, The Adkins Company

How long:
9 Years

Typical day on the job:
Jeff is a senior salesperson in the Adkins Company, so he is often called upon to help other salespersons or to cover for them, when they are not at work. The work involves answering questions for contractors and homeowners about new materials or how to solve a particular construction problem, writing purchase orders, giving information over the phone, and working with customers to determine what their needs are and what products will satisfy their needs. It is important to know what new materials and equipment are available and how to use those materials properly.

Education or training:
Jeff has attended several seminars conducted by manufacturers of building materials to teach salespeople and build-

ers how to use their products. These have been especially helpful in Jeff's training.

Previous jobs in construction:
Jeff has worked as a truck driver, warehouse worker, and counterman, and is now working half-time in the office at the Adkins Company.

Future opportunities:
Jeff plans to continue working toward more senior positions within Adkins.

Working conditions:
Jeff does a combination of office and counter work in a building materials sales company. Most of his work is carried out inside a well-lighted, air-conditioned building.

Best aspects of the job:
Jeff enjoys meeting people. You learn something new every day, either from customers or suppliers. Every job provides an opportunity to see how someone else solved a construction problem.

Disadvantages or drawbacks of the occupation:
Jeff is very happy with his occupation and cannot find any real drawbacks to mention.

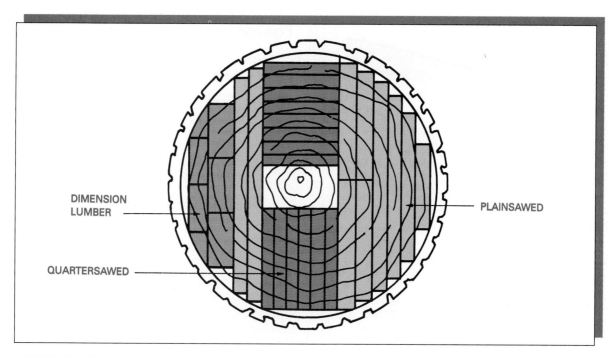

DIMENSION LUMBER

QUARTERSAWED

PLAINSAWED

FIGURE 7–12 Ways of sawing boards from a log

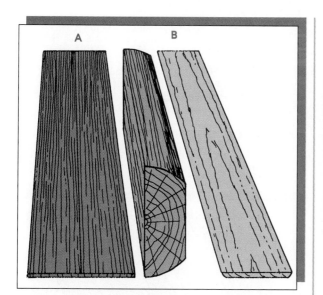

FIGURE 7–13 Grain depends on how lumber is cut from the log. Board A is quarter-sawed. Board B is plain-sawed.

as decay, knots, and cross grain. It is also affected by such physical characteristics as density and moisture content. Since some of these defects and characteristics do not cause a visible difference in the appearance of the wood, some lumber may be relatively clear but still be in a low grade or strength category.

FIGURE 7–14 Lumber on its way to steam heated drying kilns *(Courtesy of Weyerhaeuser Co.)*

Surfacing and Milling

Boards and dimension lumber for general use are planed to the proper thickness and width after sawing. This reduces the size to slightly smaller than the nominal dimensions by which it is identified, **Figure 7–16**. Some lumber is further milled to produce molding, siding, and edge-matched boards **Figure 7–17**.

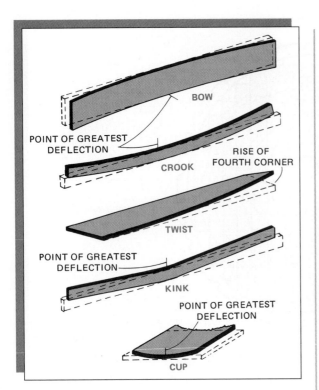

FIGURE 7–15   Various types of warp in lumber

Properties of Wood

Wood has a few characteristics that should be understood to use it to its greatest advantage. Wood has a high ratio of strength to weight in compression, **Figure 7–18**. It has relatively little strength in tension. It also has good transverse strength (resistance to forces applied to its side), but it is better in one direction than the other, **Figure 7–19**.

The ability of wood to retain its size with changes in temperature is useful in some applications. However, wood lacks **dimensional stability** with changes in moisture content. As it loses moisture it shrinks considerably in the direction of the annular rings (growth rings). It shrinks slightly less across the annular rings and very little in length. These variations in amount of shrinkage are responsible for a major part of the warpage that occurs, **Figure 7–20**. Plain-sawed lumber is affected most by warpage.

Plywood

Thin sheets of wood, called **veneer**, can be glued together at right angles to one another, to form panels. These panels have greater

Thickness		Width	
Nominal	**Dressed**	**Nominal**	**Dressed**
1	3/4	2	1 1/2
1 1/4	1	3	2 1/2
1 1/2	1 1/4	4	3 1/2
2	1 1/2	5	4 1/2
2 1/2	2	6	5 1/2
3	2 1/2	7	6 1/2
3 1/2	3	8	7 1/4
4	3 1/2	9	8 1/4
		10	9 1/4
		11	10 1/4
		12	11 1/4

Note: With nominal sizes, the inch measurement is often assumed. For instance, a 2" × 4" piece of lumber is simply called a 2 × 4 (two-by-four).

FIGURE 7–16   Nominal and minimum-dressed sizes of lumber

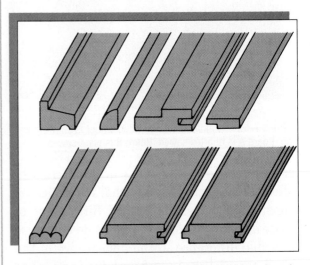

FIGURE 7–17   Some lumber is milled into plain boards and matched boards.

FIGURE 7–18 Wood is used as the framing material in many buildings because of its high compressive strength and light weight.

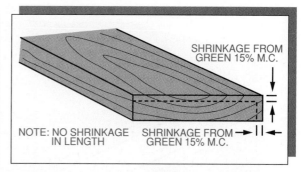

FIGURE 7–20 Plain-sawed boards shrink most in width, less in thickness, and almost none in length as they dry.

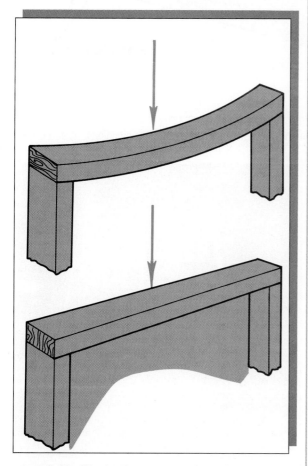

FIGURE 7–19 Wood resists transverse forces better when applied parallel to the annular rings than when applied perpendicular to the annular rings.

dimensional stability and improved strength. The material assembled in this way is **plywood**, **Figure 7–21**. Plywood has several advantages over solid lumber. It resists splitting and is dimensionally stable. Fewer pieces are required to cover a given area, and it does not warp as easily.

Veneer for plywood is either rotary cut on a veneer lathe, plain sliced, or quarter sliced. The method of cutting determines the appearance. Most plywood for general construction is made of rotary-cut softwood. This method is most economical. Plain-sliced and quarter-sliced hardwood, plywood, and premium softwood plywood are used in furniture, cabinets, and paneling. After it is sliced from the log, a veneer clipper cuts the sheets to size. A plywood patcher replaces defective areas with thin plywood patches called plugs.

Plywood is available in several types and grades for a variety of uses. It can be bonded with waterproof or nonwaterproof glue. When plywood will be exposed to moisture, such as on building exteriors or concrete forms, it must be made with waterproof glue, **Figure 7–22**. It is further graded according to the quality of its face and core veneers. Special types of plywood are also available. The American Plywood Association sets standards for most softwood plywood in construction use, **Figure 7–23**.

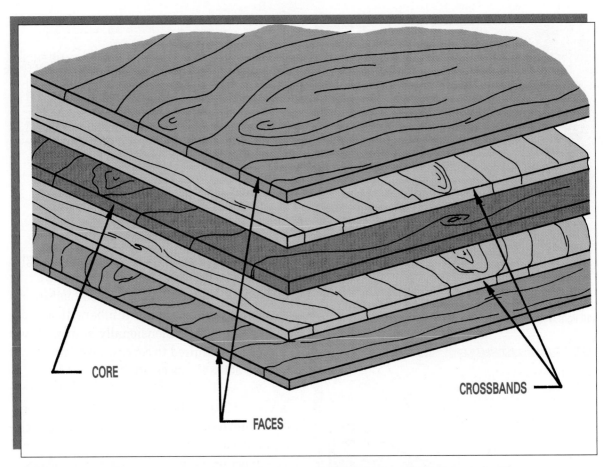

CORE

FACES

CROSSBANDS

FIGURE 7–21 Plywood consists of layers of veneer glued together with the grain running in alternate directions.

Hardboard

Hardboard is made by first exploding wood chips with high-pressure steam. The chips are loaded into a container which is then pressurized with steam. When the steam pressure is suddenly released, the chips explode into individual wood fibers. They are then pressed into thin (1/8- to 3/8-inch) sheets. Hardboard made in this manner is called standard hardboard. It is widely used for interior purposes. Tempered hardboard is treated by heat and chemicals to produce a denser, more durable surface that can be used in exterior applications, **Figure 7–24.** Hardboard is inexpensive, finishes well, and is less apt to warp than solid wood. It makes good use of wood chips and scraps from lumber and plywood production.

FIGURE 7–22 Plywood for concrete formwork is made with waterproof glue.

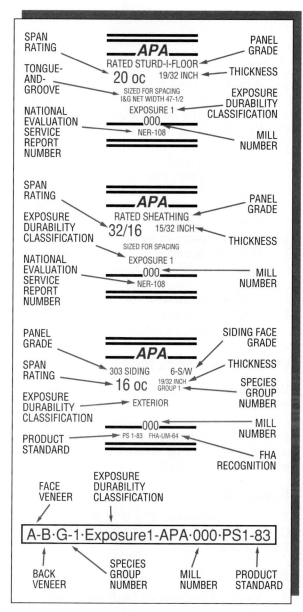

FIGURE 7–23 Typical APA trademarks, indicating the performance ratings of plywood *(Courtesy of American Plywood Association)*

Particleboard

Particleboard is another product made from wood chips. Small chips of wood are coated with synthetic resin glue and pressed into sheets under heat and pressure. Particleboard is manufactured in thicknesses ranging from 1/4 inch to 1-1/2 inches. Particleboard has a smooth surface, uniform thickness, and is dimensionally stable. These

qualities make it an excellent backing for surface materials like the plastic laminate used on countertops.

Waferboard

Waferboard is another popular manufactured wood-product sheet material. **Waferboard** is made by gluing relatively large, but thin "wafers" of wood scraps under low to medium pressure to make sheets of inexpensive building material. Waferboard is made with non-waterproof glue, so it is not good for exterior use, except where it can be protected from the weather soon after it is used. For example, it is often used as an inexpensive wall sheathing material, where weather-tight siding will be applied immediately. Waferboard has slightly better tensile strength than most particleboard products, because of the larger wafers of natural wood, but it is not nearly as strong as plywood.

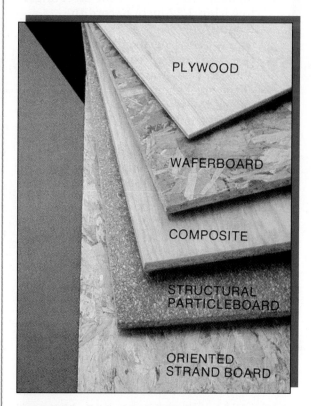

FIGURE 7–24 Forms of manufactured wood-product sheet materials *Courtesy of American Plywood Association)*

ACTIVITIES

Properties of Metals

Steel and aluminum have very different properties. In this activity you will see some of these differences.

Equipment and Materials

1 steel angle approximately 1″ × 1″ × 1/8″ × 12″ long
1 aluminum angle the same size as the steel angle
1 wax crayon
Visegrip plier
Propane or acetylene torch
Steel rule for accurate measurement at 12″
Leather gloves
2″ × 2″ pieces of thin steel and aluminum
Bucket of water

Procedure

1. Measure the length of the steel angle to within 1/32 inch. Record the length.
2. Clamp the visegrip on the midpoint of the steel angle. Fasten the visegrip in a bench vise. Approximately 6 inches of the steel should protrude on each side of the visegrip.
3. Place a small piece of crayon on one end of the steel.
4. Light the torch according to the ap proved procedure. Heat the steel angle at the end opposite the crayon.

CAUTION: Wear leather gloves when working with hot metal. Do not place the hot metal on a bench or leave it unattended where someone else might pick it up, not knowing that it is hot. Have a fire extinguisher handy at all times when using a torch.

5. Measure the amount of time between first applying the heat and the crayon melting. Record that time.

6. While the steel is still hot, measure the length to within 1/32 inch. How much did the steel expand?
7. Cool the steel in the bucket of water.
8. Repeat steps 1 through 7, using the aluminum angle.
9. Hold the sheet steel by one corner with the visegrip and heat it until it begins to get reddish orange. Record the time this took.
10. Cool the piece in water.
11. Heat the aluminum sheet in the same way, for the same length of time.
12. Cool the piece in water.
13. Fold both pieces in half. What difference is there in properties?

Effects of Moisture on Wood

Wood lacks dimensional stability with changes in moisture content. The shrinkage that occurs is responsible for a major part of the warpage in wood. In this activity, you will demonstrate the effects of moisture on wood.

Equipment

2 pieces of softwood, 1″ × 6″ × 6″
Container of water
Drying oven

Procedure

1. Write your name on both pieces of wood.
2. Compare the pieces to be sure they are of equal size in all directions.
3. Place on piece in the drying oven at a low temperature.
4. Soak the other piece in water.
5. After one day remove both pieces and compare their sizes again. Notice the direction in which the greater change took place.

Moisture Content of Wood

Dry kiln operators season wood by drying it in a kiln or oven. This removes the moisture from the wood so it will not shrink or warp. The moisture content can be measured with electronic instruments or by weighing.

Equipment

Drying oven
Balance or scientific scales
Small wood specimen

Procedure

1. Shave a handful of shavings from a piece of wood.
2. Weigh these and record their exact weight.
3. Place the shavings in a flat container in the oven at a temperature of 200°F for 30 minutes.
4. Weigh the dried wood shavings and record their weight.
5. The weight lost during drying is due to the evaporation of moisture contained in the wood cells. The moisture content is found by subtracting the oven-dried weight from the weight when cut, dividing by the oven-dried weight, then multiplying by 100.

Applying Construction Across the Curriculum

Science

What is the chemical reaction that is taking place when iron rusts? Why doesn't aluminum rust?

Science

When ingots of iron are rolled into structural steel shapes or sheet metal, do they undergo a physical change or a chemical change? What type of change occurs when iron is melted and cast into a mold?

Science and Communications

Make a display with several species of wood, indicating which are hardwoods and which are softwoods. Explain what commercial uses there are for each species.

Science and Communications

Use model making materials, such as straws, paper, and glue, to make a large-scale model showing the cellular structure of wood. Write a brief report explaining where water is found in the cellular structure of wood and what changes take place as the wood dries.

REVIEW

A. Multiple Choice. Select the best answer for each of the following questions.

1. Which of the following is *not* an ingredient of steel?
   a. Iron
   b. Carbon
   c. Slag
   d. Alloys
2. In which direction does plain-sawed lumber expand most as it gains moisture?
   a. Length
   b. Width
   c. Thickness
   d. About the same in all directions
3. Which of the following is most apt to warp?
   a. Plain-sawed lumber
   b. Particleboard
   c. Hardboard
   d. Plywood

4. Which of the following changes dimensions most with changes in temperature?
   a. Plain-sawed lumber
   b. Quarter sawed lumber
   c. Hardboard
   d. Steel
5. Which of the following changes dimensions most with changes in moisture content?
   a. Plain-sawed lumber
   b. Steel
   c. Hardboard
   d. Plywood
6. Which of the following is an important property of aluminum?
   a. Rusts quickly
   b. Does not have enough strength for structural use
   c. Is very lightweight
   d. Does not conduct electricity
7. Which of the following might be made of aluminum?
   a. Pipe
   b. Structural shapes
   c. Sheet
   d. All of these

B. Matching. Match the material in Column II with the correct term in Column I.

Column I	Column II
1. metallurgist	a. lumber
2. nonferrous	b. steel
3. drawing die	c. plywood
4. veneer lathe	d. aluminum
5. white to silver color	
6. kiln	
7. cants	
8. plugging	
9. quarter saw	
10. ferrous	

CHAPTER 8

Gypsum, Glass, and Plastics

OBJECTIVES

After completing this chapter, you should be able to:

▼ describe a variety of construction finishing materials;

▼ list the important properties of construction finishing materials; and

▼ discuss common applications of each of the materials covered.

KEY TERMS

gypsum

plaster

wallboard

lath

tempered glass

wired glass

insulating glass

thermoset

thermoplastic

foamed plastics

asphalt

caulk

adhesion

cohesion

mastic

As previously stated, almost all materials are used in some way in the construction industry. Chapters 6 and 7 discussed the most common structural materials. This chapter discusses a few nonstructural materials. Other materials are discussed in the chapters dealing with the various stages and types of construction.

Gypsum Products

Gypsum is a natural rock taken from mines throughout the world. To make a usable product, the gypsum rock is crushed, then heated in kilns. The heating drives off the water that was combined with the rock. This leaves a fine powder called plaster of paris. When plaster of paris is mixed with water, it returns to its former rock state. Plaster of paris is used for patching cracks and small areas on plastered walls.

Plaster

Gypsum **plaster** is widely used for finishing interior walls, **Figure 8–1**. It produces a durable surface and resists the spread of fire. Gypsum wall plaster is made up of plaster of paris and

FIGURE 8–1 This plasterer is finishing an interior wall with gypsum plaster. *(Courtesy of United States Gypsum Co.)*

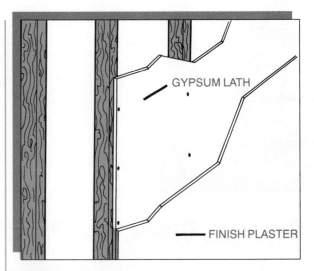

FIGURE 8–2 Gypsum lath provides a base for finished plaster.

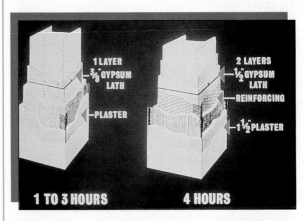

FIGURE 8–3 Gypsum lath and plaster are used to protect steel from fire. *(Courtesy of Gold Bond Building Products)*

aggregates such as sand or perlite. Finish plaster, which is used for a final coat on plastered walls, is made with plaster and lime.

Gypsum Board Products

Two common board products made with gypsum are **wallboard** and **lath**. Gypsum wallboard is a fireproof wall covering. It is made up of a gypsum plaster core, 3/8 inch to 1 inch thick, with a strong paper covering. The sheets are made in a standard width of 4 feet and in lengths from 8 feet to 16 feet. Gypsum wallboard is nailed directly to the wall frame by a lather. The lather then covers the joints with a special compound.

Gypsum lath is very similar to wallboard. It is made in sheets 16 inches wide by 48 inches long. The pieces of lath are fastened to the wall framing to provide a base over which plaster is applied, **Figure 8–2**.

Gypsum lath and plaster are used to encase structural steel and protect it from fire, **Figure 8–3**. Other shapes are used for special purposes, such as using long planks for roof decking.

Glass

Glass has been used for windows for hundreds of years. Although the equipment and some of the additives have improved, the basic ingredients are still the same.

Glass can be made with special properties. **Tempered glass** is reheated, then cooled very rapidly. This sets up stresses in the glass and produces glass that is three to five times stronger than regular glass. It is a good safety glass.

Reflective glass is made by mixing flakes of reflective metal into the glass as it is made. Reflective glass is used in buildings to reflect unwanted heat and light. It can also be used to create attractive architectural effects, **Figure**

FIGURE 8–4 Reflective glass is attractive and reduces the cooling load of buildings. *(Courtesy of PPG Industries)*

FIGURE 8–5 Wired glass is used for security purposes

8–4. Wired glass is made with high-strength steel wire embedded in it. Wired glass is used to prevent breakage and for security purposes, **Figure 8–5**.

Insulating glass is made of two sheets of glass sealed around their edges. The air space between the two sheets acts as a barrier to heat, cold, and sound, **Figure 8–6**.

Plastics

Plastics includes a large group of synthetic materials that can be formed into desired shapes at some point during their processing. Plastics can be made with nearly any property desired. Some are light enough to float on water. Some are stronger than steel. Some will not burn, and others are as transparent as

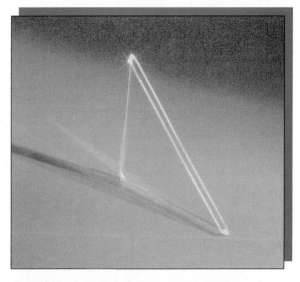

FIGURE 8–6 Insulating glass consists of two pieces of glass with an air-tight space between them. *(Courtesy of PPG Industries)*

glass. Because of the limitless properties and shapes available, plastics are an important construction material. Many paints, fabrics, and caulking materials are made with plastics.

Thermosetting and Thermoplastic Materials

There are two categories of plastics, depending on how they react to heat. Thermosetting plastic is formed by heat and pressure during its manufacture. Once **thermosetting** materials have been formed and cured, they cannot be reshaped. This can be compared to frying an egg. The raw egg can be molded into any shape, but once it is cooked, further heating will only burn it. **Thermoplastics** can be softened by heating. These materials can be reheated and shaped repeatedly. Thermoplastics can be compared to ice cubes. They can be melted and frozen over and over.

Sheets

Thermoplastic sheets are made by calendering. Calendering is the process of heating the raw material, then rolling it between polished rolls.

Transparent plastic sheets are used for unbreakable glazing in place of glass in windows and doors, **Figure 8–7**. Acrylic sheets (Plexiglass™ and Lucite™) have up to sixteen times the impact strength of glass. Polycarbonate (Lexan™ and Merlon™) has four times the impact strength of acrylics. However, plastic glazing materials are more expensive than glass, and they scratch easily.

Extrusions

In extrusion, a soft material, such as hot thermoplastics, is forced through a die to produce a continuous length of material. The extruded material is cooled and cut to length. Any thermoplastic can be extruded. Polyvinyl chloride (PVC) and acrylonitrile butadiene styrene (ABS) are extruded to make plastic

FIGURE 8–7 Plastic glazing is used to prevent safety hazards. *(Courtesy of Rohm & Hass Co.)*

FIGURE 8–8 PVC plastic pipe *(Courtesy of Larry Jeffus)*

pipe, **Figure 8–8**. Plastic pipe is inexpensive, noncorrosive, and lightweight. Vinyl is also extruded to make plastic moldings and trim.

Another important material in construction is polyethylene film. Polyethylene, a lightweight and translucent thermoplastic, is extruded into sheets 2 to 6 mil (.002″ to .006″) thick. Polyethylene film is widely used as a vapor barrier, **Figure 8–9**. Where the film will be exposed to direct sunlight, it may be colored black to filter out damaging ultraviolet rays.

Expanded Plastics

Many plastics can be **foamed** by trapping

Paula DeVaney

Occupation:
Owner/operator Seal Co, sealing and caulking company

How long:
7 years

Typical day on the job:
Paula started Seal Co as a service to do caulking and sealing of concrete. A big part of the business is sealing concrete bridge decks and road surfaces. Much of the caulking work is on commercial buildings. She started the company with only $75 of her own money. Today, seven years later, she is doing over $300,000 a year in business and has 2 employees.

The day consists of two parts: office and field. The office part of the job is answering phone calls from clients and prospective clients, lining up work and juggling the schedule. In the field, she does the sealing and caulking. Seal Co is a very small company, so it is not unusual for the owner to be involved in the actual field work.

Education or training:
Paula did not originally plan a career in construction, so she did not take construction courses in high school or college. She did, however, take a drafting course, which she recommends to anyone considering a career in construction. Her math, business, and marketing courses have also helped her develop this business.

Previous jobs in construction:
She worked summers in school as a bookkeeper for a home builder, but she has not had any experience directly related to the sealing and caulking business.

Future opportunities:
As long as the construction industry remains strong, there will be plenty of work in this field. Paula hopes to expand her business. There is no formal training for this business, so she looks for young people who are anxious to learn, then she will help them learn the business.

Working conditions:
Part of the time is spent in a comfortable office, but some is spent on construction sites. This may require working on a ladder in bitter cold or on the pavement in the heat of summer.

Best aspects of the job:
Most construction jobs are varied from one day to the next. It's never boring. Construction is a very rewarding career, because you can see what you have built when the job is done.

Disadvantages or drawbacks of the occupation:
At first she had to overcome the stereotype that women are not as capable as men in the construction field. There were some contractors who had not dealt with a women in this type of position and she had to prove herself to them. That is rarely an issue anymore.

FIGURE 8–9 Polyethylene is used to prevent moisture from passing through the floor.

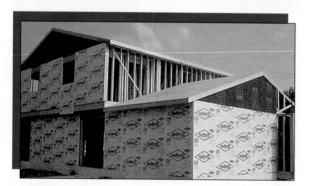

FIGURE 8–10 Foamed plastic board is sometimes used for sheating. *(Courtesy of Dow Chemical Company)*

FIGURE 8–11 Fiberglass reinforced plastic is a popular choice for bathtubs and shower enclosures. *(Courtesy of Larry Jeffus)*

air or gas bubbles in the cured plastic. This results in a very lightweight material with excellent heat insulating properties. Polyethylene, polystyrene, and polyurethane are the plastics most often foamed. When plastic is foamed and trimmed into slabs, it is called *foamed boards*, **Figure 8–10**. When it is sprayed onto a surface and foamed where it is to be used, it is said to be *foamed in place*. Foamed-in-place polyurethane is used for insulation.

Reinforced Plastics

Plastics can be reinforced with a variety of materials. A material with the desired properties is simply mixed with the raw plastic *resin* (uncured plastic). Fiberglass is one of the most familiar reinforced plastics. Glass fibers with high tensile strength are mixed with polyester, epoxy, or acrylic. Fiberglass-reinforced plastics have high strength and are relatively lightweight, **Figure 8–11**. Because reinforced plastics are made of two or more materials, they are called *composites*.

Sealants

In construction, a *sealant* is any material applied to small openings or flat surfaces to keep water out. Although there are a great number of such materials available, most construction sealants fit into one of the following groups.

Asphalt Coatings

Asphalt is a petroleum product used to make construction sealants. This black substance is either thinned with a solvent or emulsified in water to make it liquid. Asphalt sealant is applied to foundations to prevent ground water from seeping into the basement, **Figure 8–12**. It is also applied on roofs. In a liquid form, roofers use it to apply roof-covering materials. In a thick, semi-paste form, it is used to seal around chimneys, pipes, and other openings in the roof.

FIGURE 8–12 Asphalt foundation coating *(Courtesy of Cal Parlman, Inc.)*

FIGURE 8–13 Joints between construction members are sealed with caulking compound. *(Courtesy of Dow Corning Corporation)*

Liquid Sealants

Concrete and masonry materials are porous (have tiny openings). They can be improved by sealing their surfaces. On bridges, runways, and roads, linseed oil can be applied to seal the pores in the concrete. However, linseed oil has a tendency to pick up dirt and become discolored. On buildings where appearance is more important, silicone sealers may be used. Silicone is more expensive than linseed oil sealers, so it is not used for all applications. Both of these sealers are thin liquids which masons apply by spraying, rolling, or brushing.

Caulks

Caulking compounds are used to seal the joints between parts of a building. Commonly caulked joints include the joint between windows or doors and the wall, the joints in metal building panels, and expansion joints in sidewalks and roadways, **Figure 8–13**. In some applications, the **caulk** serves mainly as a filler and can be relatively hard. However, on tall buildings which sway in the wind, and in expansion joints, the caulking compound must be elastic.

The most elastic (stretchable) caulks are silicone and urethane. They are, however, more expensive than others. Oil-base caulking compound has been used longer than others, but it cures to a hard, nonelastic material and must be replaced after a few years. Butyl caulking compound is fairly elastic, lasts well, and is less expensive than silicone and urethane. The architect generally specifies the type of caulking compound to be used. Workers apply caulking compound with hand- or air-operated guns.

Adhesives

Adhesives have been used for thousands of years. The ancient Egyptians used glue to apply decorative veneers to wood objects. Until the twentieth century glues made from animal parts were the only major construction adhesives. More recently, adhesives have been made with superior strength for almost unlimited applications.

Adhesives have two important properties: adhesion and cohesion. **Adhesion** is the ability of one material to stick to another. **Cohesion** is the attraction that the molecules of a material have for one another. If an adhesive has good adhesion it will stick well to a surface. However, if that adhesive does not have good cohesion, it will pull apart when stress is placed on the adhesive joint.

Wood Glues

Animal glue is still used to a limited extent, but other glues have largely replaced it. Most modern glues for woodwork are made with a base of some kind of plastic. These glues have superior strength and other properties. **Figure 8–14** lists the most common glues and their outstanding characteristics.

Mastics

Mastics generally rely on cohesion to hold large areas of material in place. Mastics are used to apply floor coverings, roofing materials, and ceramic tiles on walls and floors. Most mastic cements have a latex or synthetic rubber base. They generally have a thick, creamy texture and are applied with a trowel, **Figure 8–15**. Most mastics are water resistant, but they do not withstand heat well.

Recent Developments in Adhesives

Modern science has developed adhesives for almost every imaginable use. Adhesives are used for assembling parts of automobiles, space vehicles, and buildings. One of these modern adhesives is epoxy. Epoxy is used for mending cracks in concrete, **Figure 8–16**. Experiments are being conducted with substituting epoxy for welding on structural steel.

ADHESIVE	COMMON NAME	PROPERTIES	USES
Animal glues	Hide glue Fish glue	Slow setting No water resistance	General interior woodwork Furniture Largely replaced by synthetics
Casein glue	Casein glue	Powder is mixed with water Can be used down to 35°F Water resistant Powder deteriorates with age	General woodwork where glue must be applied at low temperature Good on oily woods
Polyvinyl	White glue	Sets quickly Not water resistant Transparent when dry	General interior woodwork and furniture
Urea Formaldehyde	Plastic resin glue	Powder is mixed with water Slow setting Water resistant Heat resistant Low bond strength with oily woods	Laminated timbers and general woodworking where moisture resistance is desired
Resorcinol Resin	Waterproof glue	Two parts are mixed Dark color Waterproof Expensive	Exterior woodwork and laminated timbers where waterproof glue is required
Contact Cement	Contact cement	Type of rubber cement Poor heat resistance Waterproof Low strength Instant bonding	Applying plastic laminates on countertops and cabinets

FIGURE 8–14 Common wood glues and their major characteristics

FIGURE 8–15 Mastic is usually applied with a trowel.

FIGURE 8–16 Epoxy is used for mending cracks in concrete. *(Courtesy of Sika Corporation, Lyndhurst, New Jersey)*

ACTIVITIES

Glazing

When glaziers install windows, they have to cut the glass to size and secure it to the frame, using glazing compound to seal it. Glaze the frame given, following the steps in the procedure.

Equipment and Materials

Frame to be glazed
Glass, larger than frame
Glazing points
Glass cutter
Glazing compound
Steel square
Leather gloves
Putty knife

Procedure

1. Clean the frame to be glazed. Be sure that all old glazing compound and glazing points are removed.
2. Lay a piece of glass on a clean work surface and position the square to cut the glass 1/8 inch smaller than the opening.
3. Hold the glass cutter as shown in **Figure 8–17**. Score the glass in one stroke. Apply firm pressure on the glass cutter and do not go back over the scored line.

CAUTION: Wear gloves when handling glass to protect your hands from cuts.

4. Glass should be broken immediately after it is scored. The scored line will tend to repair itself if left alone.

FIGURE 8–17 The glass cutter scores the surface of the glass. (*Note:* Gloves are removed here to show hand position, but they should be worn when cutting glass.)

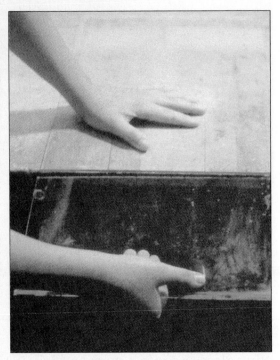

FIGURE 8–18 The glass is broken along the scored line. (*Note:* Gloves are removed to show hand positions.)

FIGURE 8–19 Glazing points are pushed into the wood frame.

Break the glass by bending it down over the straight edge of a workbench, **Figure 8–18**.

CAUTION: Dispose of broken scraps of glass in a safe place. Do not leave broken glass in the work area.

5. Follow the same procedure to cut the glass 1/8 inch smaller than the opening to be glazed.
6. Apply a thin bead of glazing compound to the inside of the frame.
7. Press the glass into the bead of compound. Secure the glass with glazing points 3 to 5 inches apart, **Figure 8–19**. Glazing points are pushed into the wood frame with the blade of a putty knife.
8. Seal the glass with glazing compound. Apply a full bead of compound. Then smooth and trim it by drawing the putty knife along the edge, **Figure 8–20**.

FIGURE 8–20 Glazing compound can be applied from a caulking gun or with a putty knife.

Plastics Identification and Applications

Plastics is the common name for materials that are made with polymers (a family of chemical substances with very big molecules). Many of the products used in construction involve polymers. These include finishing materials, wall coverings, carpets, insulation, and dozens of other familiar products. A fairly small number of polymers represents most of the plastics construction materials. To fully understand the nature of construction materials, it is valuable to be able to identify the most common plastics and to know the most important characteristics of those materials.

There are four basic tests that can be used to identify most of the plastics used in construction products: appearance, effects of heat, burning characteristics, and specific gravity (Is it heavier or lighter than water?). In this activity, refer to the Plastics Identification Chart as you test products to determine the plastic's materials and its properties.

Equipment and Materials

Assorted plastics construction products, such as:

siding
pipe
sheeting used for vapor barrier
plastic glazing (glass substitute)
electrical outlet cover plate
foamed boards (used for sheathing and
   foundation insulation)
electrical cable insulation

Procedure

Test several samples of construction plastics and design a chart to record the name of the plastics, your test results, the type of construction product you tested, and the important properties of the material.

1. Determine if the material is a thermoplastic or thermosetting material. Most thermoplastics will soften easily with a hot soldering iron. Most ther-

mosetting plastics will not soften as readily.

CAUTION: Never leave a hot soldering iron unattended. The person who uses the soldering iron is responsible for protecting others from accidental burns. Be careful to place the hot soldering iron on a proper rest to avoid damage to the work surface.

When testing foamed plastics, it may be necessary to crush a sample to eliminate as much air space as possible. The thin walls of expanded (foamed) plastics cells are much easier to melt or crush than are most solid plastics.

2. Examine the appearance of the material, test it for floating, and burn a small sample over a candle flame, as necessary to identify it according to the plastics identification charts, **Figures 8–21** and **8–22**.

CAUTION: Flame testing should only be done in area that is approved for the use of open flame. Samples to be burned should be as small as possible and must be immersed in water immediately after conducting the test. Wear protective clothing while conducting flame tests, as many plastics drip flames and sputter. Do not inhale plastics fumes deeply. A small sniff is all that is necessary to identify the material.

Applications of Materials

Inspect your school, home, or another structure in your community to find several uses of each of the materials listed below. Draw a cross-section sketch showing how each of the materials is used.

1. Gypsum board
2. Gypsum plaster
3. Wired glass
4. Insulating glass
5. Plastic sheet

Plastics	Appearance	Float/ Sink	Burning Characteristics	Properties and Uses
ABS	Tough, metallic sound when struck	Sink	Black smoke, yellow flame, drips	Fairly hard, tough, easy to mold into intricate shapes. Used for tool handles, knobs, some pipes.
Acrylic	Transparent, can be colored, brittle	Sink	Blue flame with yellow top, sweet fruity smell popping sound.	Transparent. Used for safety glazing, but scratches easily.
Polycarbonate	Transparent, not usually colored, hard	Sink	Dense black smoke, spurts orange flame, self extinguishing	Very tough (difficult to break). Expensive. Used for lenses on lamps.
Polyproplyene	Waxy feel, translucent, soft texture	Float	Burns rapidly with blue flame, drippings burn, hot area transparent	Decomposes in ultraviolet light (sunlight), may be molded or rolled into thin sheeting. Commonly used for plastic bags and vapor barriers.
Polyvinyl Chloride (PVC)	Tough, semi flexible soft to semi-hard texture	Sink	Yellow flame with green base, self extinguishing, black or gray smoke	Flexible, easily colored, resists ultraviolet light. Used for pipes, siding, and moisture barrier where toughness is required.
Styrene	Brittle, shiny surface, easily molded	Sink	Yellow flame, much black smoke and soot	Hard and brittle. Inexpensive. Used for inexpensive molded parts and foamed as small (usually white) beads.

FIGURE 8–21 Thermosetting Plastics (soften without charring under soldering iron)

Plastics	Appearance	Burning Characteristics	Properties and Uses
Epoxy	Hard, usually filled with metallic or glass fibers, unless used as an adhesive or coating	Yellow flame, some spits black smoke, chars, but continues to burn when flame sources removed.	Tough, strong adhesive. Used for high-strength electrical castings as an adhesive, and as a coating.
Melamine	Hard, brittle, shiny surface	Difficult to burn, cracks, turns white yellow flame, self extinguishing, fish-like smell	Easy to mold, good electrical insulating properties. Used for counter tops and electrical components.
Phenolic	Hard, brittle, sometimes illed with reinforcing fibers	Cracks, deforms, difficult to burn, yellow flame, little black smoke, distinctive smell	Excellent electrical insulating properties, difficult to color white. Used for molded handles, knobs and electrical components.
Polyester	Hard, brittle, usually filled or reinforced	Yellow flame with blue edge, ash and black beads, dense black smoke, no dripping, sweet smell	Can be cast crystal clear or colored. Most common plastics for reinforced castings, often referred to simply as fiberglass.
Urethane	Flexible, soft rubber-like	Yellow flame with blue base, thick black smoke, acid smell	Very flexible, and stretches. Often foamed or cushions and insulating boards. and insulating boards.

FIGURE 8–22 Thermosetting Plastics (Char under heat, but do not soften readily)

6. Plastic film
7. Extruded plastic
8. Asphalt coating
9. Caulking compound
10. Mastic adhesive

Applying Construction Across the Curriculum

Science

Look up the word *hydration*, then explain the manufacture of plaster, incorporating the principles of hydration and dehydration. What chemical reaction takes place when plaster is exposed to heat, such as in a fire? Why is gypsum plaster a good fire protection material?

Science and Communications

Obtain at least three plastic products used in construction. Find out what companies manufacture those products. Write to one company for each of the three products you selected and get information about the raw materials used to manufacture that type of plastic, the manufacturing process, and any health or environmental hazards involved. Make a report to your class on one of the three products.

REVIEW

Material Descriptions. For each of the materials listed, give (a) one important property, and (b) one application of the material in construction.

1. Polyethylene
2. Gypsum wallboard
3. Mastic
4. Gypsum lath
5. Tempered glass
6. Insulating glass
7. Extruded PVC
8. Foamed plastics
9. Asphalt sealant
10. Silicone sealant
11. Silicone caulking compound
12. Oil-base caulking compound

Section Three

Tools

Tools are especially important resources in the construction industry. Construction workers rely on the use of tools more than the workers in most industries. Building a structure requires the measurement, cutting, forming, and assembly of materials.

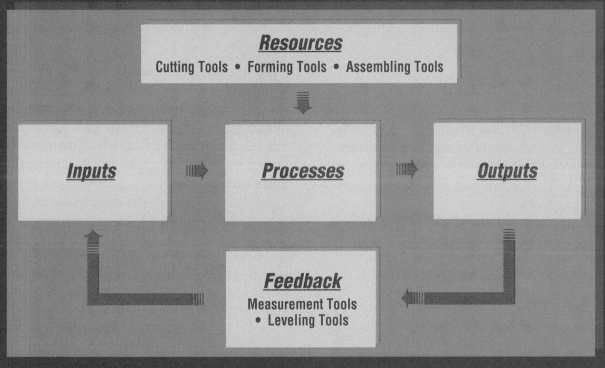

Chapter 9
Measurement and Layout Tools
covers systems of measurement, accuracy, tools for linear measurement, tools for measurement angles, and levels.

Chapter 10
Cutting, Sawing and Drilling Tools
covers tools for shearing, sawing, drilling, and boring.

Chapter 11
Fasteners, Fastening Tools, and Forming Tools
explains the most common fasteners for construction along with the tools for using those fasteners. This unit also explains the tools used for forming and smoothing construction materials.

Chapter 12
Power Tools
gives instructions on the use of all types of portable power tools, including basic information on the construction of these tools. Safety is stressed throughout this unit.

CHAPTER 9

Measurement and Layout Tools

OBJECTIVES

After completing this chapter, you should be able to:

▼ describe the measurement and layout tools commonly used in the construction industry;

▼ take accurate measurements to the smallest graduations indicated on the measuring tool being used;

▼ lay out square corners and angles; and

▼ determine the levelness or plumbness of any line.

KEY TERMS

linear measure	*right angle*
tape measure	*spirit level*
rafter square	*plumb bob*
stair gauge	*builder's level*
speed square	*target rod*
combination square	*laser*
sliding T bevel	
marking gauge	
divider	
trammel point	
chalk line	

A major factor in the quality of any construction project is the degree of accuracy in its layout and measurements. Absolute accuracy can never be achieved, only a relative degree of accuracy. It is always possible to take any measurement more accurately.

Persons in the construction trades should never allow preventable errors in their measurements. Some errors occur naturally due to expansion and contraction during heating and cooling and inaccuracy of equipment. However, with proper equipment, errors can be kept to a minimum.

Systems of Measurement

The *English system of linear measure* is used to measure most distances in building construction. In this system the *yard* is considered the base unit, **Figure 9–1**. The yard is divided into three *feet*. The foot is divided into twelve *inches*. The inch can be further divided into halves, eighths, sixteenths, and so on.

These divisions are called the *fractional parts of an inch*. The fractional parts of an inch

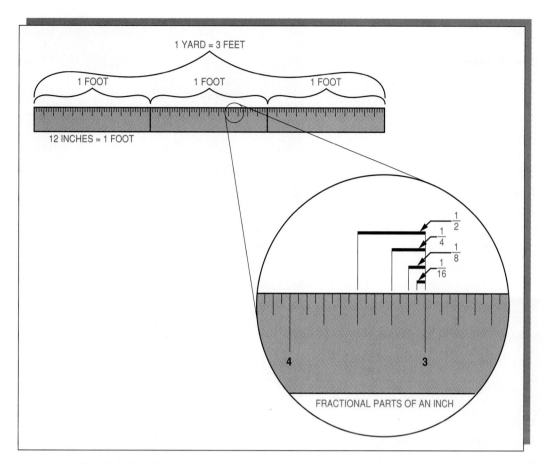

FIGURE 9–1 The English system of linear measure is based on the yard. The yard is divided into three feet. Each foot is divided into twelve inches. Each inch is further divided into fractional parts.

are named according to the number of these parts that make up an inch. It is possible to continue dividing these fractional parts until the desired degree of accuracy is reached.

Dimensions for construction are normally specified in feet, inches, and fractional parts of an inch. It is customary to reduce fractions to their simplest terms. For example, 26 and 6/16 inches is expressed as 2 feet, 2-3/8 inches. This is commonly written 2'-2-3/8".

Construction involves the coordination of a large number of materials and parts. The standards for this coordination have been established using the English system. Also, all existing structures are designed and built using the English system. Maintenance and additions to these structures would be more difficult if another system of measurement were used. Tools are available with metric scales and markings for use where necessary.

Measurement Tools

The measuring of distances is called **linear measure**. In the construction of a building, linear measurement is done with either steel tape measures or folding rules. Measurement of area and volume are only adaptations of linear measure. This can also be accomplished with these tools. Area is found by measuring width and length. Volume is found by measuring width, length, and height or depth. All of these involve measuring distances.

Tape Measures

Steel **tape measures**, **Figure 9–2**, are available in several lengths ranging from 6 feet to 100 feet. The 8- to 20-foot lengths are most often used for measuring the sizes and position of building parts. The longer lengths are used to establish building lines and to lay out the building. The shorter sizes usually have a siliding hook on the end so that both inside and outside measurements can be taken, **Figure 9–3**. The size of the case is marked on these tape measures so that inside measurements can be taken, **Figure 9–4**. The longer

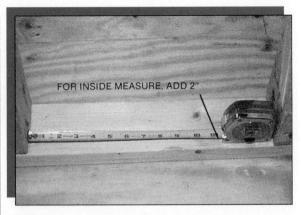

FIGURE 9–4   For inside measurement the width of the case is added to the tape reading.

FIGURE 9–2   Steel tape rules are available in lengths from 6 feet to 100 feet.

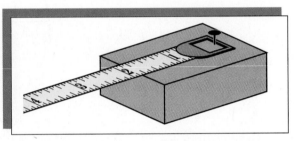

FIGURE 9–5   The fitting on a long tape can be hooked on a nail.

steel tape measures generally have a fitting that can be hooked over the outside of a corner or slipped over a nail, **Figure 9–5**.

Layout Tools

Layout tools include all those used for laying out lines and checking angles.

Rafter Square

One of the most commonly used tools for layout work is the **rafter square**, or framing square, **Figure 9–6**. It includes tables and scales that are used to lay out octagons, and rafters of several different types. It contains linear measure scales, as well. The variety of uses of the rafter square is beyond the scope of this text.

Although the rafter square is also used as a measurement tool, it is listed here with

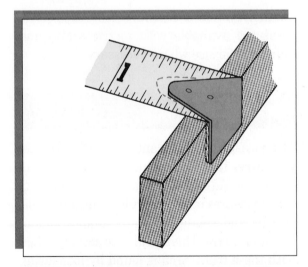

FIGURE 9–3   Most 6 to 25 foot tape rules have a sliding hook so they can take both inside and outside measurements. As the hook is pushed against a surface for taking inside measurements, it slides back to adjust for its thickness.

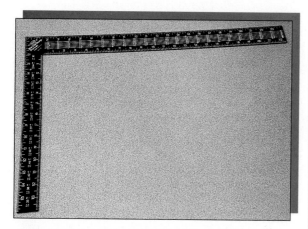

FIGURE 9–6 The rafter square is one of the most versatile measurement and layout tools.

FIGURE 9–7 Stair gauges are fastened on the square to lay out repeated angles.

layout tools because of its great value in layout work. Nearly all building-trade workers rely on it for checking the squareness of corners and laying out parts.

Stair Gauges

Stair gauges are used with the rafter square when several parts must be laid out with the same dimensions. These small fittings are clamped to the square to maintain the desired setting, **Figure 9–7**.

Speed Square

A fairly recent new tool, which has become widely used is the **speed square**, **Figure 9–8**. The speed square is a triangular piece of metal, with several scales for marking the cuts most often made on rafters and angles. An adjustable arm can be set to the desired cut or angle, making it easy to position the speed square quickly and mark the desired cut or angle. Although the speed square is very different from other squares in appearance, it performs the same functions and its sides are clearly marked, making it easy to understand.

Combination Square

The **combination square**, **Figure 9–9**, has a movable head on a 12-inch blade. The

FIGURE 9–8 Speed square

head of the combination square has a right-angle surface, a 45-degree angle surface, and a spirit level. The combination square can be substituted for a try square in most cases. It can also be used for laying out 45-degree angles, **Figure 9–10**.

Sliding T Bevel

The **sliding T bevel**, **Figure 9–11**, consists of a handle with movable blade. This blade can be locked in any position, making it useful for laying out various angles. It can be adjusted for 90-degree and 45-degree angles, but rafter squares, try squares, and combination squares are generally preferred for this.

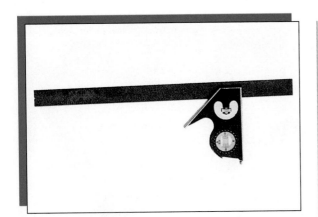

FIGURE 9–9 Combination square *(Courtesy of Sears, Roebuck & Co.)*

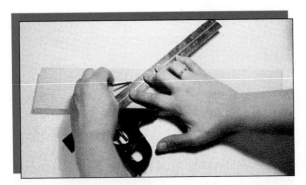

FIGURE 9–10 Marking a 45-degree angle with a combination square

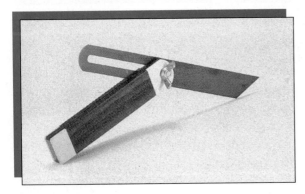

FIGURE 9–11 The sliding T-bevel can be adjusted for any angle. *(Courtesy of Sears, Roebuck & Co.)*

Marking Gauge

The **marking gauge** is used for marking a line parallel with an edge. **Figure 9–12** shows the parts of a marking gauge. In use it is held by the head with the pin pointing slightly toward the user. With the face against the edge of the material, it is pushed away from the user so the pin scribes the desired line, **Figure 9–13**. The scale on the beam of the marking gauge should not be relied upon for accuracy. Adjust the marking gauge with another measuring device.

Dividers and Trammel Points

Circles and arcs are laid out with **dividers**, **Figure 9–14**. For a small radius, the dividers are set with a separate measuring device. With the desired radius set, one leg is held on the

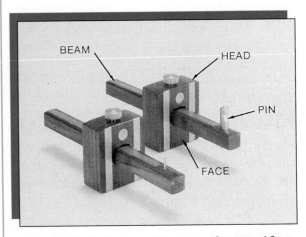

FIGURE 9–12 Parts of a marking gauge *(Courtesy of Sears, Roebuck & Co.)*

FIGURE 9–13 Scribing a line with a marking gauge

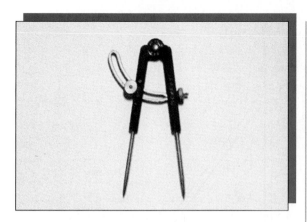

FIGURE 9–14 Wing dividers are used to scribe circles and arcs with small radii.

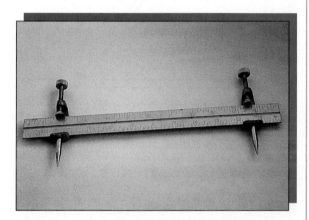

FIGURE 9–15 Trammel points can be attached to a beam of any length to scribe large circles and arcs.

center of the circle or arc and the other scribes a curved line. For larger radii, **trammel points** are attached to a bench rule or piece of wood. Once the trammel points are set, they can be used in the same manner as dividers. The radius possible with trammel points is limited only by the length of the piece to which they are attached, **Figure 9–15**.

Chalk Line Reel

A **chalk line** reel is used to mark long straight lines, such as a partition on a floor, **Figure 9–16**. The reel case contains powdered chalk that coats the line as it is pulled from the reel. The chalk-covered line is then stretched

tight while a point near its midpoint is pulled away from the surface to be marked. When the line is released it snaps against the surface to be marked and the chalk is deposited on the surface. This marks a straight line, **Figure 9–17**.

Lines similar to that used in a chalk line reel are also used by masons to indicate the desired height of a *course* (row) of bricks or blocks. The line is stretched along the wall. The masonry units are leveled with this line as they are put in place, **Figure 9–18**. The important characteristics of this line are (1) its ability to be pulled tight without stretching and (2) its strength.

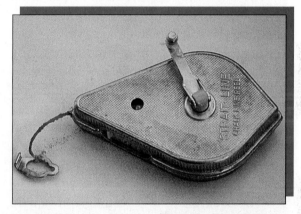

FIGURE 9–16 The chalkline reel applies powdered chalk to the line as it is pulled from the case.

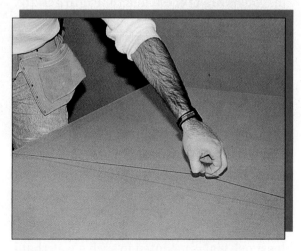

FIGURE 9–17 Snapping a chalk line

FIGURE 9-18 This mason is laying blocks to a line. *(Courtesy of Richard T. Kreh, Sr.)*

6-8-10 Method and Checking Diagonals

Although they are not tools in the usual sense, these are valuable techniques for checking the squareness of large corners and rectangles. It can be proven mathematically that any triangle with sides of 6, 8, and 10 units of length includes a **right angle**. This principle is used in building construction by measuring 6 feet along one side of a corner, and 8 feet along the other side. If the corner is a 90-degree angle the distance between these points is 10 feet, **Figure 9–19**.

The squareness of a rectangle or square can also be checked by measuring the diagonals. When all of the corners are 90 degrees, the diagonals are of equal length, **Figure 9–20**.

Levels

Spirit Level

The **spirit level**, **Figure 9–21**, makes use of the fact that an air bubble rises to the top of a container filled with liquid. A small, slightly curved glass tube (*vial*) is nearly filled with alcohol (*spirit*). The vial is mounted in an aluminum, magnesium, or wood straightedge. The level position is marked on the vial. At least one vial is mounted in such a way that it

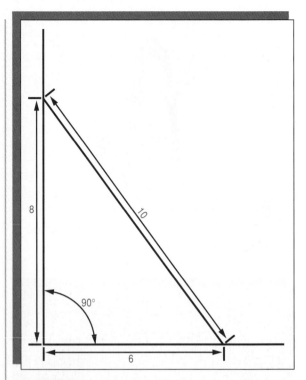

FIGURE 9-19 A triangle with sides of 6, 8, and 10 units of length includes a right angle.

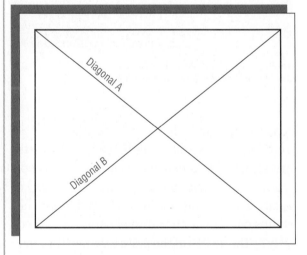

FIGURE 9-20 When the length of diagonal A equals the length of diagonal B, the rectangle is square.

indicates when the straightedge is *plumb*, or in a perfectly vertical position, **Figure 9–22**.

Spirit levels are available in sizes ranging from less than one foot to six feet long. Workers in many of the building trades use spirit levels.

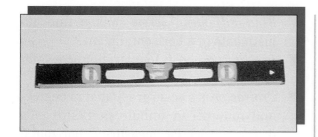

FIGURE 9–21   The spirit level uses the fact that an air bubble rises to the top of a container of liquid.

Carpenters usually use 2-foot levels to install rough and finished woodwork, while masons often use longer levels to check their work.

Line Level

The line level, **Figure 9–23**, is a small spirit level with fittings so that it can be hung on a tightly stretched line. The line level is used to check the levelness of a line between two points that are too far apart to be checked with a conventional 2- to 6-foot spirit level.

NOTE: All levels must be treated with great care. A sudden jolt may move the vial within its mounting, making the level inaccurate.

Plumb Bob

The **plumb bob**, **Figure 9–24**, is a pointed weight with some means for attaching a string. When the plumb bob is suspended on its string, the force of gravity causes the string to be perfectly plumb or vertical. In this way, a plumb bob can be used to check the plumbness of a line, much the same as a spirit level can be used. A plumb bob is most often used to locate a point directly below another, such as a point on a floor directly below a point on a ceiling.

Builder's Level

The **builder's level**, like a spirit level, is a device for checking the levelness between two or more points. In addition, it can be used for

FIGURE 9–22   This carpenter uses a spirit level to make sure that the woodwork is plumb.

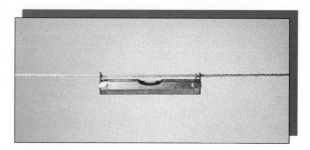

FIGURE 9–23   The line level is a small spirit level with special fittings so it can be hung on a tightly stretched line.

measuring angles on a horizontal plane. Unlike the transit level, it cannot be used for measuring angles on a vertical plane. The basic parts of the

FIGURE 9–24 A plumb bob causes a line to hang in a perfectly vertical position.

builder's level are shown in **Figure 9–25**. The functions of these parts are as follows:

▼ *Telescope* contains the lens, focusing adjustment, and cross hairs for sighting.
▼ *Telescope level* is a spirit level used for leveling the instrument prior to use.
▼ *Clamp screw* locks the instrument in position horizontally.

▼ *Find adjusting screw* makes fine adjustments in a horizontal plane.
▼ *Leveling base* holds four leveling screws for leveling the instrument prior to use.
▼ *Protractor* is a scale graduated in degrees and minutes (a minute is 1/60th of a degree), used for measuring horizontal angles.

Two accessories are required for most operations performed with the builder's level. The *tripod* is a separate 3-legged stand that provides a stable mounting base for the level. A **target rod** is a separate device with a scale graduated in inches and fractional parts of an inch. This is what one actually focuses on when sighting through the telescope.

The builder's level is used by workers in several of the building trades during many phases of construction. It is used to check the depth and levelness of *excavations* (holes dug or earth moved for foundations or grading), the levelness of footings and foundation walls, the

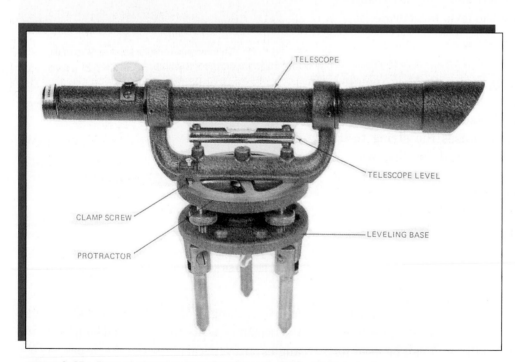

FIGURE 9–25 Parts of a builder's level *(Courtesy of L. S. Starrett Company)*

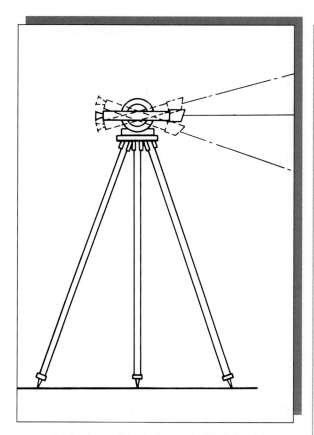

FIGURE 9–26 A transit is similar to a builder's level, but can be tilted to measure vertical angles.

levelness of wood and steel framing, and the positioning of forms for concrete work.

NOTE: The builder's level is a delicate instrument and must be handled with care. When not in use, the lens should be protected with the lens cap. It should always be stored and transported in a case designed for that purpose. Even when secured in its case, the builder's level should not be subjected to rough handling. Do not rub dust or dirt off the lens. Blow it off or brush it off lightly.

Surveyor's Transit

A surveyor's transit is like a builder's level, except that it works in one more direction. The **transit** can be tilted up and down to read vertical angles, **Figure 9–26**. By using trigonometry, surveyors can calculate vertical dimensions if they know the vertical angle. There are also electronic instruments for taking surveying measurements.

Any leveling job that can be done with a builder's level can also be done with a transit. However, the reverse is not always true. Some of the jobs done with a transit cannot be done with a builder's level.

Laser Measurement

Laser technology is quite new. The first laser device was built in a laboratory in 1960. For nearly twenty years, lasers were mainly laboratory curiosities. However, in recent years laser technology has been applied to manufacturing, medicine, retailing, home entertainment, and construction.

How a Laser Works

The word **laser** stands for *L*ight *A*mplification by *S*timulated *E*mission of *R*adiation. A laser device is made up of a gas-filled tube with mirrors at each end and a power supply, **Figure 9–27**. There are four classes of lasers. The class depends on the kind of gas in the tube and the intensity of input and output energy. Lasers used in construction use a mixture of helium and neon gas. The reflective mirror at one end is designed to reflect nearly 100 percent of the light striking its surface. The mirror at the opposite end has a small opening which allows about 2 percent of the light to pass and reflects 98 percent of the light. As a voltage from the power supply is applied to the gas, the gas molecules have absorbed all the energy they can hold. They *radiate* (give off) energy in the form of light. The light striking the 100 percent mirror is reflected back through the gas, further stimulating the gas molecules, and strikes the mirror with the small opening. That mirror reflects most of the light back to the gas again. Thus the amount of light energy inside the tube continues to grow.

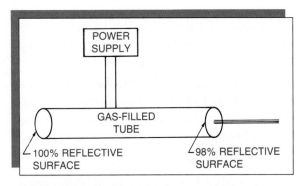

FIGURE 9–27   Simplified laser

The light energy that passes through the small opening in one of the mirrors is the laser beam.

Laser light has two important properties. Laser light is *monochromatic*—it is purely made up of one color. The light from an incandescent lamp contains all of the colors of the visible spectrum. Most of the colors we see are a result of a surface reflecting one color more than others, but the other colors are still present. The light radiated by stimulating gas in a laser is all the same color. Because laser light rays are all the same color, they are not scattered by lenses and mirrors.

Laser light rays are parallel to one another. Most of the light sources we commonly use radiate *divergent* light. This means that the farther the light travels, the more its rays spread, **Figure 9–28**. Laser light rays diverge very little.

The monochromatic and parallel characteristics of laser light make it very controllable. It is possible to send a very precise beam of laser light long distances through the atmosphere without losing its precision. The accu-

racy of laser beams over distances makes them very useful as references in construction. Laser reference beams are used for aligning pipes, erecting walls, controlling tunneling equipment, and many other construction operations, **Figure 9–29**.

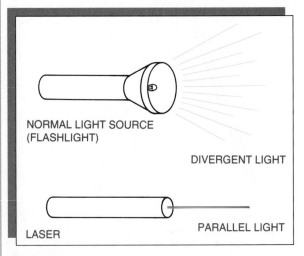

FIGURE 9–28   Most light diverges (spreads), but laser light is parallel.

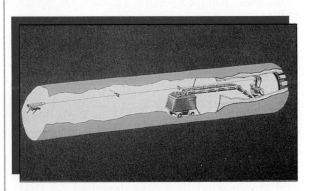

FIGURE 9–29   A laser reference beam can be used to control the machine that digs a tunnel. *(Courtesy of Spectra-Physics)*

ACTIVITIES

Marking Stock

Accurate measurement and layout are important factors in the construction industry. In this activity you will practice measuring with some of the tools discussed in this chapter.

Equipment and Materials

1" × 12" × 24" piece of softwood
Tape measure or folding rule
Combination square or rafter square
Compass or wing dividers

Procedure

1. Two duplicate pieces of wood stock will be laid out on the ends of a tote carrier that will be constructed in another unit.
2. Using the square, check to see that one end of the lumber is square with one long edge. If it is not square, mark a square line near the end.
3. With the tape measure or folding rule, and using the blade of the square as a straightedge, mark a line 9-1/2 inches from the square edge.
4. Working from the square edge, mark a square line across the piece 10 inches from the square end. Mark the center of this line (point A, **Figure 9–30**).
5. Using point A as the center, scribe a circle with a 1-inch radius.
6. Mark point B and C, 6 inches from the square end.
7. Using the blade of the square as a straightedge, draw lines from points B and C to the edge of the circle.
8. Repeat steps 2 through 7 to lay out the second part on the opposite end of the lumber.

Using a Spirit Level

Construction workers use spirit levels to determine the plumbness or levelness of work they are installing. Work carefully while following the procedures in this activity.

Equipment and Materials

2-foot spirit level
Line level

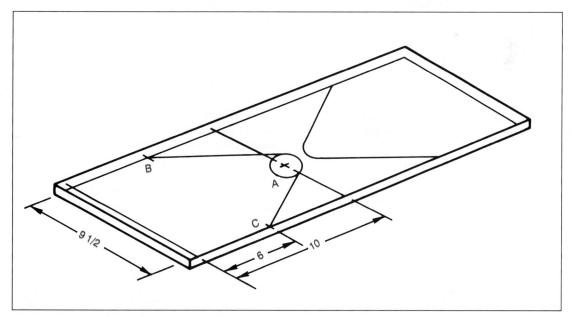

FIGURE 9–30   Layout of stock for "Marking Stock" Activity

Chalk line
Tape measure

Procedure

1. Mark a point near one end of a vertical surface. The surface should be at least eight feet long.
2. Use a line level to find a level point on the opposite end of the surface. Keep the line tight. Move one end of it up and down slightly until the bubble in the level is centered between the marks on the vial.
3. Snap a chalk line between these two points.
4. Measure to find the midpoint of the chalk line snapped in step 3. At this point, draw a plumb line using the 2-foot spirit level and chalk.

Laying Out Corners

The 6-8-10 method is a quick and accurate way to check the squareness of a rectangle or square. Carpenters sometimes use this method to check the squareness of a wall they are erecting. In this activity you will lay out four corners using the 6-8-10 method.

Equipment and Materials

Four wooden stakes approximately 3 feet long
Hatchet or heavy hammer
100-foot tape measure
200 feet of mason's line

NOTE: Four chalk marks on a paved surface can be substituted for the wooden stakes.

Procedure

1. Drive a stake in the ground at a point designated by the instructor. About 2 feet of the stake should be left above ground.
2. Drive a second stake exactly 50 feet from the first. Stretch a line between these two stakes.
3. Using the 6-8-10 method, stretch another line at right angles to the first

one. Mark a point on one line 6 feet from the stake, and on the other line 8 feet from the stake. When these two points are 10 feet apart the lines are at right angles.
4. Drive a third stake 30 feet from the second one and along the right-angle line.
5. Repeat steps 3 and 4 to drive the fourth stake 30 feet from the first one.
6. Check the accuracy of the lines by measuring the diagonals of the rect angle. If the diagonals are equal, the stakes are properly located; if not, correct the error.
7. Leave these stakes in place for the next activity.

Leveling Corners With a Builder's Level

A cement mason may want to check the levelness of a concrete form. To do this, the mason measures the height at each corner of the form using a builder's level. In this activity you will check the levelness of the rectangle measured in the third activity using a builder's level.

Equipment and Materials

Builder's level
Tripod
Target rod
Marking crayon

NOTE: The builder's level is a delicate instrument and should be handled with care. When not in use, it should be stored in the case designed for that purpose. All manufacturers supply instructions for setting up and using their instruments. The instructions given here are only general guides. The manufacturer's instructions should be consulted for more details.

Procedure

1. Set up the tripod in the approximate center of the rectangle laid out in the third activity. The legs of the tripod should be about three feet apart and

firmly set in the ground. Remove the protective cover from the head of the tripod. The tripod should be nearly level.

2. Remove the level from its case and set it in place on top of the tripod. Hand tighten the clamp screw.
3. Turn the leveling screws down so they contact the tripod plate.
4. Turn the telescope so that it is over one pair of leveling screws. Adjust these two screws so that the telescope level indicates that the telescope is level, **Figure 9–31**.
5. Turn the telescope so that it is over the other pair of leveling screws. Adjust these two screws so the telescope is level.

6. Repeat this over each pair of leveling screws alternately, until the telescope is level in all positions.

NOTE: Once the telescope has been leveled, be careful not to kick or move the tripod legs. If the tripod is accidentally moved or jarred, repeat the leveling procedure.

7. Remove the lens cap from the telescope. Have a partner hold the target rod on top of one of the stakes while the telescope is aimed and focused. The markings on the target rod should be in sharp focus.
8. Record the reading on the target rod at this stake.
9. Repeat steps 7 and 8 at each stake.

NOTE: Care must be taken not to move the tripod when aiming the telescope.

10. Mark the three highest stakes at the level of the lowest one. For example, if the target rod reads 41-1/2 inches at the lowest stake and 40 inches at another stake, the higher stake should be marked 1-1/2 inches from the top, **Figure 9–32**.

Applying Construction Across the Curriculum

Social Studies

Look up information on an ancient measuring system. Make a poster describing the measuring system. Try to include interesting information about problems with the system, why it was eventually replaced, or how it satisfied the civilization of its time.

Science

Look up information in the library and explain what system we use to standardize our systems of measure throughout the world today. What role do national governments play in ensuring the accuracy of our measuring systems?

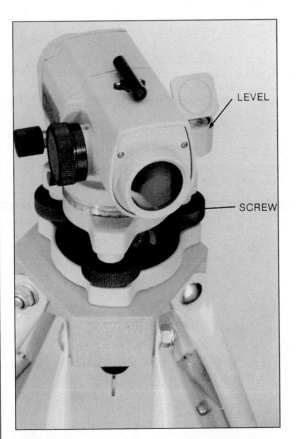

FIGURE 9–31 Adjust the leveling screws so the telescope level indicates that the instrument is level. (*Courtesy of Sears, Roebuck & Co.*)

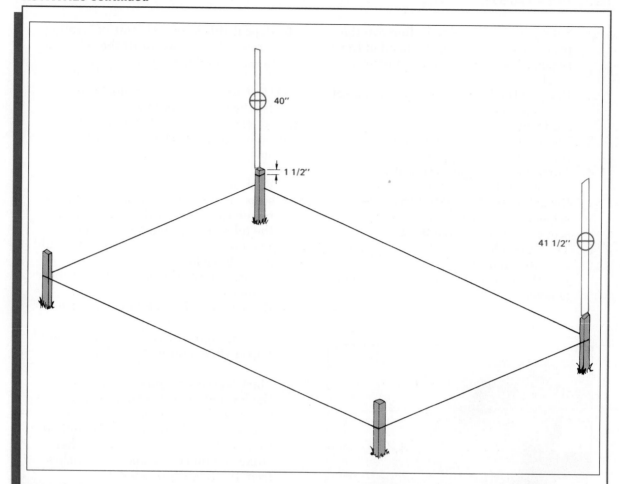

FIGURE 9–32   The numbers on a target rod increase from bottom to top. The higher the reading, the lower the stake. The target on the rod on the left is actually 41-1/2 inches above the mark on the stake.

REVIEW

1. List the measurement and layout tools that would be used by workers in each of the following trades:
   a. Mason
   b. Carpenter
   c. Excavator
   d. Electrician
   e. Sheet metal worker

2. Refer to a catalog provided by the instructor and list a complete set of measurement and layout tools. Include a brief description of each, indicating important features and price.

3. Give the dimension indicated in each of the following:

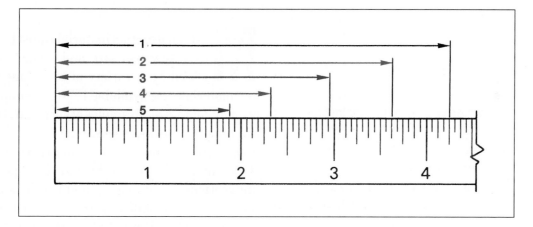

CHAPTER 10

Cutting, Sawing, and Drilling Tools

OBJECTIVES

After completing this chapter, you should be able to:

▼ list and identify the cutting, sawing, and drilling tools used most often by construction workers;

▼ name the proper tool to use for cutting, sawing, and drilling operations; and

▼ demonstrate the proper way to perform basic operations with cutting, sawing, and drilling tools.

KEY TERMS

wood chisel	miter
cold chisel	miter box
brick set	hacksaw
bench plane	hand drill
aviation snips	twist drill
utility knife	carbide
kerf	countersink
crosscut saw	center punch
rip saw	
coping saw	
backsaw	

Most construction materials must be cut to a specified size and shape, then joined by a variety of techniques. A few materials are available from the supplier already cut to size. Examples of such materials are concrete blocks, bricks, pipe fittings, and precut studs. However, in some cases, even these materials must be cut for special applications or joining techniques. It is also important to be skilled in the use of cutting, sawing, and drilling tools. Any accuracy achieved through careful measurement and layout will be lost if the materials are not cut to size accurately.

Cutting Tools

Cutting is an operation that is familiar to everyone. However, cutting is often confused with sawing. The operations are quite different. In sawing, small chips of the material are cut free, then pushed out of the work area or saw cut. In cutting, a sharp edge parts the material without removing chips.

The cutting principle is basically the same regardless of the material being cut. Cutting tools vary according to the materials because of the differences in the materials. For example, a thin, knife-like edge may be suitable for cutting soft plastic. To cut hard materials, such as steel, a thicker, stronger cutting edge is needed.

Chisels

Chisels are used for making small cuts in wood, metal, and masonry materials. **Wood chisels**, **Figure 10–1**, are fitted with a wood or plastic handle. The thin, sharp cutting edge of a wood chisel is ground with a bevel on one side only. Wood chisels are available in assorted sizes from 1/4 inch wide to 2-1/2 inches wide.

Wood chisels should never be struck with a hammer. Often, hand pressure is all that is needed. A soft-faced mallet can be used when added force is needed. The depth of cut can be controlled by placing the beveled edge of the chisel against the surface of the wood, **Figure 10–2**.

CAUTION: Wood chisels in proper condition are very sharp. Never place either hand in front of the chisel. Always cut away from the body.

Cold chisels, **Figure 10–3**, are used for cutting metal and masonry materials. They are made of a single piece of hardened and tempered steel. Cold chisels have a blunter bevel than wood chisels, but they are still ground to a sharp edge. Cold chisels are struck with a hammer and, for this reason, do not have handles.

CAUTION: If the head of a cold chisel becomes mushroomed, the sharp edges should be ground off before the chisel is used. The sharp edges could cause an injury.

A **brick set**, **Figure 10–4**, is quite similar to a cold chisel. Bricklayers use it to cut bricks and concrete blocks to size. The cutting edge of the brick set is the width of a standard brick. Bricks and blocks are cut by first scoring each side with one sharp blow on the brick set. Then the unit is broken with a final blow on this

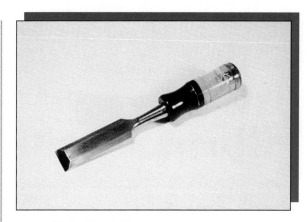

FIGURE 10–1   Wood chisel

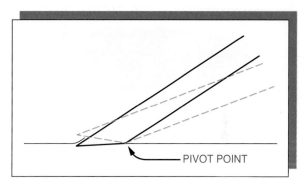

FIGURE 10–2   The depth of cut with a chisel can be controlled by placing the beveled side down and raising or lowering the handle.

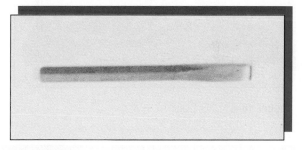

FIGURE 10–3   Cold chisel

score line. Bricks and blocks can be cut with a cold chisel or a mason's hammer. The wide edge of the brick set results in a straighter cut.

Planes

Carpenters use planes to remove a small amount of wood from the surface of a piece of lumber or millwork. **Bench planes** come in

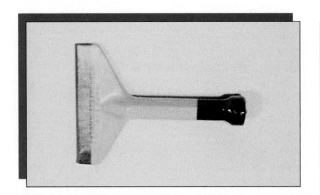

FIGURE 10–4 Brick set chisel

FIGURE 10–5 Block plane

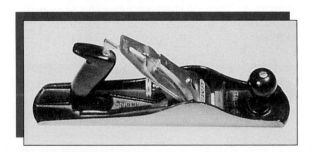

FIGURE 10–6 Jack plane

several sizes and each size is named differently. In order of their sizes, from largest to smallest, they are: jointer, fore, jack, smooth, and block. The largest planes, *jointer planes* and *fore planes*, are used mainly for producing flat surfaces. The smallest, the *block plane*, **Figure 10–5**, is used for planing end grain and where it is convenient to hold the plane in only one hand. The *jack plane*, **Figure 10–6**, is the most used, all-around plane.

When planing the face or edge of a piece of wood, hold the piece securely in a vise. Always plane with the grain, never across the grain. Place the front of the plane bottom on the wood and push forward while applying pressure downward on the knob, **Figure 10–7A**. As the front of the plane passes over the far end of the wood, relax the pressure on the knob but maintain pressure downward on the handle, **Figure 10–7C**. If the plane is sharp, properly aligned, and properly used, a long and smooth curled shaving should form.

To plane the end grain, precautions must be taken to prevent the far edge from splitting. This can be done in one of three ways: (1) by placing a piece of scrap against the far edge; (2) chamfering the edge in the waste portion; or (3) by planing from one edge toward the center, then from the other edge toward the center, **Figure 10–8**.

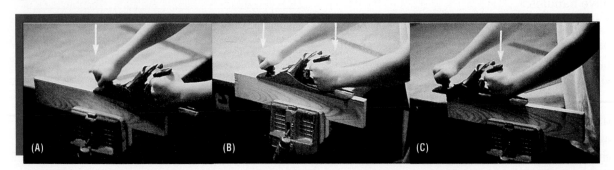

FIGURE 10–7 (A) At the beginning of the stroke, pressure is applied to the knob. (B) At midstroke, pressure is applied to both the handle and the knob. (C) At the end of the stroke, pressure is applied to the handle.

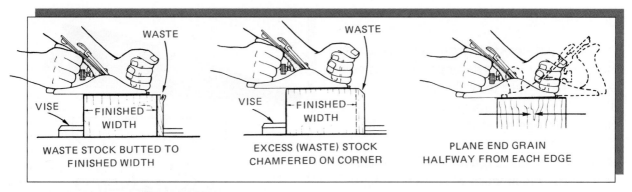

FIGURE 10–8   Three methods of planing end grain

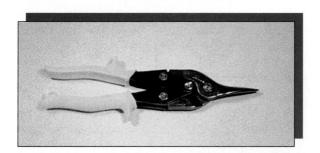

FIGURE 10–9   Aviation snips

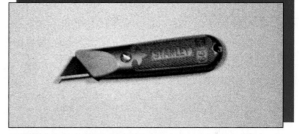

FIGURE 10–10   Utility knife

Snips

It is often necessary for construction workers to cut sheet metal. Carpenters and masons cut sheet metal flashing which is used to seal the joints between walls and roofs, chimneys and roofs, and other features. Sheet metal workers use large amounts of sheet metal in the ducts for heating, ventilating, and air-conditioning systems. The most common hand tool for cutting sheet metal is **aviation snips**, **Figure 10–9**. Three types of aviation snips are available for cutting straight lines and gradual curves, cutting left-hand curves, and cutting right-hand curves.

Utility Knife

The **utility knife**, **Figure 10–10** is also used by workers in several of the building trades. It is often called a *Sheetrock knife* because it is used extensively for cutting *gypsum wallboard* (Sheetrock™).

Sawing Tools

As mentioned earlier, sawing removes small chips of material from the saw cut (**kerf**). The sawing principle is the same regardless of the material being sawed. However, different arrangements of teeth are needed for different materials. Saw blades also vary depending on the kind of cut to be made. The coarseness of a saw is called its *pitch*. Saw pitch is measured in points per inch. This is the number of points of saw teeth in one inch of the blade, **Figure 10–11**

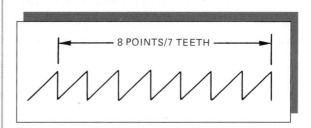

FIGURE 10–11   Saw pitch is measured in points per inch.

Crosscut Saws

Crosscut saws are used to saw wood across the grain. Crosscut saws for general purposes usually have eight to twelve points per inch. The teeth are filed at a slight angle so that they come to sharp points at alternate sides. The top third of each tooth is *set* (bent slightly) to the sharp-pointed side, **Figure 10–12** Set is needed in most saws to produce a kerf that is slightly wider than the thickness of the blade. This prevents the saw from binding in the kerf.

As the crosscut saw is pushed forward, the first parts to contact the wood are the sharp points at the sides of the kerf. This cuts the wood fibers off before they are removed. As the teeth cut deeper, the body of the tooth removes the sawdust from the kerf.

CAUTION: A properly sharpened saw should be handled with care. Keep hands away from the saw teeth. Never allow the teeth to contact metal or other hard materials.

Rip Saw

A **rip saw** is used to cut with the grain of wood. This is called *ripping*. Rip saws generally have four to eight points per inch. Rip saw teeth are filed straight across the blade, so that each tooth is chisel shaped, **Figure 10–13** To prevent binding, the teeth are set much like crosscut teeth.

As the rip saw teeth contact the wood, the full thickness of each tooth chisels out a small amount of wood. Because the sawing is in the direction of the wood grain, it is not necessary to cut the fibers off before removing them. Rip saw teeth are usually larger than crosscut teeth because they must remove larger amounts of material in making long rip cuts.

Coping Saw

Sometimes in fitting molding to inside corners, carpenters cut the profile of one piece

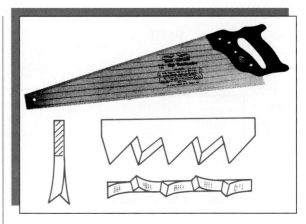

FIGURE 10–12   Crosscut saw *(Courtesy of Sears, Roebuck & Co.)*

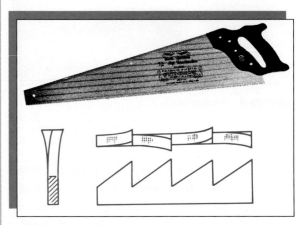

FIGURE 10–13   Rip saw *(Courtesy of Sears, Roebuck & Co.)*

of molding on the end of the other. This is called *coping*. To make these intricate cuts, a coping saw is used. A **coping saw** is a small frame that holds a thin, fine-pitch crosscut blade, **Figure 10–14** Coping saws are useful for all kinds of sawing where intricate shapes are involved. The blade is installed so the teeth point toward the handle. The cutting action occurs as the saw is pulled, not pushed.

Compass Saw

The *compass saw* is another crosscut saw for sawing curves. A compass saw is constructed more like an ordinary handsaw, but the blade

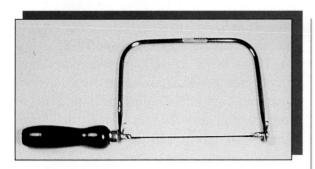

FIGURE 10–14 Coping saw

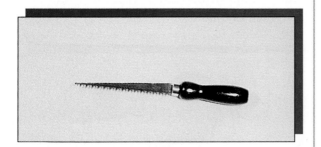

FIGURE 10–15 Compass saw for wallboard

is narrower and shorter, allowing it to saw around curves, **Figure 10–15**

Backsaw

The **backsaw** is a short, fine-toothed crosscut saw with a reinforced back, **Figure 10–16** Because of their rigidity and fine teeth, backsaws are used for sawing joints in wood and other sawing where a high degree of accuracy and a smooth cut are important.

Miter Box

A **miter** is a cut at a 45-degree angle, often used to join two pieces at a corner. To guide the saw in making miter joints, carpenters use a **miter box**. A wooden miter box can be made of three pieces of hardwood, forming a trough into which accurate saw kerfs are presawed.

Most carpenters and cabinetmakers use mechanical miter boxes. This is a metal frame holding a long backsaw, **Figure 10–17** The frame can be adjusted for any angle from 90 degrees to approximately 30 degrees. Usually a stop is provided to help locate the 90-degree, 45-degree right, and 45-degree left positions.

Hacksaw

Metal can be sawed, much like wood, with a **hacksaw**. A hacksaw is made up of a metal frame, usually adjustable in size, with some means for attaching and tightening the blade, **Figure 10–18** Hacksaw blades are made of high-carbon steel with fine-pitch teeth. The teeth on a hacksaw should point away from the handle so that the cutting action occurs on the forward stroke.

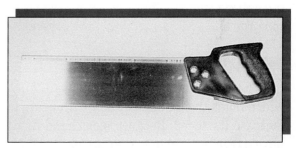

FIGURE 10–16 Backsaw

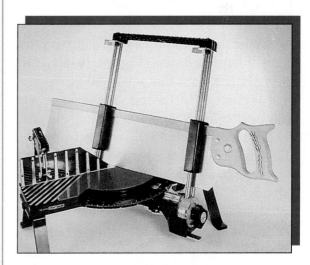

FIGURE 10–17 Miter box *(Courtesy of Sears, Roebuck & Co.)*

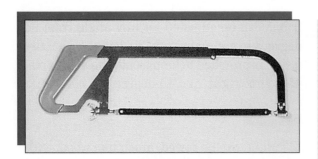

FIGURE 10–18 Hacksaw

Drilling Tools

The principle used in drilling tools is similar to the cutting and sawing principles. The greatest difference is that the cutting surfaces are moved in a circle instead of a straight line. All drilling operations require two devices: the tool with the cutting edges, called a *bit*, and a device to turn the bit.

Hand Drill

The **hand drill, Figure 10–19,** has a *chuck* which holds the bit and a crank with gears to increase its speed. Because most hand drills have a three-saw chuck, they are used to turn bits with a round shank.

Twist Drills

Twist drills are formed of a solid, straight piece of hardened steel, **Figure 10–20.** They have a round shank that is turned with a hand drill or electric drill having a three-jaw chuck. The cutting lips are ground at an angle. The center starts the hole and, as the bit penetrates the surface, a quarter portion of the cutting edge contacts the surface of the hole. Most twist drills are ground with a 118-degree angle between the two cutting lips. These are suitable for drilling in wood and most metals. By altering the angles of the cutting lips, twist drills can be manufactured for drilling in a variety of materials. For drilling in concrete and masonry materials, special twist drills are made. They have **carbide** (a very hard, man-made material) cutting edges, **Figure 10–21.**

Twist drills are sized by four systems: number sizes, letter sizes, metric sizes, and fractional-inch sizes. Most drilling done in construction uses fractional-inch drill bits. These range from 1/16 inch to 1/2 inch with straight shank, and larger with the shank portion reduced in size to fit normal chucks. Most hand drills cannot hold drill bits over 3/8 inch in diameter.

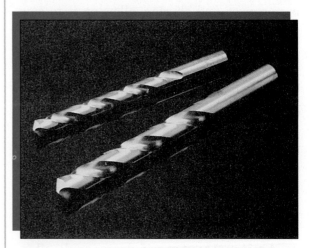

FIGURE 10–19 Hand drill

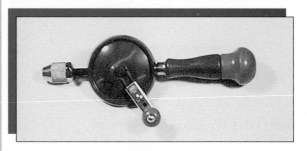

FIGURE 10–20 High speed steel twist drill bit (*Courtesy of the Irwin Company*)

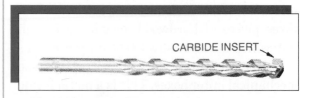

CARBIDE INSERT

FIGURE 10–21 Carbide-tipped masonry drill bit (*Courtesy of Sears, Roebuck & Co.*)

Countersink

One operation that is often performed with drills is making holes for screws. A common type of screw, called a flathead screw, requires an enlarged hole near the surface of the material being fastened. This part of the hole is called a countersink and is formed with a tool called a **countersink**, **Figure 10–22**. Countersinks can be purchased with a round shank and made of tool steel for countersinking metal. They can also be purchased with a square tang for use in a bit brace. Countersinks that are intended for use in wood should never be used in metal.

Center Punch

Although they are not a type of drill, center punches are discussed here because they are often used in drilling operations. A **center punch** is a solid piece of tool steel, usually round, with a point ground on one end, **Figure 10–23**. To help start a drill at a precise point, a small dimple is first made with a center punch. This holds the point of the drill in place as the hole is started.

There is also a spring-loaded type of center punch, called a *centering punch*. It can be used to locate the starting points for holes to fasten hardware to wood. The centering punch has a beveled end that fits into the countersink of the hardware. The punch is then struck with a hammer, leaving a dimple at the starting point.

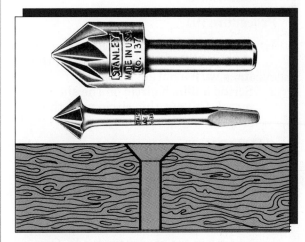

FIGURE 10–22   Countersinks and a countersunk hole *(Courtesy of Stanley Tool Co.)*

FIGURE 10–23   Center punch

ACTIVITIES

Cutting Operations

Drywall installers and sheet metal workers use a variety of cutting and measuring tools. Be very careful when working with sharp-edged instruments.

Equipment and Materials

Piece of gypsum wallboard, approximately $2' \times 2'$
Scraps of aluminum flashing or other sheet metal
Utility knife
Aviation snips
Rafter square
Dividers
Pencil

Procedure (I)

1. Lay out an 18-inch square on the face of the wallboard. The lines should extend to the edge of the wallboard.
2. Using the utility knife, cut the paper on the lines marked in step 1.

CAUTION: Do not place your hands in front of the knife.

3. Bend the wallboard away from the cutting side to break the gypsum core.
4. Cut the paper on the back of the wallboard.

Procedure (II)

1. Lay out several square corners on the sheet metal. Practice cutting on these lines with aviation snips.
2. Scribe a 6-inch circle on the sheet metal. Cut out this circle with the snips.

CAUTION: The edges of sheet metal are sharp. Wear gloves when handling sheet metal.

Sawing and Planing

Sawing and planing tools are important equipment in the carpentry trade. Carpenters must know which tools will best complete a task and give the desired surface.

Equipment and Materials

Piece laid out in Chapter 9, first activity
Combination square
Crosscut saw
Rip saw
Coping saw
Block plane or jack plane

Procedure (Refer to Figure 10–24)

1. Use the crosscut saw to cut the ends square.
2. Rip the piece to within 1/4 inch of the marked width.
3. Cut the angled edges and the curved portion with a coping saw.
4. With a properly adjusted jack plane, plane the ripped edge to the line. Check often with a combination square to be sure the edge is square.
5. Use either a block plane or a jack plane to smooth the end grain where it was sawed with the crosscut saw.

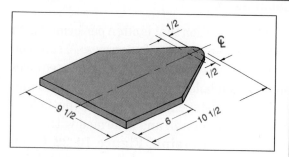

FIGURE 10–24 End of tote carrier

Drilling Holes

Equipment and Materials

Pieces from second activity
Combination square
Tape measure
Hand drill and 3/8-inch twist drill bit

Procedure

1. Lay out the centerlines of the end pieces as shown in **Figure 10–24.**
2. On this centerline mark off two points: one 1/2 inch from the top and the other 1 inch from the top.
3. Drill a 3/16-inch hole at each of these points. Remember to take precautions to protect the back from splintering as the bit passes through.

Cutting Masonry

Masons use brick sets and hammers to cut bricks and blocks to size. The cut must be neat and square if the course is to be accurate and attractive.

Equipment and Materials

Several bricks
Brick set
Hammer
Chalk

Procedure

1. Make a line around a brick.
2. Cut the brick on this line with a brick set. Practice until a neat, square cut can be made.

Applying Construction Across the Curriculum

Social Studies

What were some of the first cutting, sawing, and drilling tools used by early humans? How did the development of cutting tools affect the progress of civilization?

Science

Most hand tools rely on one or more of the simple machines (e.g., lever, inclined plane, etc.). What simple machines are employed by each of the tools discussed in this chapter?

REVIEW

A. Matching. Match the use or description in Column II with the correct tool in Column I.

Column I	Column II
1. Bit brace	a. Used for sawing metal
2. Twist drill bits	b. Used for drilling holes in metal
3. Crosscut saw	c. Has a square tang
4. Hacksaw	d. Used for sawing with the grain
5. Combination drill	e. Used for sawing across the grain
6. Hand drill	f. Used for cutting wallboard
7. Auger bits	g. Used for sawing intricate shapes
8. Rip saw	h. Used for drilling screw holes
9. Coping saw	i. Has a two-jack chuck
10. Utility knife	j. Has a three-jaw chuck

B. Matching. Match the use or description in Column II with the correct tool in Column I.

Column I	Column II
1. Crosscut saw	a. 10 points per inch
2. Hacksaw	b. For drilling in concrete
3. Cold chisel	c. Thin, sharp edge
4. Wood chisel	d. Very fine pitch
5. Rip saw	e. Turned with a hand drill
6. Carbide drill bit	f. Turned with a bit brace
7. Steel drill bit	g. For cutting masonry
8. Round shank	h. For drilling wood or metal
9. Square tang	i. 6 points per inch
10. Brick set	j. Blunt-angled cutting edge

C. Tool Descriptions. Refer to a catalog provided by the instructor and list a complete set of cutting, sawing, and drilling tools used by a carpenter. Include a brief description of each, indicating important features and price.

CHAPTER 11

Fasteners, Fastening Tools, and Forming Tools

OBJECTIVES

After completing this chapter, you should be able to:

▼ describe common types of nails and screws and the systems for sizing them;

▼ identify and demonstrate the proper use of the fastening and forming tools used most often by construction workers; and

▼ explain the markings on coated abrasives.

KEY TERMS

common nail	*drywall screw*
box nail	*deck screw*
finishing nail	*plastic*
penny size	*float*
face nailing	*trowel*
toenailing	*masonry jointer*
clinching	*coated abrasive*
nail hammer	
nail set	
mason's hammer	
gauge size	
flathead screw	
round head screw	

The range of fasteners used in the construction industry is almost limitless. Hundreds of kinds of special fastening devices are available from a large number of suppliers. However, a few types of nails and screws account for most of the fasteners used.

Nails

Many kinds of nails can be used for a variety of purposes. Several of the most common nails are shown in **Figure 11–1** Most of these are designed for a specific purpose, but a few are used for many different purposes.

Common Nails

Common nails are the most often used of all nails. Common nails have a flat head and smooth shank. They are used for most applications where the special features of another type of nail are not needed.

Box Nails

Box nails are similar to common nails except they have a thinner shank and thinner head. Because of their thin shank, box nails are less apt to split the wood used for boxes and

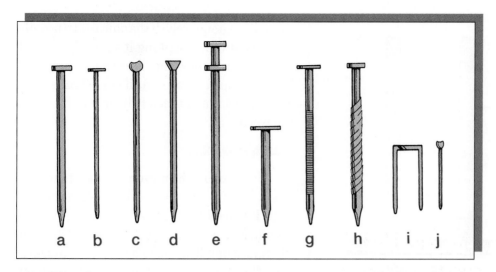

FIGURE 11–1   Some of the most common kinds of nails: (a) common nail; (b) box nail; (c) finishing nail; (d) casing nail; (e) duplex-head nail; (f) roofing nail; (g) ring-shank drywall nail; (h) screw-shank nail; (i) staple; (j) wire brad

crates. They are also useful where paint will be applied over the nailed surface. Due to their thin head, box nails do not show up as readily under paint. Box nails are often coated with a chemical that resists rusting and makes the box nail more difficult to withdraw.

Finishing Nails

Finishing nails have very small heads and somewhat thinner shanks than common nails. The small head of a finishing nail can be driven below the surface of the wood and concealed with putty so that it is completely hidden. Finishing nails are used for installing trim and millwork where appearance is important. Because of their small heads, finishing nails do not have as much holding power as common nails and box nails.

Nail Lengths.   Nail lengths are specified by **penny size**. The term penny (abbreviated *d*) was adopted in the early days of nailmaking. The penny size indicated the size of nails that could be purchased at the rate of 100 for a given number of pennies. Today, the penny size indicates the length of nails, regardless of the type, **Figure 11–2** For example, 8d common

nails, finishing nails, and box nails are different diameters, but they are all 2-1/2 inches long.

Nailing Techniques.   Nails can be driven in basically three ways: face nailing, toenailing, and clinching. In **face nailing**, nails are driven straight through one member and into the other. **Toenailing** is used where it is not possible to use face nailing and where resistance to withdrawal is needed. A nail is driven into the side of one member at an angle, penetrating the adjacent surface and the other member,

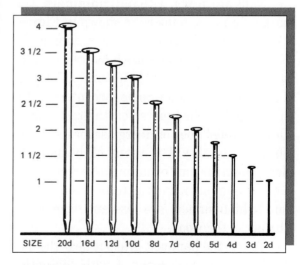

FIGURE 11–2   Penny sizes of nails

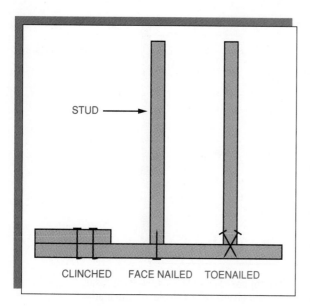

FIGURE 11–3 Nailing techniques

Figure 11–3 By toenailing from both sides, the members are less apt to pull apart. **Clinching** refers to driving nails through both members and bending over the points to prevent withdrawal.

Hammers

Nail Hammers

Nails are driven with **nail hammers**. The parts of a nail hammer are shown in **Figure 11–4** The *claw* is used for withdrawing nails. Nail hammers come with handles of wood, steel, and fiberglass, and in 13-ounce, 16-ounce, and 20-ounce weights. Some carpenters prefer steel-handled hammers for their strength. Others prefer wood or fiberglass for its shock-absorbing quality. The weight of the hammer depends on the kind of nailing to be done. With a heavy hammer it is easier to drive large nails, but light hammers are less apt to mar the surface of the wood.

To use the hammer, grasp it near the end of the handle and swing it with the entire forearm. The face of the hammer should strike the nail head squarely. If the nail bends (usually as a result of not striking it squarely), pull it out and start a new one. To *set* a finishing nail below the surface, use a **nail set**, **Figure 11–5** This drives the nail head about 1/16 inch into the wood.

Mason's Hammer

The **mason's hammer** (or brick hammer) has a square face and a chisel-like cutting edge, **Figure 11–6** Like nail hammers, mason's hammers are available with wood, steel, or fiberglass handles and in a variety of weights. The flat face of the hammer is used for

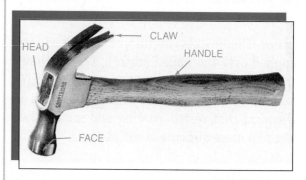

FIGURE 11–4 Parts of a nail hammer *(Courtesy of Sears, Roebuck & Co.)*

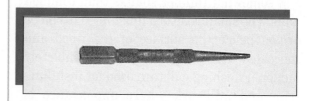

FIGURE 11–5 Nail set

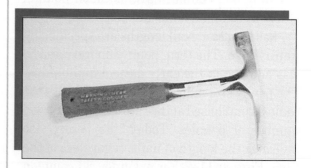

FIGURE 11–6 Mason's hammer

occasional nailing and for striking cold chisels and brick-set chisels. The other end of the hammer can cut bricks and concrete blocks. The cutting edge of the mason's hammer is used like the brick-set chisel. The brick or block is struck on all four sides to score it. It is then broken with a final, sharp blow.

Screws

Screws are a common kind of fastener in construction. A variety of sizes and types of screws are available for fastening several kinds of material in a wide range of situations. Wood screws are listed according to the following:

▼ *Material.* Brass screws are used for exterior application, but steel screws are stronger and much less expensive.

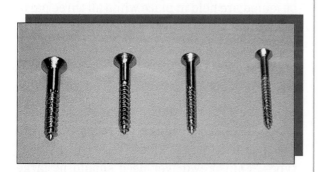

FIGURE 11–7 These screws are all the same length, but different gauges.

FIGURE 11–8 These screws are all the same gauge, but different lengths.

▼ *Gauge size.* The **gauge size** indicates the diameter of the unthreaded part of the screw. The higher the number, the larger the screw, **Figure 11-7.**

▼ *Length.* Wood screws are available in lengths from 1/4" to 4", **Figure 11-8**

▼ *Head type,* **Figure 11-9 Flathead screws** are generally used where a flat surface is desired. **Round head screws** or oval head screws are more difficult to conceal.

Another type of wood screws that has become quite popular is the type that was first developed for fastening gypsum wallboard, **Figure 11-10 Drywall screws** which are intended for interior use have finer threads and are not generally coated to resist corrosion. **Deck screws** are similar, but have coarser threads to drive faster and are galvanized or

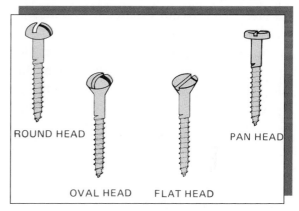

FIGURE 11–9 Common screw head shapes

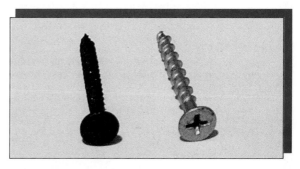

FIGURE 11–10 Drywall screws are intended for interior use and deck screws are intended for exterior use.

Screw Gauge	0	1	2	3	4	5	6	8	10	12	14	16	18
Shank Hole	1/16	5/64	3/32	7/64	7/64	1/8	9/64	11/64	3/16	7/32	1/4	17/64	19/64
Pilot Hole	1/64	1/32	1/32	3/64	3/64	1/16	1/16	5/64	3/32	7/64	7/64	9/64	9/64

FIGURE 11–11   Chart of drill sizes for wood screws

otherwise coated to resist corrosion. Both of these types of screws are available only with Phillips heads and their heads flare out to cover a wider surface than conventional wood screws. Their wider heads improve their holding qualities, but make them more difficult to conceal in finish work.

Drilling Screw Holes

Fastening two pieces of wood together with flathead screws requires three operations with a drill. A shank hole just large enough to fit the unthreaded portion of the screw must be drilled in one piece. A pilot hole slightly smaller than the threaded portion of the screw must be

drilled in the other piece. The chart in **Figure 11–11** shows suggested sizes for shank holes and pilot holes for most screw sizes. The top surface of the shank hole must be countersunk to accept the head of the screw, **Figure 11–12**

These operations can be combined into one operation by using a *combination drill*, **Figure 11–13** Combination drills are made for a specific screw length and gauge. The pieces of wood to be joined are held in place while all three parts of the hole are drilled the proper size.

Screwdrivers

Most screws have either straight-slotted heads or Phillips heads, **Figure 11–14** Screwdrivers are available in a range of sizes to fit both types. The screwdriver used should be large enough so the blade fills the slot in the screw head. Screwdrivers are not made of hardened steel and should not be used for prying.

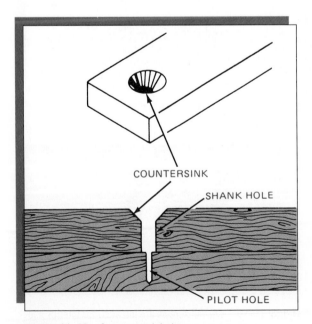

FIGURE 11–12   Countersunk hole

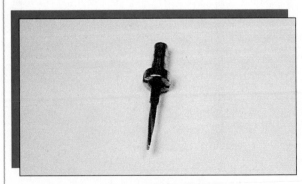

FIGURE 11–13   Combination drill makes the pilot hole, shank hole, and countersink in one operation.

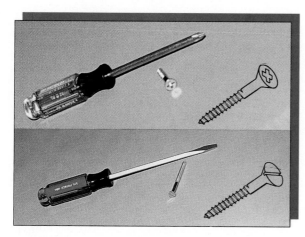

FIGURE 11–14 Phillips and straight-slot screwdrivers *(Courtesy of Sears, Roebuck & Co.)*

Wrenches

Although many of the fasteners used in construction do not require wrenches, most construction workers carry one or two adjustable wrenches in their tool kits. Adjustable wrenches are available in various lengths from 4 inches to 18 inches. These have one movable jaw that can be adjusted to fit any size fastener. When using an adjustable wrench, the force should be applied toward the movable jaw, **Figure 11–15**

Plumbers and pipefitters use *pipe wrenches* to turn pipes and pipe fittings. Pipe wrenches

have a series of teeth on their jaws, **Figure 11–16** so they can grip the round surface. Pipe wrenches are available in lengths from 6 inches to 2 feet. Usually two pipe wrenches are needed—one to turn the fitting and one to hold the pipe.

Clamps

Clamps are used by welders to hold parts during welding. Cabinetmakers and carpenters use clamps when gluing parts together. Plumbers, electricians, and other building tradesworkers use them to hold pieces in place temporarily during assembly or installation. There are many types of clamps, but most construction workers are familiar with two or three types.

C clamps, **Figure 11–17** are lightweight and easy to handle. Construction workers often use them to hold parts in place during construction. *Bar clamps* are mostly used by carpenters and cabinetmakers for clamping glued assemblies, **Figure 11–18** They are available in lengths ranging up to eight feet.

FIGURE 11–15 When using an adjustable wrench, the pressure should be applied toward the adjustable jaw.

FIGURE 11–16 Pipe wrenches are used by plumbers to turn pipes and fittings.

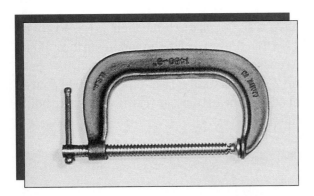

FIGURE 11–17　C clamp

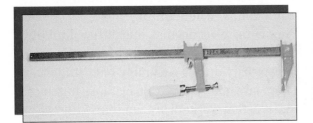

FIGURE 11–18　Bar clamps are used to glue up stock.

Hand screw clamps are also used mainly by woodworkers. These clamps are fast-acting and can be adjusted for nonparallel surfaces.

When gluing up wooden assemblies, clamps should be placed every few inches along the glue line. Clamps with steel clamping surfaces should be padded with scraps of softwood. This prevents marring the parts being glued.

Forming Tools

Many construction materials are applied while they are in a plastic state. **Plastic** means able to be formed or shaped. These materials are shaped or smoothed with *forming tools*. Most of the forming tools used in the building trades are characterized by a smooth working surface used to smooth the material.

Floats

Concrete workers use **floats** to smooth out some of the unevenness in the surface of freshly placed concrete, **Figure 11–19** As soon as possible after the concrete is placed in the forms, it is floated. This leaves a textured but uniform surface. Floats can be made of wood, aluminum, or magnesium.

Concrete Trowels

When concrete is firm, but still plastic enough to be worked a little, it can be troweled for a smoother surface. **Trowels**, **Figure 11–20**, are made of high-carbon steel and have a

FIGURE 11–19　Floating concrete *(Courtesy of Portland Cement Association)*

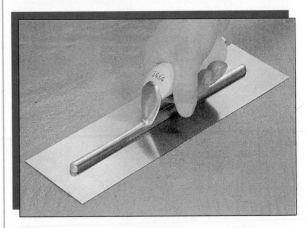

FIGURE 11–20　Cement trowel *(Courtesy of Goldblatt Tool Co.)*

FIGURE 11–21 Concrete masons using power trowels to finish a large concrete floor *(Courtesy of Marshalltown Trowel Co.)*

smoother working surface than floats. Trowels have a much smoother surface. Power trowels with rotary blades are often used to trowel large areas, **Figure 11–21** Trowels similar to those used by concrete workers are used by plasterers in smoothing plaster on walls and ceilings.

Mason's Trowel

Masons use another type of trowel, **Figure 11–22**, for spreading mortar. The pointed shape of the *mason's trowel* is suited for applying mortar to masonry units and removing excess mortar as it is squeezed out of the joint. Mason's trowels are available in a range of sizes. Masons who work with bricks (bricklayers) often select a small size. A mason working with concrete blocks may select a larger size. In either case the quality of the steel and the balance of the

FIGURE 11–22 Mason's trowel *(Courtesy of Goldblatt Tool Co.)*

trowel are important considerations in selecting a trowel.

Masonry Jointers

When the mortar is just hard enough to show a thumbprint, the mason smoothes the joints with the **masonry jointer**, **Figure 11–23**. This improves the appearance of the masonry job and helps make the joints watertight.

Joint Knife

After drywall mechanics install gypsum wallboard, the joints are concealed with a special compound and paper tape. The compound for this joint system is applied either with a plaster trowel or a joint knife, **Figure 11–24** A *joint knife* is similar to a putty knife, except that joint knives are usually 5 to 6 inches wide.

FIGURE 11–23 Before the mortar is completely set, the mason smoothes the joint with a jointer.

FIGURE 11–24 Joint knife

Coated Abrasives

Although these are not tools in the common sense, coated abrasives are included here because they are important for smoothing construction materials. Coated abrasives are often referred to as sandpaper, although only one type is actually sandpaper. **Coated abrasives** include all products made of a paper or cloth backing coated with sharp particles for smoothing materials.

The grit may be natural or synthetic. The most common abrasive materials and their outstanding characteristics are listed in **Figure 11-25** Garnet and aluminum oxide are the most common abrasives for smoothing wood. Emery, silicon carbide, and aluminum oxide are commonly used on metals.

The coarseness of the abrasive grains is marked on the back of the sheet with a number, **Figure 11-26** The number of the grit indicates the number of wires per inch in a screen through which the grains will pass. For example, if the abrasive sheet is marked with number 120, the grains will just pass through a screen with 120 wires per inch running in each direction. Coarse abrasives, such as numbers 32 and 40, are used for sanding wood floors. Very fine abrasives, such as number 400, are used for smoothing painted surfaces. The letter marking on the back of the coated abrasive indicates the weight of the backing material. A-weight paper is used for smoothing finished woodwork. D-weight paper is generally used with coarser abrasives for heavier work.

Coated abrasives are either open coat or closed coat. *Closed-coated abrasives* have all available space covered with abrasive grains. *Open-coated abrasives* have some space between the abrasive grains. Closed-coated abrasives cut faster, but open-coated abrasives have less tendency to "plug up" when smoothing paint, varnish, or resinous materials.

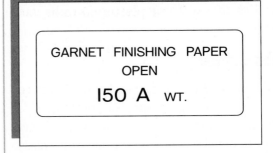

FIGURE 11-26 Markings on the back of coated abrasive

Abrasive	Color	Description
Flint	Light tan	Relatively soft natural abrasive sed for some stonework
Garnet	Reddish brown to orange	Natural abrasive used extensively for woodwork
Aluminum oxide	Brownish gray	Very hard man-made abrasive used for metals, wood, and other materials
Emery	Dark brown	Natural abrasive commonly used for polishing metal
Silicon carbide	Black	Hard man-made abrasive used to smooth and polish metals and for wet sanding finishes

FIGURE 11-25 Common abrasive materials

ACTIVITIES

Nailing

Carpenters must be familiar with the different types of nails and nailing techniques. How parts are fastened together greatly influences the finished product.

Equipment and Materials

Nail hammer
Four pieces of wood, $2'' \times 4'' \times 2'$
Six pieces of wood, $1'' \times 3'' \times 2'$
Supply of nails—12d common, 6d common, 6d finish

Procedure

1. Face nail one pair of $2'' \times 4''$ pieces and toenail one pair of $2'' \times 4''$ pieces as shown in **Figure 11–27**.
2. Lap and nail one pair of $1'' \times 3''$ pieces with 6d common nails as shown in **Figure 11–28** Clinch the nails.
3. Lap and nail one pair of $1'' \times 3''$ pieces with 6d finishing nails as shown in **Figure 11–28** Clinch the nails.
4. Hold one part of each assembly in a vise. Pull the assemblies apart, noting the differences in the nails' holding power.

CAUTION: As the assemblies are pulled apart, remove the nails. Exposed nail points cause injuries.

5. Lap and nail one pair of $1'' \times 3''$ pieces with 6d common nails, as in step 2, but drive the nail 3/4 inch from the end of the top piece.
6. Reverse the ends of the pieces and repeat step 5. Dull the nail points first by holding the heads on a hard surface and tapping on the points with a hammer.

NOTE: This is a technique often used by carpenters to prevent splitting.

Fastening with Screws

Carpenters and other construction workers must know which drill and screw combination to use to fasten objects

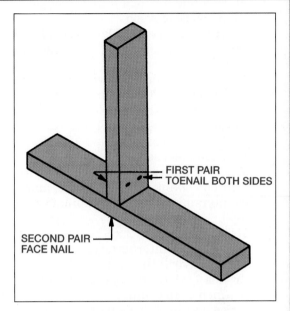

FIGURE 11–27 Assembling for "Nailing" activity

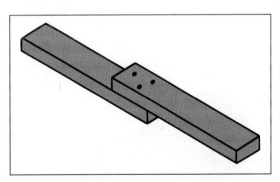

FIGURE 11–28 Assembly for "Nailing" activity, steps 2 and 3

properly. This activity demonstrates how specific screws should be installed.

Equipment and Materials

Two pieces of wood, $1'' \times 3'' \times 2'$
Hand drill and set of twist drill bits
$1\text{-}1/2 \times 8$ combination drill—countersink type
1×8 combination drill—counterbore type
$1\text{-}1/2 \times 8$ flathead steel screws
1×8 round head steel screws
Screwdriver
Marking gauge

Procedure

1. Mark a line 3/8 inch from the edge of one piece of wood.
2. Refer to **Figure 11–11**and drill the proper size shank hole for a #8 screw on this line.
3. Hold the two pieces of wood together as in **Figure 11–29** Mark the location, and drill the pilot hole.
4. Countersink the shank hole.
5. Fasten the pieces with a 1-1/2 × 8 flathead steel screw.
6. With the pieces still assembled, drill a second shank hole, pilot hole, and countersink using the combination drill.
7. Drive another 1-1/2 × 8 screw.
8. Using the counterbore-type combina tion drill, drill the proper holes to drive a 1 × 8 round head screw with a 3/8-inch deep counterbore.
9. Drive a screw in this hole.
10. Examine the assembly. All screw heads should be tightly seated. The pieces of wood should be in tight contact along their entire length.

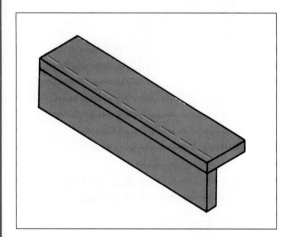

FIGURE 11–29   Assembly for "Fastening with Screws"

Trowel Practice

Plasterers finish interior walls and ceilings with plaster coatings that form fire-resistant and relatively soundproof surfaces. The first step in plastering is learning how to use the trowel to produce a smooth coating.

Equipment and Materials

Gypsum wallboard or plywood, approximately 24″ × 24″
Two pounds of patching plaster
Container for mixing
Plaster or concrete trowel

Procedure

1. Mix the plaster according to instructions. Patching plaster is used because it sets quickly. This allows completion of the activity in one class period. To extend the drying time, add a very small amount of vinegar.

CAUTION: Plaster sets underwater. Do not dispose of unused plaster in a sink drain as it will clog the drain.

2. Dampen the surface of the wallboard or plywood. This prevents the wallboard from drawing the water out of the plaster.
3. Trowel an even coat of plaster onto the surface of the wallboard or plywood. Try to obtain a smooth coat without making ridges with the edges of the trowel.

Applying Construction Across the Curriculum

Social Studies

The early settlers in North America placed a very high value on their fasteners—nails and screws. They even sorted through the ruins after a home was destroyed by fire, so they could save the fasteners. Why did they place such a high value on their fasteners? How were nails made in colonial America?

Science

Most hand tools and fasteners rely on one or more of the simple machines (e.g., lever, inclined plane, etc.). What simple machines are employed by each of the tools and fasteners discussed in this chapter?

REVIEW

A. Multiple Choice. Select the best answer for each of the following questions.

1. What kind of nail has a small diameter and a thin, flat head?
   a. Box
   b. Finish
   c. Common
   d. Casing

2. Which kind of nail is most often used to install wood molding?
   a. Box
   b. Finish
   c. Common
   d. Duplex

3. Which method for nailing would be best suited for fastening the studs in **Figure 11–30**?
   a. Face nailing
   b. Toenailing
   c. Clinching
   d. Nail Setting

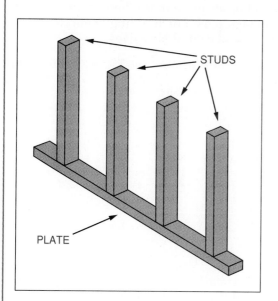

FIGURE 11–30

4. What is the length of a 2-1/2 × 6 screw?
   a. 6 inches
   b. 1/6 inch
   c. 6/4 (1-1/2) inches
   d. 2-1/2 inches

5. Which screw has the largest diameter?
   a. 1-1/4 × 4
   b. 1-1/4 × 8
   c. 2-1/2 × 6
   d. 3 × 6

6. What should be the diameter of the shank hole for a 1-1/2 × 6 screw?
   a. 1/16 inch
   b. 6/32 inch
   c. 9/64 inch
   d. 3/32 inch

7. Which of the following is the coarsest coated abrasive?
   a. 150A
   b. 80D
   c. 120D
   d. 120A

8. Which of the following tools is least likely to be used by a mason?
   a. Combination drill
   b. Jointer
   c. Wood float
   d. Pointed trowel

9. Which of the following tools is least likely to be used by a carpenter?
   a. Nail set
   b. Screwdriver
   c. Aluminum float
   d. Bar clamp

10. Which nail is longest?
    a. 6d box nail
    b. 10d casing nail
    c. 8d common nail
    d. 4d finishing nail

B. Tool Descriptions. Refer to a catalog provided by the instructor to list a complete set of fastening and forming tools. Include a brief description of the important characeristics of each, including price. Also, name at least one building trade that uses each tool. It is not necessary to list fasteners and coated abrasives.

CHAPTER 12

Power Tools

OBJECTIVES

After completing this chapter, you should be able to:

▼ safely perform basic operations with each of the tools listed; and

▼ identify the most outstanding features of the power tools listed.

KEY TERMS

ampere

double insulated

plain bearing

roller bearing

ball bearing

reciprocating saw

portable circular saw

crosscut blade

rip blade

combination blade

electric drill

spade bit

orbital sander

straight-line sander

router

router bit

pneumatic

Power Tool Construction

Power tools perform the same basic operations as hand tools. Work can be done more quickly with power tools because of the speed and power their motors or engines deliver to the working parts. This increased speed of operation requires special features in the construction of power tools.

Motors

The size of the motor on most electrical power tools is specified by the amount of electrical current it uses. Electric current is measured in **amperes** (abbreviated *amps*). A typical motor size for an electric saw, for example, is 8 amps, **Figure 12–1**. The higher the amperage, the higher the power output of the tool. Of course, higher amperage generally means more weight and greater cost.

Battery-Operated Tools

In recent years, several manufacturers have introduced portable power tools that operate on rechargeable batteries, **Figure 12–2**. When not in use, these tools are kept

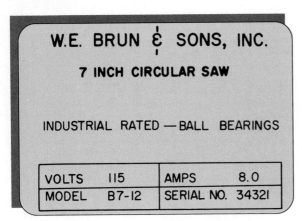

FIGURE 12–1   Power tool identification plate

plugged into a normal house-current electrical outlet. This keeps the batteries charged. In normal use, these tools can be used for several hours without recharging. Although they lack the peak power of conventional power tools, they have plenty of power for most jobs. Battery-operated tools have two advantages over conventional power tools. They are more portable, because they can be used without an outside power source. Also, there is little chance of serious electrical shock, as they operate on low-voltage batteries.

Insulation and Grounding

To protect portable power tool users from electrical shock, some power tools are double insulated.

FIGURE 12–2   Battery operated drill *(Courtesy of Porter-Cable Corporation)*

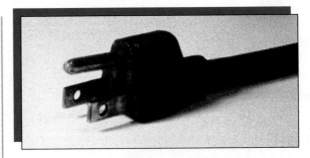

FIGURE 12–3   Three-prong grounding-type plug

Double-insulated tools have two chassis. The inner one contains all of the electrical parts. The outer one is completely insulated from the inner one. In case of an electrical malfunction, no voltage can reach the operator.

Tools that are not double insulated should be connected to an electrical ground. Such tools have a three-wire power cord. The third wire is attached to a grounding prong on the plug, **Figure 12–3**. Grounded electrical outlets are connected to a ground within the system.

Bearings

There are three kinds of bearings used in power tools. **Plain** (or *sleeve*) **bearings** consist of a smooth surface on the inside of a short sleeve, **Figure 12–4**. These bearings have a shorter life expectancy than others. They are not widely used in top-quality tools. *Needle* or **roller bearings** are free-rolling, straight rollers which allow a shaft to roll in its support. **Ball bearings** are another kind of rolling bearings. Ball bearings are generally found where bearings must resist end-to-end as well as side-to-side motion.

CAUTION: PORTABLE POWER TOOL SAFETY

1. Tools that are not double insulated should be connected to an electrical ground.
2. Wear eye protection when operating power tools.
3. Keep all guards and protective devices in place.

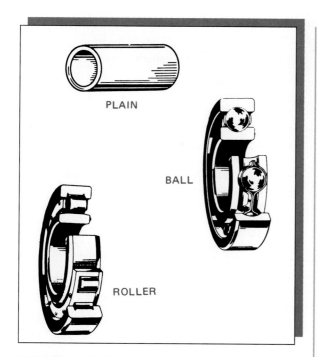

FIGURE 12–4　Bearings used in tools

PLAIN

BALL

ROLLER

4. Do not use defective tools.
5. Unplug the tool when changing bits, blades, and attachments.
6. Unplug the tool when left unattended.
7. Check to see that the cutting edge of the tool will have a clear path as it penetrates the workpiece.
8. Use power tools only after you have received instruction on their use.

Reciprocating Saw

The **reciprocating saw**, **Figure 12–5**, works on the same sawing principles as hand saws. An electric motor drives a gear mechanism, which in turn causes the blade to move up and down. The *shoe*, or base of the saw, adjusts to tilt the saw up to 45 degrees to either side. The blade is held in a chuck, which makes it easier to change.

Blades of various sizes and pitches are available for a variety of sawing jobs, **Figure 12–6**. Some saws also have variable-speed controls to allow even greater flexibility.

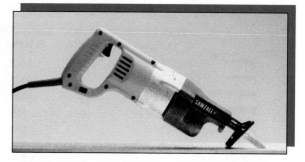

FIGURE 12–5　Reciprocating saw (*Courtesy of Milwaukee Electric Tool Corporation*)

To use the reciprocating saw, follow these steps:

1. Check to see that the proper blade is securely installed.
2. Hold the piece to be cut in a vise, or clamp it securely in place.
3. Plug in the saw.
4. Turn on the saw and start the cut.
5. Avoid trying to cut too sharp a curve as this will break the blade.

Portable Circular Saw

The one power tool used most often by carpenters is the **portable circular saw**, **Figure 12–7**. It consists of a motor, a handle with a trigger switch, a shoe that tilts up to 45 degrees and moves up and down to vary the depth of cut, a blade, and a blade guard. Various blades are available for different operations, **Figure 12–8**. The blade used most often is the **combination blade**. This blade can be used for crosscutting and ripping, the two most commonly performed operations.

Many serious injuries are caused by improper use of circular saws and by use of defective saws. To crosscut with the circular saw, follow these steps:

1. Check to see that a combination or **crosscut blade** is installed securely with the teeth pointing up in front.

	Blade Length in Inches	Teeth Per inch	Blade Width In Inches	Applications
	3	10	5/16	Fast Cutting Set Tooth-Wood
	3	6	5/16	Faster Cutting Shark Tooth
	4 1/4	10	1/2	Fast Cutting Set Tooth
	4 1/4	6	1/2	Faster Cutting Shark Tooth
	4 1/4	10	1/4	Scroll and General Smooth Cutting Soft and Hard Grain Woods-Plywood-Masonite
	4 1/4	6	1/4	Fast Scroll and Rough Cutting Soft and Hard Grain Woods
	4 1/4	10	3/8	General Wood Cutting-Asphalt Tile Fiber-Paper-Plastic-Laminates-Lucite-Plexiglass
	4 1/4	6	3/8	General Rough Cutting Roof Rafters and General Frame Cutting-Plunge Cutting
	3 1/2	10	1/4	Smooth Scroll and Circular Cutting Masonite-Plywood-Soft and Hard Grain Trim Stock-Plastics
	3 1/2	6	1/4	Fast Scroll and Circular Cutting Solid Grain Wood-Masonite-Plastics
	3 1/2	10	3/8	General Straight and Large Curvature Cutting-Solid Grain Wood Plywood-Masonite-Plastics-Soft Aluminum Extrusions
	3	10	1/4	Smooth Scroll and Circular Cutting-Plywood, Straight Grain Wood, Masonite-Plastics-Plunge Cutting
	3	6	1/4	Fast Scroll and Circular Cutting-Plunge Cutting Straight Grain Woods-Hard Board
	3	10	3/8	General Wood Cutting, Fiber, Paper and Plastic Laminates Plexiglass, Rubber Linoleum
	3	6	3/8	Rough Cutting Wood
	2	10	13/64	Smooth Finish Cutting of Straight, Curvature, Round Finish and Trim Materials and Plunge Cutting
	2	—	5/16	Cutting-Cardboard-Cloth-Leather Rubber and Sponge Type Plastics

High-Speed Steel-Fiberglass Cutting

	Blade Length in Inches	Teeth Per inch	Blade Width In Inches	Applications
	4 1/4	6	3/8	Cutting Fiberglass-Fiberglass Bonded to Plywood Sheet Rock-Asphalt Tile-Plastics-Plaster
	2 5/8	6	3/8	Cutting Fiberglass-Fiberglass Bonded to Plywood Sheet Rock-Asphalt Tile-Plaster

Carbide Tip-Problem Material Cutting

	Blade Length in Inches	Teeth Per inch	Blade Width In Inches	Applications
	3 1/2	6	3/8	Cutting Fiberglass, Asphalt Tile, Plastics, Sheet Rock, Plaster and General Wood Cutting

High Speed Steel-Metal Cutting

	Blade Length in Inches	Teeth Per inch	Blade Width In Inches	Applications
	3 5/8	14	3/8	Cutting Brass-Bronze-Copper and Non-Ferrous Metals 5/32 to 1/4" Thick Angle Iron-Mild Steel Sheets and Tubing 5/32 to 1/8" Wall Thickness
	3 1/2	24	1/2	Cutting Window Openings in Steel Core Fire Doors-Copper-Brass and Steel Tubing to 7/8" in Diameter
	3	10	3/8	Cutting Brass-Bronze-Copper-Aluminum to 1/2" in Thickness Steel-Cast Iron to 3/16" in Thickness
	3	14	3/8	Cutting Non-Ferrous Metals to 1/4" in Thickness-Cutting Angle Iron Mild Steel Sheets and Tubing to 1/8" Wall Thickness
	3	24	3/8	Cutting Steel Sheets to 1/8" Cutting Tubing Thin Wall to 1 3/8" Diameter
	3	24	1/4	Cutting Steel Sheets and Tubing 3/32 to 1/8" Wall Thickness
	1 3/4	14	1/4	Cutting Steel Sheets and Tubing 3/32 to 1/8" Wall Thickness
	1 3/4	24	1/4	Cutting Steel Sheets and Tubing 3/32 to 1/8" Wall Thickness

FIGURE 12–6   Reciprocating saw blades

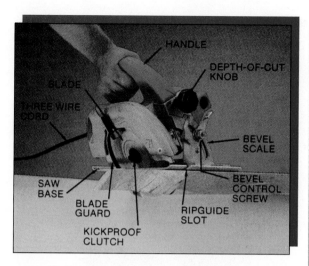

FIGURE 12–7 Circular saw *(Courtesy of Milwaukee Electric Tool Corporation)*

CAUTION: Do not overtighten the blade screw. Most saws have a special washer that allows the blade to slip if it binds. This prevents the saw from kicking back toward the operator.

2. Adjust the shoe so that the blade protrudes through the bottom of the stock about 1/2 inch.
3. Check the blade guard to see that it works easily.
4. Rest the stock on saw horses so that none of the portion to be cut off is supported.

CAUTION: The stock to be cut off should not be held or supported. This may cause the blade to bind.

5. Mark a line on the stock where it is to be cut.
6. Plug in the saw.
7. Grasp the saw by the two handles provided and turn it on.
8. Rest the forward end of the shoe on the stock and push the blade into the stock slowly, but firmly. Avoid turning the saw from side to side.
9. After the cut is completed, release the trigger but hold the saw until the blade stops.
10. Unplug the saw.

The **power miter saw**, sometimes called a power miter box or chop saw, is a circular saw specially designed for making accurate miters and cutoffs, **Figure 12–9**. It is a circular saw with a small table to hold the work. The saw can be adjusted from side to side for cuts at any angle up to 45 degres. A very important feature of a power miter saw is the precision of the bearings and other moving parts. Because these parts are manufactured to very close tolerances, there is very little side-to-side play in the action of the saw or wobble in the blade. The power miter saw can accommodate the same blades as any other circular saw, but carbide tipped blades are used most often, because they produce the finest cuts.

CAUTION: It is generally necessary to hold the work with one hand fairly close to the path of the saw blade. Before turning on the saw, check to see that the blade is completely covered by the proper guards and that all guards work properly. Check twice to see that your hand is at least several inches from the path of the blade. Do not attempt to cut a piece that is too small to be held easily.

Electric Drills

Electric drills are used by workers in most of the building trades, **Figure 12–10** Electric drill sizes are listed according to the maximum size bit that can be held in the chuck of the drill. The most common sizes are 1/4 inch, 3/8 inch, and 1/2 inch. Some electric drills have variable-speed controls and reversing switches. For drilling in concrete and masonry, a low speed is used. For drilling in softwood, a high speed is used. The reversing switch allows the drill to be backed out of a hole.

Any of the bits that can be used in a hand drill can also be used in an electric drill. **Spade bits** can also be used in electric drills, **Figure 12–11** Spade bits are available in larger sizes than bits used in hand drills.

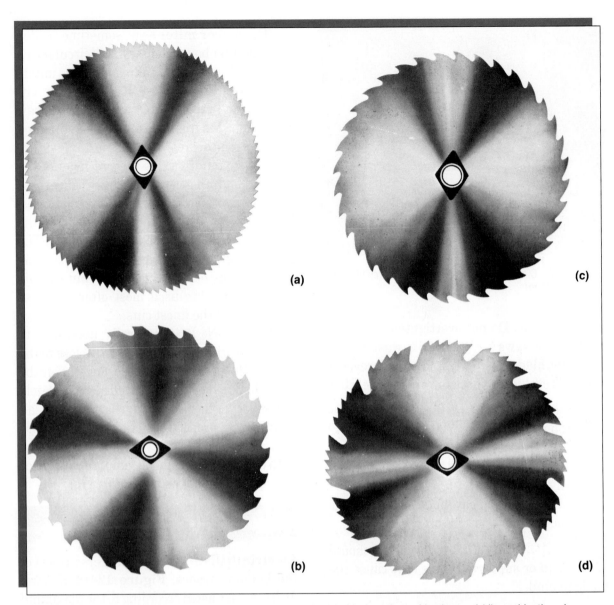

FIGURE 12–8 Types of circular saw blades: (a) crosscut, (b) rip, (c) chisel-tooth combination, and (d) combination planer

FIGURE 12–9 Power miter saw

FIGURE 12–10 Portable electric drill *(Courtesy of Milwaukee Electric Tool Corporation)*

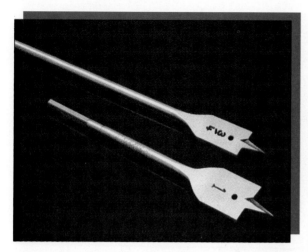

FIGURE 12–11 Spade bits *(Courtesy of the Irwin Company)*

Finishing Sanders

The finishing sander, **Figure 12–12** is a lightweight power tool for moving coated abrasives. The abrasive is moved in one of two ways, depending on the type of sander. The **orbital sander** moves the abrasive in a small circular path. The **straight-line sander** moves the abrasive in a straight, back-and-forth motion, **Figure 12–13**

Most finishing sanders hold either 1/3 or 1/4 of a standard sheet of coated abrasive. The abrasive paper can be torn accurately by creasing it, then tearing it over the corner of a

FIGURE 12–12 Finishing sander

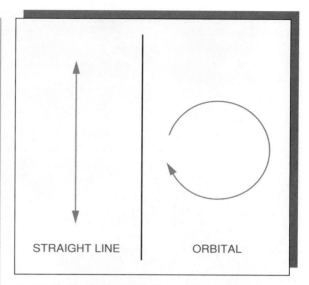

FIGURE 12–13 Action of finishing sanders

workbench or board. The pad on the bottom of the sander has two clamps to hold the edges of the abrasive. To use the sander, it is simply turned on and moved slowly around the surface to be smoothed.

Routers

Carpenters and cabinetmakers use routers to shape the edges of woodwork, cut out openings in finished millwork, cut recesses for door hinges (called *gains*), and trim plastic facing material. The **router** is a high-speed motor with a chuck to hold various cutters and a base to control the router, **Figure 12–14** The base is moved up and down on the motor to control the amount the **router bit** protrudes into the work. Some type of fine-adjustment device is also included.

Router Bits

One manufacturer makes over 40 types of router bits, **Figure 12–15** Several shapes are available for producing a decorative edge on wood. Others are used on the surface of the work for decorative or functional cuts. Some

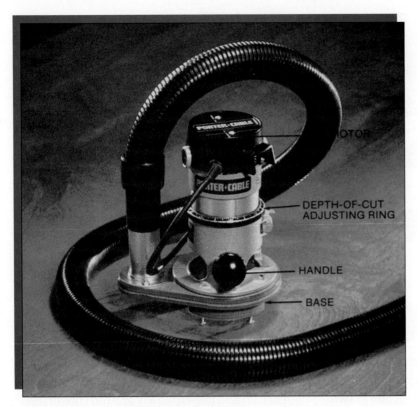

FIGURE 12–14   Router *(Courtesy of Porter-Cable Corporation)*

router bits are designed for piercing and cutting panels. Some trim the edge of plastic laminates used on counter tops.

Router bits used on plastic have carbide cutting edges. Bits used on wood are available with either steel or carbide cutting edges. The cost of carbide bits is higher, but they stay sharp much longer and produce smoother cuts. Some router bits have a steel (**Figure 12–16**) or ball-bearing pilot (**Figure 12–17**) to guide the router. To use other types, a guide is attached to the router.

Shaping an Edge with the Router

1. Insert a bit with a pilot in the router chuck.
2. Adjust the base of the router so the desired portion of the bit is exposed.
3. Clamp the workpiece to a bench so the edge to be shaped extends beyond the edge of the bench.
4. Plug the router in and turn it on.

5. Place the router base on the left end of the workpiece and move the router from left to right. If the bit does not take a full cut on the first pass, make a second pass to finish the cut.
6. Turn off and unplug the router.

Router Edge Guide

1. Insert the desired bit in the router chuck.
2. Adjust the router base for the desired depth of cut.
3. Mount the edge guide on the router base. Adjust the edge guide so that the distance from the bit to the guide is equal to the desired distance from the edge of the workpiece to the cut.
4. Clamp the workpiece to a bench. Check to see that the clamps are positioned where they will not interfere with the router.
5. Plug in the router and turn it on.
6. Place the router on the end of the work

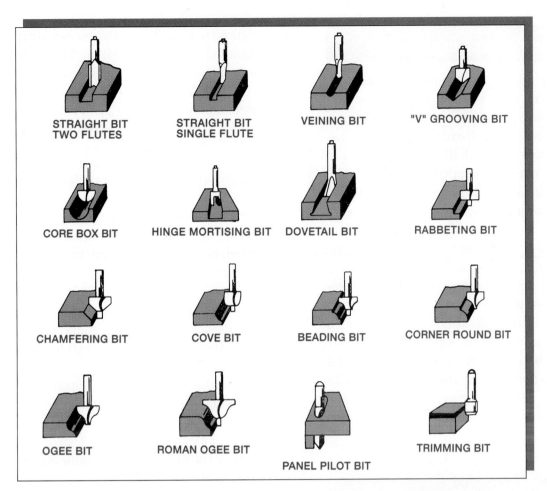

STRAIGHT BIT TWO FLUTES

STRAIGHT BIT SINGLE FLUTE

VEINING BIT

"V" GROOVING BIT

CORE BOX BIT

HINGE MORTISING BIT

DOVETAIL BIT

RABBETING BIT

CHAMFERING BIT

COVE BIT

BEADING BIT

CORNER ROUND BIT

OGEE BIT

ROMAN OGEE BIT

PANEL PILOT BIT

TRIMMING BIT

FIGURE 12–15   Router bits

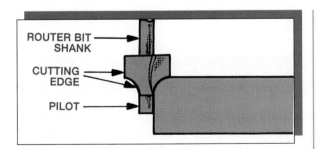

ROUTER BIT SHANK

CUTTING EDGE

PILOT

FIGURE 12–16   Router bit, with pilot to follow edge of work

with the edge guide against the edge of the stock.

7. As the router is pushed through the stock, be careful to keep the edge guide against the edge of the stock.

8. Turn off and unplug the router.

FIGURE 12–17   Router and bit with ball-bearing pilot

Gasoline-Powered Equipment

Gasoline-powered equipment is common on a construction site, **Figure 12–18** Gasoline-powered water pumps are used to *dewater* (take the water out of) excavations. Gasoline-powered carts are used, like wheelbarrows, to transport concrete and other heavy materials. Power trowels are used in concrete finishing. Gasoline-powered electrical generators are very common.

Most of these devices use a small gasoline engine similar to those found on lawn tractors. Operating principles and maintenance procedures for gasoline engines would be beyond the scope of this book. Manufacturers provide instructions for using and servicing their engines. Those instructions must be followed if the engine and the equipment it drives are to give good service. Many of the problems operators have with small gasoline engines can be traced to lack of required service, dirty or contaminated fuel, and attempts by untrained operators to perform maintenance.

CAUTION: Many injuries and deaths result from improper use of gasoline-powered equipment. These safety rules must be observed:

▼ Read and understand the manufactur-er's instructions before you operate any power equipment. Your teacher may prefer to go over the operating procedures with you instead of having you read the manufacturer's instructions.

▼ Never operate a gasoline engine indoors or in any area that is not well ventilated.

▼ Store gasoline in properly marked containers that have been approved for that purpose.

▼ Do not strike a match or hold an open flame within 50 feet of gasoline.

▼ Do not try to override the governor or speed control on a gasoline engine.

▼ Do not fill the fuel tank of a gasoline engine while the engine is running.

▼ Before attempting any adjustments or service, stop the engine and disconnect the sparkplug wire.

FIGURE 12–18 Gasoline-powered generator *(Courtesy of Multiquip Inc.)*

Air-Powered Tools

Air-powered tools are often great timesavers. Many of the nailing jobs in construction can be done much more quickly with **pneumatic** nailers, **Figure 12–19** (*Pneumatic* means air-powered.) The pneumatic nailer holds up to 300 special nails that can be purchased in lengths from 1-1/2 inches to 3-1/4 inches. With small nails, a pneumatic nailer can be used for finish carpentry, such as installing molding. With longer nails, a pneumatic nailer can be

FIGURE 12–19   This carpenter is using a pneumatic nailer to build a wall frame. *(Courtesy of Duo-Fast Corporation)*

used for applying sheathing and subflooring, or even framing.

CAUTION: Air may seem harmless, but it can be dangerous. Under the pressure normally used for pneumatic tools, air from an air hose can penetrate the skin or do permanent damage to the eyes. Never point a high-pressure air hose at a person or use it to blow dust from your clothes.

ACTIVITIES

Reciprocating Saw and Electric Drill

Equipment and Materials

Reciprocating saw with wood-cutting blade
Portable electric drill
1-inch spade bit
Piece of wood, 3/4″ × 2-1/2″ × 24″ (finished size)
80-grit coated abrasive

Procedure

1. Lay out the tote carrier handle as shown in **Figure 12–20**.

2. Saw the outside shape of the handle with a reciprocating saw.
3. Mark the locations of the two holes for the handle, **Figure 12–20**.
4. Drill a 1-inch hole at each of the locations marked in step 3.
5. Connect the outside edges of these holes with straight pencil lines, **Figure 12–21**.
6. Saw out the remainder of the handle with the reciprocating saw.
7. Smooth all sawed surfaces with a file, then 80-grit abrasive.

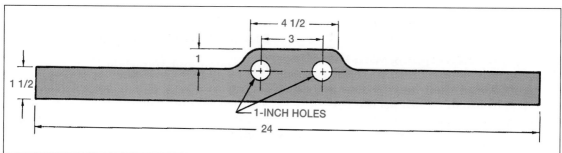

FIGURE 12–20   Tote carrier handle

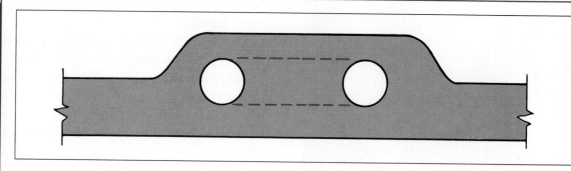

FIGURE 12–21 Saw out handle along dotted lines.

Portable Circular Saw

Equipment and Materials

Portable circular saw
Piece of wood, 3/4″ × 9-1/2″ × 24″ (finished size)
Square
Tape measure
Straightedge and pencil
Jack plane

Procedure

1. Cut one end of the stock square with the circular saw.

CAUTION: Adjust the blade of the saw to extend through the stock a maximum of 1/2 inch.

2. Mark a second square line 24 inches from the square end.
3. Cut the stock off at this point.
4. Plane one edge of the stock if it has not already been planed.
5. Mark a line 9-1/2 inches away from the planed edge and parallel to it.
6. Rip the stock about 1/4 inch wider than the line.

CAUTION: When ripping with the circular saw, be careful not to let the saw kerf pinch the blade. Also, be sure the blade does not strike the workbench or saw horse.

7. Plane the sawed edge to the line.

NOTE: This piece is the bottom of the tote carrier.

8. Repeat steps 1 through 7, or use hand tools, to cut two pieces 3/4″ × 6″ × 25-1/2″. These will be used for the sides of the tote carrier. When all pieces are cut to size, the carrier can be assembled as shown in **Figure 12–22**.

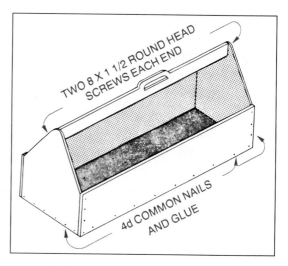

FIGURE 12–22 Tote carrier

Shaping an Edge with a Router

Equipment and Materials

Router
1/4-inch rounding-over bit
Handle for tote carrier (cut out in the first activity)

Procedure

1. Insert the bit in the router.

2. Adjust the router base so the top of the curved cutting edge is flush with the router base.
3. Clamp the tote carrier handle to a work surface.
4. Plug in the router. Shape all edges of the stock except the ends.

NOTE: It will be necessary to reposition the clamps after part of the work is shaped.

5. Turn the stock over and shape the edges on the second side.

Applying Construction Across the Curriculum

Social Studies

What were the discoveries or developments that led to the invention and growth in the use of power tools? Try to find more than just the discovery of electricity. For example, why were power circular saws slow to be accepted after their invention?

Communications

Make a safety poster about the use of power tools. It is more important to design a poster that will be noticed and read by all and send a powerful message than to try to include everything you know on one poster. Make your poster as tasteful and attractive as possible so that it can serve a real function in your school.

REVIEW

A. Multiple Choice. Select the best answer for each of the following questions.

1. What is the main advantage of power tools over hand tools?
   a. Power tools normally outlast hand tools.
   b. Greater accuracy is possible with power tools.
   c. Power tools can perform an operation more quickly.
   d. Power tools can perform operations that cannot be performed with hand tools.
2. How is the size of the electric motor on a power tool usually specified?
   a. Torque
   b. Speed
   c. Volts
   d. Amperes
3. What is the purpose of the third wire on the electrical cord of a power tool?
   a. To connect the tool to an electrical ground
   b. To allow the tool to be used with 220-volt electricity
   c. To carry the electrical current if one wire breaks
   d. The third wire is the switch leg.
4. Which of the following materials cannot be cut with a reciprocating saw?
   a. Wood
   b. Plastics
   c. Metals
   d. All of these can be cut with a reciprocating saw.

5. When using a portable circular saw, how much should the blade extend through the stock?
   a. 1 inch
   b. 1/2 inch
   c. It should be adjusted so the teeth just reach the far side of the stock.
   d. The depth of cut cannot be controlled on a circular saw.

6. In which direction should the teeth of a portable circular saw blade point?
   a. Up in front
   b. Down in front
   c. Opposite to the direction of rotation
   d. It depends on the type of blade being used.

7. Which bit cannot be used in an electric drill?
   a. Auger bits with a square tang
   b. Twist drill bits
   c. Spade bits
   d. Combination drill bits

8. What is the advantage of carbide router bits?
   a. They last longer.
   b. They produce smoother cuts.
   c. They can be used for wood and plastics.
   d. All of these.

9. Which operation cannot be done with a router?
   a. Trim plastic laminates
   b. Cut hinge gains in doors
   c. Cut openings in millwork
   d. Trim sheet metal.

10. Which portable power tool is most apt to be used by carpenters, electricians, and plumbers?
    a. Router
    b. Finishing sander
    c. Circular saw
    d. Electric drill

B. Tool Descriptions. Refer to catalogs provided by the instructor to list a set of portable power tools. Include one of each of the tools discussed in this unit. Also include outstanding features, size, price, and attachments available. Indicate which building trades are most likely to use each of the tools listed.

Section Four

Light Construction Systems

This section focuses on the processes of light construction. In this section, the resources from earlier sections come together in the processes that result in constructed output. It is through these processes that structures take shape.

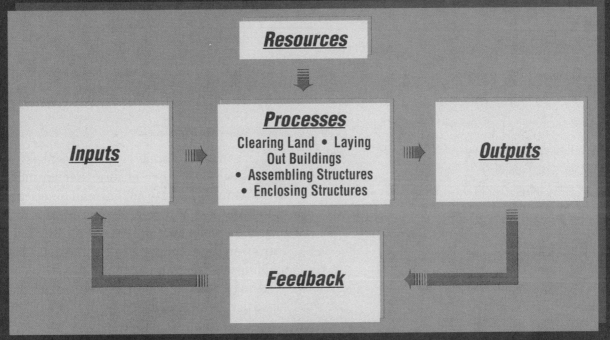

Chapter 13
Site Work
prepares the site for work to begin on the structure. This unit includes laying out the building lines, excavation, and rough grading.

Chapter 14
Foundation Systems
begins the structure. Two major classes of building foundations are explained: slab-on-grade, or mat, and foundation walls.

Chapter 15
Floor Systems
begins with an explanation of the 4-inch module upon which all building dimensions are based. The unit then explains the construction of a typical joist-framed floor.

Chapter 16
Wall Framing Systems
includes the framing of openings for windows and doors, sheathing, and special considerations for energy efficiency.

Chapter 17
Roof Systems
explains how to read a rafter table and develops an understanding of simple roof frames and roof coverings.

Chapter 18
Enclosing the Structure
covers the processes necessary to enclose the framed building completely: cornices, windows, doors, siding, masonry veneer, and stucco.

CHAPTER 13

Site Work

OBJECTIVES

After completing this chapter, you should be able to:

▼ interpret a plot plan;
▼ lay out building lines; and
▼ erect batter boards.

KEY TERMS

environmental impact study

setback

plot plan

finished grade

existing grade

solar orientation

batter boards

excavation

frost line

existing grade

bench mark

fill

grade stake

cut

Government Regulation of Building Sites

An important role of government is to give us rules to ensure that an individual's acts are not harmful to society. What one person constructs can have a big impact on many others. Local, state, and federal governments all have rules to control how building sites are developed.

A major area of concern is the impact the construction will have on the environment. If there is any chance that the construction project will affect the surrounding environment, the developer or owner might be required to complete an environmental impact study. An **environmental impact study** examines such things as how the drainage of a site will affect the area, **Figure 13–1**. If the site work changes the slope of the land or if wetlands are involved, there will probably be some environmental impact. The developer will probably be required to have methods designed to ensure that the project will not harm the environment. Sometimes drainage pipes will solve the problem, or the slope of the site might be altered. In extreme cases, such as where the

14-16-4 (2/87)—Text 12

PROJECT I.D. NUMBER

617.21
Appendix C
State Environmental Quality Review

SHORT ENVIRONMENTAL ASSESSMENT FORM
For UNLISTED ACTIONS Only

SEQR

PART I—PROJECT INFORMATION (To be completed by Applicant or Project sponsor)

1. APPLICANT /SPONSOR	2. PROJECT NAME

3. PROJECT LOCATION:

   Municipality   County

4. PRECISE LOCATION (Street address and road intersections, prominent landmarks, etc., or provide map)

5. IS PROPOSED ACTION:

☐ New ☐ Expansion ☐ Modification/alteration

6. DESCRIBE PROJECT BRIEFLY:

7. AMOUNT OF LAND AFFECTED:

   Initially _____ acres   Ultimately _____ acres

8. WILL PROPOSED ACTION COMPLY WITH EXISTING ZONING OR OTHER EXISTING LAND USE RESTRICTIONS?

☐ Yes ☐ No   If No, describe briefly

9. WHAT IS PRESENT LAND USE IN VICINITY OF PROJECT?

☐ Residential ☐ Industrial ☐ Commercial ☐ Agriculture ☐ Park/Forest/Open space ☐ Other
Describe:

10. DOES ACTION INVOLVE A PERMIT APPROVAL, OR FUNDING, NOW OR ULTIMATELY FROM ANY OTHER GOVERNMENTAL AGENCY (FEDERAL, STATE OR LOCAL)?

☐ Yes ☐ No   If yes, list agency(s) and permit/approvals

11. DOES ANY ASPECT OF THE ACTION HAVE A CURRENTLY VALID PERMIT OR APPROVAL?

☐ Yes ☐ No   If yes, list agency name and permit/approval

12. AS A RESULT OF PROPOSED ACTION WILL EXISTING PERMIT/APPROVAL REQUIRE MODIFICATION?

☐ Yes ☐ No

I CERTIFY THAT THE INFORMATION PROVIDED ABOVE IS TRUE TO THE BEST OF MY KNOWLEDGE

Applicant/sponsor name: _____   Date: _____

Signature: _____

If the action is in the Coastal Area, and you are a state agency, complete the Coastal Assessment Form before proceeding with this assessment

OVER

FIGURE 13–1 Environmental impact assessment form

PART II—ENVIRONMENTAL ASSESSMENT (To be completed by Agency)

A. DOES ACTION EXCEED ANY TYPE I THRESHOLD IN 6 NYCRR, PART 617.12?   If yes, coordinate the review process and use the FULL EAF.
☐ Yes   ☐ No

B. WILL ACTION RECEIVE COORDINATED REVIEW AS PROVIDED FOR UNLISTED ACTIONS IN 6 NYCRR, PART 617.6?   If No, a negative declaration may be superseded by another involved agency.
☐ Yes   ☐ No

C. COULD ACTION RESULT IN **ANY** ADVERSE EFFECTS ASSOCIATED WITH THE FOLLOWING: (Answers may be handwritten, if legible)

C1. Existing air quality, surface or groundwater quality or quantity, noise levels, existing traffic patterns, solid waste production or disposal, potential for erosion, drainage or flooding problems? Explain briefly:

C2. Aesthetic, agricultural, archaeological, historic, or other natural or cultural resources; or community or neighborhood character? Explain briefly:

C3. Vegetation or fauna, fish, shellfish or wildlife species, significant habitats, or threatened or endangered species? Explain briefly:

C4. A community's existing plans or goals as officially adopted, or a change in use or intensity of use of land or other natural resources? Explain briefly.

C5. Growth, subsequent development, or related activities likely to be induced by the proposed action? Explain briefly.

C6. Long term, short term, cumulative, or other effects not identified in C1-C5? Explain briefly.

C7. Other impacts (including changes in use of either quantity or type of energy)? Explain briefly.

D. IS THERE, OR IS THERE LIKELY TO BE, CONTROVERSY RELATED TO POTENTIAL ADVERSE ENVIRONMENTAL IMPACTS?
☐ Yes   ☐ No   If Yes, explain briefly

PART III—DETERMINATION OF SIGNIFICANCE (To be completed by Agency)

INSTRUCTIONS: For each adverse effect identified above, determine whether it is substantial, large, important or otherwise significant. Each effect should be assessed in connection with its (a) setting (i.e. urban or rural); (b) probability of occurring; (c) duration; (d) irreversibility; (e) geographic scope; and (f) magnitude. If necessary, add attachments or reference supporting materials. Ensure that explanations contain sufficient detail to show that all relevant adverse impacts have been identified and adequately addressed.

☐ Check this box if you have identified one or more potentially large or significant adverse impacts which **MAY** occur. Then proceed directly to the FULL EAF and/or prepare a positive declaration.

☐ Check this box if you have determined, based on the information and analysis above and any supporting documentation, that the proposed action **WILL NOT** result in any significant adverse environmental impacts **AND** provide on attachments as necessary, the reasons supporting this determination:

_____
Name of Lead Agency

_____   _____
Print or Type Name of Responsible Officer in Lead Agency   Title of Responsible Officer

_____   _____
Signature of Responsible Officer in Lead Agency   Signature of Preparer (If different from responsible officer)

_____
Date

2

FIGURE 13–1   (Continued)

habitat of an endangered species will be destroyed, the construction project may not be permitted on the desired site.

Site Plans

The building and zoning ordinances of most communities regulate the placement of buildings on the site. Usually the side and front setbacks are regulated. **Setback** is a term to describe the distance a building is allowed to be from the property lines or street. By specifying minimum setbacks, an ordinance can ensure that no single building is closer to the street than its neighbors. Side setback limits prevent a building from crowding its neighbors.

Local governments usually require that a **plot plan**, **Figure 13–2**, be included as a part of the set of working drawings. The *plot plan* indicates the location of the proposed building, location of trees to remain on the site, and the finished grade. The **finished grade** is the elevation above sea level of the site after construction is completed. The plot plan, or site plan, for large construction projects is prepared by the architect or surveyor. In residential construction, working drawings are often obtained from a planning service and the architect never sees the construction site. In this case, the general contractor often prepares a plot plan.

Solar Orientation

The location of a building in relation to its site is important. Often the site serves a functional role in the total design. Orientation to the sun, usual wind direction, position of trees, and the effects of neighboring buildings can be used to achieve maximum energy efficiency. This is called **solar orientation**.

Deciduous trees, which lose their leaves in the winter, can be used to control the solar energy striking a building. In the winter, the sun shines through the deciduous trees on the south side of a building. In the summer, the trees shade the south side of the building, **Figure 13–3**.

On some sites air currents of normal wind direction affect the best orientation of the buildings. Moving the house in **Figure 13–4** farther up the hill could raise the outside temperature 10°F in the summer. This would increase the cost of air conditioning.

Laying Out Building Lines

After the plot plan is submitted and a building permit is obtained, the position of the new building is staked out. Stakes are driven into the ground where the corners of the building will be.

The front line of the building is usually established first. If the building is to be parallel with the street, this line can be found by measuring from the curb or centerline of the street. A line is stretched between two stakes to indicate where the front building line will be. The stakes are driven at the locations of the front corners of the building, **Figure 13–5**.

The sides of the building are found by laying out lines at right angles to the front line. Right angles can be measured by the 6-8-10 method explained in Chapter 9. However, corners are usually laid out by using a builder's level or transit. The instrument is set up over a corner stake. If the builder's level is used, it is then focused on a target rod held over the other front corner stake, **Figure 13–6**. The advantage of a transit for laying out corners is that it can be aimed directly at the stake, eliminating the target rod. With the level or transit in this position, the 360-degree scale is

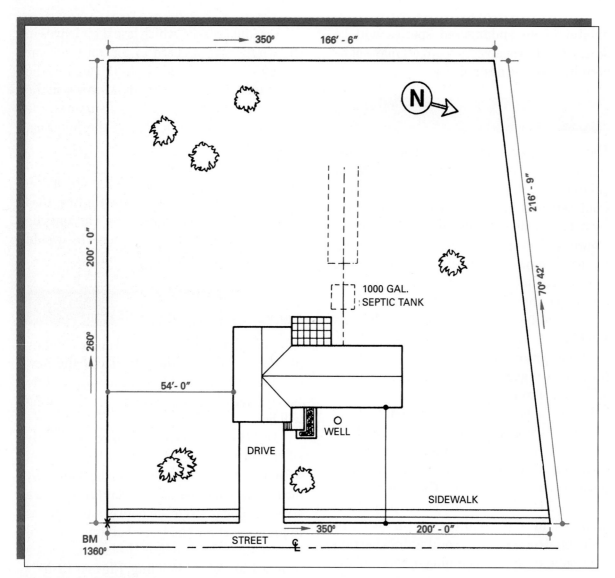

FIGURE 13–2 Typical plot plan

adjusted to zero. Then the telescope is turned so the scale reads 90 degrees. The telescope is now aimed at the side building line.

When both side lines have been established, the back corners are found by measuring from the front corners. If the building is irregular in shape, it is divided into several smaller rectangles and laid out in the same way, **Figure 13–7**.

The squareness of the layout should be checked by measuring the diagonals. If all of the corners are 90 degrees, the diagonals are equal, **Figure 13–8**. It is very important for the building lines to be accurate. If the lines are not correct, many materials will have to be cut during construction, wasting a great amount of labor.

Erecting Batter Boards

Batter boards are usually erected under the direct supervision of the general contractor. Batter boards allow the corner stakes to be removed without losing the building lines.

Batter boards are erected at each corner about 5 feet outside the building lines. By placing them away from the building lines, the batter boards are not disturbed when the excavation (earth moving) is done. Four 2″ × 4″ stakes are driven into the ground and two pieces of 1″ × 6″ or 1″ × 8″ lumber are nailed to them, **Figure 13–9**. A line stretched over the

FIGURE 13–3   The winter sun passes through deciduous trees. The summer sun is blocked by deciduous trees.

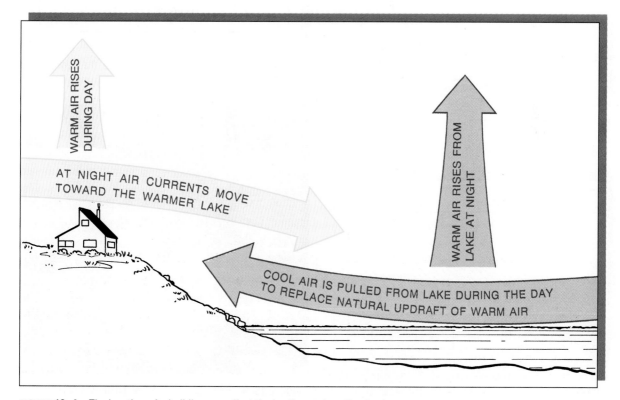

FIGURE 13–4   The location of a building can affect the heating and cooling load.

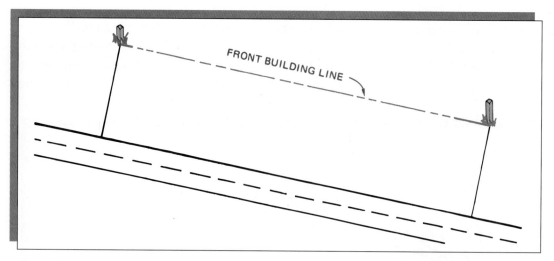

FIGURE 13–5 The first step in laying out a building is to drive stakes for the front building line.

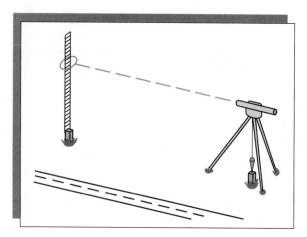

FIGURE 13–6 The level or transit is first lined up on the front building-line stakes.

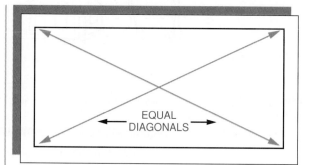

FIGURE 13–8 When all corners are square, the diagonals are equal.

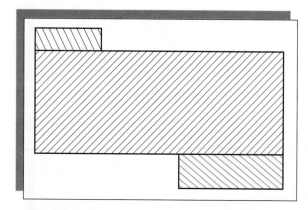

FIGURE 13–7 Irregularly shaped buildings are divided into smaller rectangles, then each part is laid out separately.

FIGURE 13–9 Batter boards are erected at each corner.

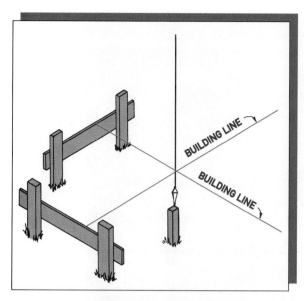

FIGURE 13–10 A plumb bob is suspended over the corner stake to find the point where the building lines cross.

corner stakes is used to mark the building lines on the batter boards. To insure that the line is directly over the corner stake, a plumb bob is suspended over the stake, **Figure 13–10** When the building line has been accurately marked on the batter boards, a saw kerf about 1/4 inch deep marks it permanently. The batter boards will be used to guide the excavation and construction of the foundation.

Excavation

Even small buildings have great weight. The weight of the building must be supported by the earth or rock directly beneath it. Because of this fact, it is important to consider the soil on which the building will rest.

If the area has been a landfill site, the earth and fill may not be fully compacted. Landfill sites also often contain *organic materials* (materials that will decompose, such as wood and garbage) that will eventually cause the area to settle. Some kinds of earth compact more easily than others.

Most soil contains some moisture. In many parts of the country, this moisture freezes during the winter. As the ground freezes, it swells just as water does. When the ground thaws in the spring, it contracts. The depth to which earth freezes in winter is called the **frost line**. This swelling and contracting creates severe problems if it occurs under the building.

To overcome these problems of unstable soil, the building rests on a base below the frost line and on solid earth. To construct this base, the earth must be *excavated* (dug out), **Figure 13–11** Often, a cellar area is excavated and included in the structure. Buildings that do not have cellars either have a concrete floor directly on the soil or a crawl space. A *crawl space* is an area big enough to permit access to electrical and mechanical systems under the floor.

The excavation is usually done by an excavating contractor. Excavating contractors use a variety of heavy equipment to complete the excavation. If the earth must be removed, the excavating contractor generally hauls it away in dump trucks. Excavators must know how to interpret plot plans and use a builder's level to accurately measure the excavation.

Grading

The excavator usually grades the land surrounding the building lines to the rough grade. The *rough grade* is the elevation and slope of the finished grade minus a few inches for top soil. If the finished grade is substantially different from the **existing grade** (grade before construction), a surveyor indicates where the finished grade will be. The surveyor uses a bench mark as a starting point. **Bench marks** are fixed points indicating the elevation above sea level at that point, **Figure 13–12** The surveyor determines grade lines by adding to

FIGURE 13-11 Excavating for a slab-on-grade foundation

FIGURE 13-12 A bench mark can be an official marker or any stationary object of known elevation.

or subtracting from the elevation at the bench mark.

Sometimes added **fill**, or earth, is needed to bring the area up to finished grade. The surveyor marks the desired level on **grade stakes**, **Figure 13-13** When earth must be removed, or **cut**, the surveyor indicates the required depth of excavation on the grade stakes.

FIGURE 13-13 Grade stakes tell the excavator where the finished grade should be.

ACTIVITIES

Establishing Building Lines

Before excavation begins, the corners of the proposed structure must be staked. A surveyor usually marks the building lines of large commercial projects. On lighter commercial buildings and residential construction, the general contractor often does this work.

Establishing these points requires accurate measurements. Work carefully to lay out the building in this activity.

Equipment and Materials

100-foot tape measure
Ball of line
4 wood stakes, 2″ × 2″ × 2′
16 wood stakes, 2″ × 4″ × 4′
8 boards, 1″ × 6″ × 4′
4-pound hammer
Plumb bob
Hand crosscut saw

Procedure

1. Drive two stakes 20 feet apart and 10 feet back from the property line. The front property line will be designated by the instructor.
2. Use the 6-8-10 method to lay out the side building lines.
3. Drive stakes on the side building lines and 30 feet from the front line. These stakes should be 20 feet apart.
4. Check the diagonals to be sure that the building lines are square.
5. Drive four 2″ × 4″ stakes about five feet outside the building lines at each corner. Refer to **Figure 13–9**.
6. Nail a 1″ × 6″ board to each pair of stakes to complete the batter boards. The 1″ × 6″ pieces should be nearly level to make it easier to locate the building lines.
7. Attach a line to one batter board and stretch it over the corner stakes.
8. Hold a plumb bob over the corner

stakes to check the position of the lines.
9. Repeat steps 7 and 8 for all four sides of the building.
10. Saw a kerf approximately 1/4 inch deep to mark the location of the building lines on the batter boards.

Plot Plan

When planning a project, the architect or surveyor prepares a plot plan. This shows where the building and other features are located on the site. It also indicates the finish grade for each corner of the building so the excavator knows how much to fill or cut. The general contractor also uses the plot plan to stake out the house and locate such features as the sidewalks, wells, gas and water lines, and landscaping.

Equipment and Materials

Architect's scale
Paper and pencil
30-60-90° triangle

Procedure

Draw a plot plan for a house, 26 feet wide (front and back) and 48 feet long. The house is to be situated on a 200 foot by 200 foot lot. The house is to face a street on the south side of the lot. The setback is to be 50 feet. The house is to be centered between the east and west boundaries. The elevation at a bench mark on the edge of the street is 474 feet. The finished grade is to be 478 feet at the front of the house and 477 feet at the back of the house. The northeast corner of the lot is at 470 feet and the northwest corner is at 467 feet. Include a 3-foot sidewalk beside the street and a 3-foot walk leading to the front entrance, 24 feet from the west end of the house. A 9-foot driveway leads to a garage in front of the west end. Draw the plot plan to a scale of 1/16″ = 1′-0″.

Applying Construction Across the Curriculum

Social Studies

Obtain a copy of a short environmental impact assessment form. Your teacher will tell you what form to use. It might be a copy of the one in **Figure 13–1** or it might be the one that is used in your area. Choose any site where you would like to build a teen activity center. If there are no open sites in your area, you may assume that existing structures will be demolished. Complete the owner/developer part of the short environmental assessment review form for your teen center.

Organize a group of two to four students to act as an Environmental Review Board. The Environmental Review Board will review the application for environmental approval to build a teen center.

Science and Mathematics

The position of the sun relative to the Earth changes constantly and is different for any point on the Earth. Architects and solar designers need to consider this angle in their designs. The sun's position can be calculated by formulae that are published in reference books or it can be calculated by software based on these formulae. Using library references or software, which might be provided by your teacher, determine the angle of elevation of the sun above the horizon in your city at noon on the first day of summer. (This is called the *angle of elevation.*) If you have difficulty finding information on the sun's angle of elevation write to the U.S. Naval Observatory, 3450 Massachusetts Ave., N.W., Washington, DC 20392-5420, and ask for Circular #171. The latitude and longitude of your city or town can be found on most maps or in an atlas.

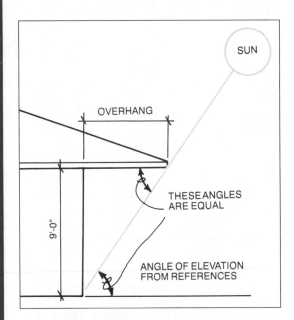

FIGURE 13–14 One side (the height of the wall) and one angle (the angle of elevation) are known. Solve for the other side (the overhang).

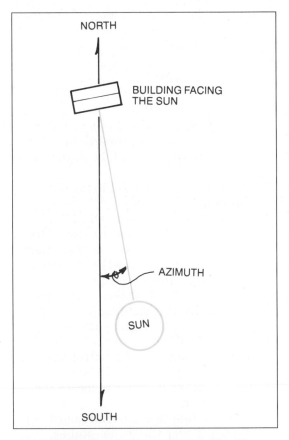

FIGURE 13–15 The sun rises on the east side of the north-south line and crosses the line (0° azimuth) during the day to set on the west side of the line.

Using this information calculate the amount of roof overhang necessary to completely shade a 9 foot high wall from the sun on the first day of summer. Hint: Begin by drawing a vertical line to represent the wall. Then draw a line at the angle of elevation from the bottom of the wall toward the sun. Draw a horizontal line to represent the overhang of the roof, **Figure 13–14** Now solve the right triangle, using trigonometry.

In your search for the angle of elevation you will probably see the terms *azimuth* and *angle of declination*. Azimuth refers to how far to one side or the other of a true north-south line the sun is positioned. To be exposed to the greatest amount of the sun's energy a surface should face the azimuth of the sun at the angle of elevation of the sun, **Figure 13–15** The angle of declination refers to the angle of the sun relative to the surface that would be created by slicing the Earth in half at the equator, **Figure 13–16** For any position not on the equator the angle of elevation will be different from the angle of declination.

Mathematics

If the surface of a concrete slab is to be at an elevation of 1441.62 feet and the nearest bench mark is at an elevation of 1435.12 feet, how far above or below the bench mark is the surface of the slab? If a second slab is to be placed 14 inches above the first, how far above or below the bench mark is the second slab?

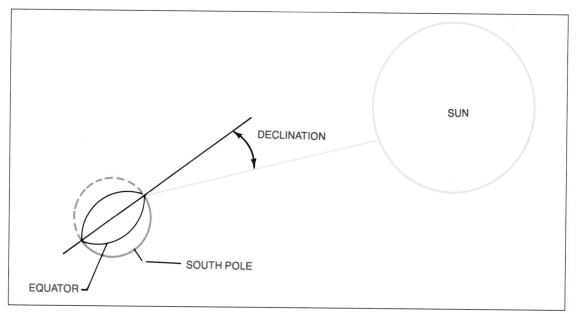

FIGURE 13–16 The angle of declination is the angle at which the sun's rays strike the equator, but as you move away from the equator, the angle of elevation becomes different from the angle of declination.

REVIEW

1. Refer to **Figure 13–17** to determine the amount of cut (earth removed) and fill (additional earth) needed for the new building in **Figure 13–18** Write the number of feet and the word "cut" or the word "fill" for each of the lettered points.

2. List all of the occupations that were probably involved with the site planning, site preparation, and earthwork for your school.

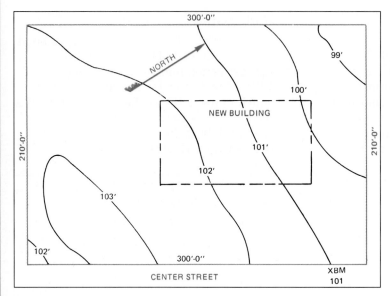

FIGURE 13–17   Topographical map

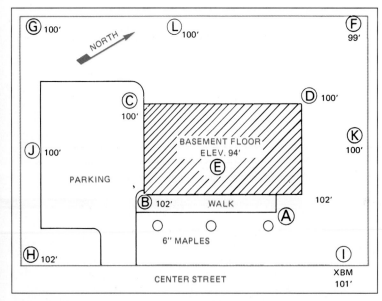

FIGURE 13–18   Plot plan

CHAPTER 14

Foundation Systems

OBJECTIVES

After completing this chapter, you should be able to:

▼ explain the most important design considerations for the foundation of a small building;

▼ list the major steps in the placement of a concrete slab foundation;

▼ explain the function of spread footings; and

▼ construct a concrete block wall.

KEY TERMS

superstructure

slab-on-grade

seismic

monolithic

frost line

spread footing

screed

placing

running bond

dampproof

parging

Foundation Design

The foundation is the base upon which the **superstructure** of a building rests, **Figure 14–1** (The *superstructure* is all of the structure above the foundation.) When the great weight of a building is placed on soil, the soil tends to flow around the walls of the build-

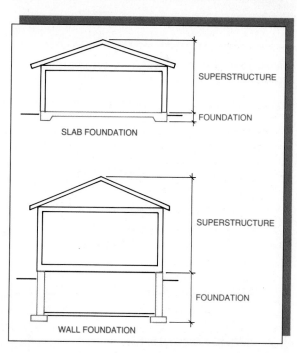

FIGURE 14–1 The structure consists of the foundation and the superstructure.

ing. The foundation spreads the weight of the building over a greater area. The soil is thus less apt to move and the building does not settle.

The foundation also provides a strong base for the superstructure. This eliminates shifting of the structural parts. Without a strong, rigid foundation the superstructure would flex and crack.

In some buildings, the foundation encloses a basement. The basement can be planned as living space, **Figure 14–2**. There must be enough headroom and the area must be dry. It is also desirable to have good ventilation and windows for natural light. These features are included on the architect's foundation plan.

Slab-on-Grade Foundations

In **slab-on-grade** construction, the first floor of the building serves as the foundation, **Figure 14–3**. The earth is prepared without extensive excavation. Then, the concrete slab is placed directly on the approximate finished grade of the site. Slab-on-grade construction is popular in warmer climates for three main reasons: (1) The ground does not freeze, so there is no concern for the action of freezing and thawing; (2) the cost of excavation is saved; (3) often in warmer climates there is a great deal of ground water which makes the soil less stable. The slab spreads the weight of the building over the largest possible area without the need for excavation, which would collect ground water.

Constructing a Slab

The actual slab construction is simpler than other types of foundation construction. There are several steps that must be taken before the concrete is *placed* (since concrete is not considered liquid, it is not poured).

Concrete forms for a slab are relatively simple. Often wood 2 × 10s or 2 × 12s are all

FIGURE 14–2 The basement area may be made into living spaces. *(Courtesy of American Wood Council)*

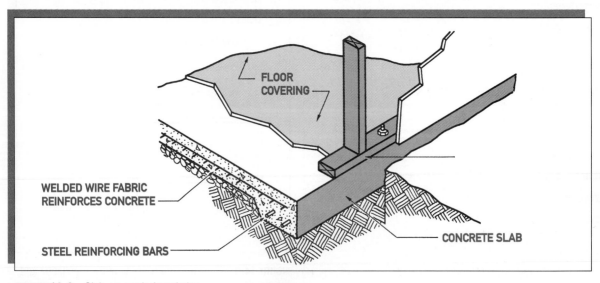

FLOOR COVERING

WELDED WIRE FABRIC REINFORCES CONCRETE

STEEL REINFORCING BARS

CONCRETE SLAB

FIGURE 14–3 Slab-on-grade foundation

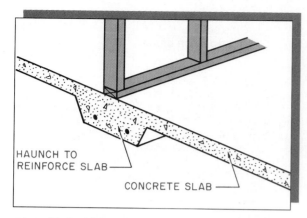

FIGURE 14–4 A haunch is a thickened part of a slab to reinforce the slab under a bearing wall.

FIGURE 14–5 Piping sleeves in place under slab area *(Courtesy of Larry Jeffus)*

that are needed. The forms will contain the concrete around the outside edge. The forms also control the height and levelness of the top of the slab. The forms are placed according to the building lines which were established with batter boards. The height of the forms is measured with a builder's level or transit.

With the forms in place, depressions in the soil are shoveled out to form haunches. A haunch is a thickened portion of the slab to support extra loads, **Figure 14–4**.

Any pipes that will run under or through the slab are positioned next. It may be necessary to add piping or repair it after the concrete is in place. Therefore, large-diameter pipes are normally installed as sleeves. Piping is installed in these sleeves, **Figure 14–5**.

Plastic sheeting is laid over the entire area to keep the finished slab dry. Welded wire mesh is put in place to reinforce the finished slab. The actual placement and finishing of the concrete is the quickest part of the job.

Inverted-T Foundations

In the western part of the United States, where **seismic** (earthquake) activity requires extra strong foundations, but there is little danger of frost damage, inverted-T foundations are common. As its name suggests, the inverted-T foundation design resembles the letter T turned upside down, **Figure 14–6**. Form work for a

very deep foundation of this type would be very difficult, so this type is not practical where the footing must be deep to get below the frost line. However, the inverted-T allows for good placement of reinforcing steel in a monolithic foundation system, so it is a strong design. **Monolithic** is a term used to describe something that is all one piece. A monolithic foundation combines the wall or vertical section in one piece with the footing portion.

Wall Foundations

In areas where the earth's surface freezes in the winter, the foundation must reach below the frozen earth. The maximum depth to which

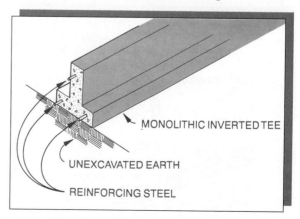

FIGURE 14–6 The inverted-T foundation design combines the footing and wall in one piece.

the earth freezes is called the **frost line**. Foundations in areas of frost usually are made up of footings and walls.

Footings

The base upon which a wall-type foundation is supported is called a **spread footing**. It is called this because it "spreads" the weight of the house over a larger area. All of the weight of a house is supported by the foundation walls. These walls vary in thickness from 6 inches to 12 inches. The pressure these walls apply to the ground is greatly reduced by resting the walls on a concrete footing that is wider than the thickness of the wall, **Figure 14–7**. Similarly, where posts support the weight of beams in the interior of the basement, footings spread their pressure over a greater area, **Figure 14–8.**

Footing Design

In general practice, the excavation for the foundation is done to a depth below any loose fill, soft organic material, and the frost line. Excavation must extend below the frost line because the forces of freezing and thawing cannot be practically controlled. However, in some cases, it may not be practical to excavate to the depth at which the soil can support the building on conventional sized footings. This is particularly true in wet areas, such as many

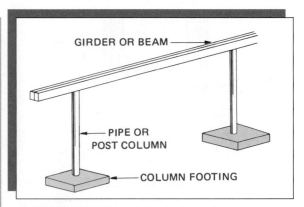

FIGURE 14–8 Pipe columns or posts which support girders or beams rest on concrete footings.

parts of the southeastern United States. For this reason, footing designs must be varied to suit the conditions. Usually footing design is specified by building codes, **Figure 14–9.**

Where footing design is not specified by a building code, it is up to the architect to indicate the size of the footing. The footing is shown on the foundation plan by a broken line, **Figure 14–10.** Usually a note is included to indicate

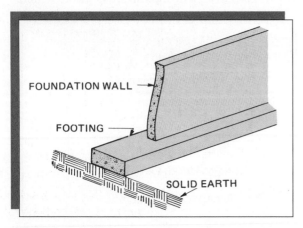

FIGURE 14–7 Spread footings distribute the weight of the structure over a larger area than the foundation wall.

Footings

Sec. 2907. (a) **General.** Footings and foundations, unless otherwise specifically provided, shall be constructed of masonry or concrete and in all cases extend below the frost line. Footings shall be constructed of solid masonry or concrete. Foundations supporting wood shall extend at least 6 inches above the adjacent finish grade. Footings shall have a minimum depth below finished grade as indicated in Table No. 29-A unless another depth is recommended by a foundation investigation.

(b) **Bearing Walls.** Bearing walls shall be supported on masonry or concrete foundations or piles or other approved foundation system which shall be of sufficient size to support all loads. Where a design is not provided, the minimum foundation requirements for stud bearing walls shall be as set forth in Table No. 29-A.

TABLE NO. 29-A—FOUNDATIONS FOR STUD BEARING WALLS MINIMUM REQUIREMENTS

NUMBER OF STORIES	THICKNESS OF FOUNDATION WALL (Inches)		WIDTH OF FOOTING (Inches)	THICKNESS OF FOOTING (Inches)	DEPTH OF FOUNDATION BELOW NATURAL SURFACE OF GROUND AND FINISH GRADE (Inches)
	CONCRETE	UNIT MASONRY			
1	6	6	12	6	12
2	8	8	15	7	18
3	10	10	18	8	24

NOTES:

Where unusual conditions or frost conditions are found, footings and foundations shall be as required in Section 2907 (a).

The ground under the floor may be excavated to the elevation of the top of the footing.

FIGURE 14–9 Most building codes include some regulations of footing design. *(Courtesy of the International Conference of Building Officials)*

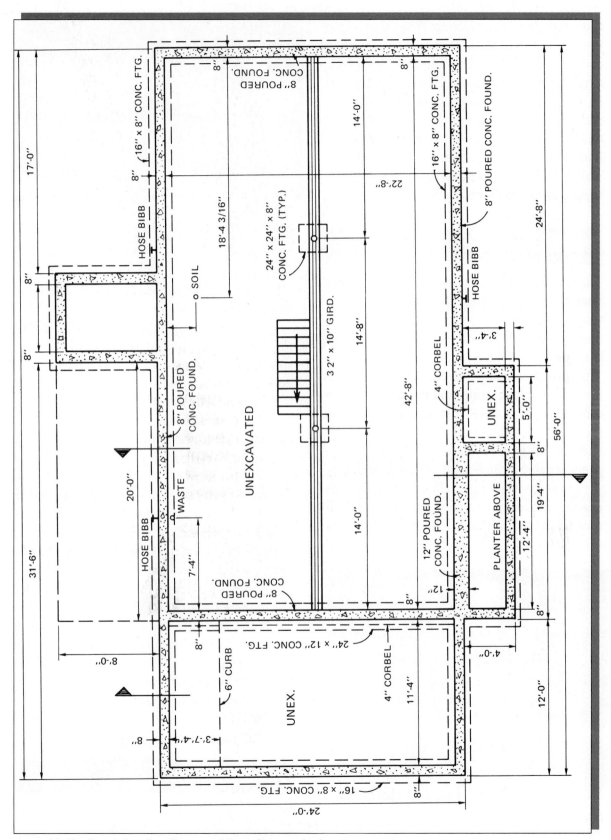

FIGURE 14–10 Foundation plan *(Courtesy of Home Planners, Inc.)*

the dimensions of the footing. When the footing is not designed by the architect and no building code is available, the builder relies on experience. Typical footings for residential and light commercial construction are twice as wide as the thickness of the foundation wall and the same depth as the thickness of the foundation wall. For example, if the foundation wall is 10 inches thick, the footing is built 20 inches wide and 10 inches high.

Placing Footings

The excavator digs to the depth of the basement floor if one is included in the plans. Otherwise, the excavation is dug to the bottom of the foundation walls. In either case, the excavation extends about 2 feet outside the building lines. This provides space for carpenters to erect forms for concrete work and for masons to lay concrete blocks. Construction laborers finish the excavation for the footings by hand with shovels, **Figure 14–11** It is done by hand to prevent going beyond the depth of the footings.

On some jobs, the ditches for the footings are dug carefully and the concrete is placed directly in them without the need for forms. On

FIGURE 14–12 This worker is placing screed boards which will help level the surface of the concrete. *(Courtesy of Cal Parlman, Inc.)*

other jobs, lumber is used to construct forms in the ditches.

If the concrete is placed directly in the ditch, grade stakes are driven every few feet in the center of the ditch. The height of these stakes is measured with a builder's level. When the concrete is flush with their tops, the footing is level. If side forms are used, they are leveled as they are built. As the concrete is placed it is leveled with the top of the forms, **Figure 14–12**

As the concrete is placed, it is screeded off. A **screed** is a straightedge used to push excess concrete into low spots. The concrete is not smoothed beyond screeding. The rough surface left by screeding provides for better bonding of the mortar if concrete blocks will be used for the foundation wall.

If the foundation walls are to be made of concrete, reinforcing bars are usually placed in the footing. The reinforcing bars are bent upwards, so that they will also be imbedded in the foundation wall when it is placed, **Figure 14–13** (Putting concrete into the forms is called **placing** it, not pouring it.)

Foundation Wall Design

When a concrete foundation is used, carpenters erect forms of plywood or metal. *Form carpentry*

FIGURE 14–11 The bottom of the footing excavation should be carefully leveled.

FIGURE 14–13 Reinforcing bars might be used to secure the foundation wall to the footing.

FIGURE 14–14 Forms for a concrete foundation *(Courtesy of Portland Cement Association)*

FIGURE 14–15 Mortar is usually mixed in power mixers. *(Courtesy of Stone Construction Equipment Inc.)*

is an important branch of the carpentry trade, **Figure 14–14** As soon as possible after the concrete cures, the forms are *stripped* (removed), cleaned, and stored for use on the next job.

Concrete blocks are also a popular material for foundations. Although a concrete block foundation takes somewhat longer to build, the materials are less expensive. Concrete blocks are laid by masons.

Laying a Concrete Block Foundation

When the concrete blocks are delivered to the construction site, they are stacked at several points around the footings. The lime, sand, and cement (or masonry cement) are stacked near where they will be mixed. The cement is protected from rain by a tarp.

Before any mortar is mixed or blocks are laid, the masons mark the location of the wall on the footing with a chalk line. The wall is then measured to determine if any blocks will have to be cut. The most common length of concrete blocks is 16 inches. It may be necessary to cut one block in every course to build the planned wall.

Masonry laborers use a power mixer to mix the mortar, **Figure 14–15** The mixed mortar is transferred to mortar pans located next to the footing. The masons take their mortar from these pans. A full bed of mortar is spread along the footing for a length of several blocks. The blocks are set in the mortar bed with mortar spread on the head joints (end joints). As each block is laid, it is checked for levelness and plumbness with a spirit level.

The corners are constructed first, **Figure 14–16,** so the main portion of each wall can be laid to a line. As each course of blocks is laid, a length of mason's line is stretched between the

FIGURE 14–16 The corners are built first, then the wall is filled in.

FIGURE 14–17 This mason is laying blocks to a line.

tops of the corner blocks for that course, **Figure 14–17**. The masons use this line as a guide in setting that course to the right height. The plumbness is still checked with a spirit level.

The blocks for a foundation are usually laid in **running bond**. This means that the head joints are located over the center of the block below, **Figure 14–18** With 8″×8″×16″ blocks, this is done by alternating the blocks at the corner. If blocks of any other size are used, the corner blocks must be cut.

When the mortar is just hard enough so that it leaves a thumbprint when pressed with the thumb, the joints are tooled. The most common jointing tool is the *convex jointer*; it produces a concave joint. Portions of the foundation that will be hidden by earth on the outside do not have to be tooled.

Dampproofing and Drainage

To help prevent moisture from seeping through the foundation walls, they are **dampproofed**. This often consists of **parging** (coating) the wall with a 1/2-inch coat of portland cement and lime plaster and covering this with asphalt waterproofing, **Figure 14–19** Dampproofing

FIGURE 14–18 (A) Stacked bond, (B) Running bond

does not extend above the finished grade line. For a more complete waterproof job, some foundations are covered with a *waterproof membrane* (thin sheet of material) and hot tar.

FIGURE 14–19 Parging the foundation wall with portland cement mortar

FIGURE 14–20 Flexible perforated plastic drain pipe

As ground water reaches the coated foundation, it runs down to the bottom of the foundation. Plastic or clay drain pipes are placed around the footing to carry this water away from the building. Plastic drain pipe has holes to allow the water to enter, **Figure 14–20**. Clay drain tiles have loose joints that allow the water to enter. These pipes are buried in crushed stone to permit a free flow of water.

Wood Foundations

Foundation walls are usually made of concrete or concrete block. However, a type of specially treated wood foundation is gaining popularity. The wood for these foundations is treated with preservative under pressure to prevent it from decaying. It is further protected by covering it with a thin sheet of plastic during construction, **Figure 14–21**.

FIGURE 14–21 Wood foundation *(Courtesy of the American Wood Council)*

ACTIVITIES

Footing Design

Structural engineers design the load-carrying parts of a building. They must know how each type of footing will settle on a particular type of soil. To do this, they work closely with the architect to find the amount of weight on the building. They also consult soil engineers for facts about the soil on the site. Using this data, they then decide the size and depth of the footings for the building.

Equipment and Materials

Container at least $1' \times 1' \times 1'$
Sand
Peat moss
Lumber, $2'' \times 6'' \times 10'$
Lumber, $2'' \times 4'' \times 10'$
Tape measure or folding rule
Concrete block, $8'' \times 8'' \times 16''$

Procedure

1. Fill the container with sand to a depth of at least one foot.
2. Place the $2'' \times 6'' \times 10'$ piece of lumber on edge on the sand.
3. Place the concrete block on the $2'' \times 6''$ piece as a weight.
4. Measure the depth to which the $2'' \times 6''$ piece settles in the sand.
5. Repeat steps 1 through 4, but place the $2'' \times 4''$ lumber on its 4-inch surface under the $2'' \times 6''$ lumber.
6. Repeat steps 1 through 5 using peat moss instead of sand.
7. What effect does the footing have in each type of soil?

Laying Concrete Blocks

(This activity should be done by a team of four to eight students.) A sound concrete block foundation depends upon a mason's skill and accuracy. Masons must be able to lay blocks in a straight line and in proportion. Their work must be uniform. All blocks must have a 3/8-inch joint and be firmly set in place. Good construction depends on a sound foundation.

Equipment and Materials

Tape measure
Chalk line
112 concrete blocks, $2'' \times 8'' \times 16''$
Sand-line training mortar (see Masonry Activity, Chapter 6)
Trowels
Spirit level
Mason's line and line blocks
Convex jointer

Procedure

1. On the floor or pavement, lay out a square with 9'-4" sides.
2. While some students are laying out the activity, others should mix the mortar.
3. With the students working alone or in pairs, spread mortar inside the square at each corner and lay the first course of blocks.
4. Level and plumb each block leaving 3/8-inch joints. Any excess mortar should be cut off with a trowel and returned to the supply.
5. Build each corner four courses high, **Figure 14–22.** Care should be taken to see that all joints are 3/8 inch and that all blocks are level and plumb.
6. When all corners are built, stretch the mason's line between two corner blocks of the same course. **Figure 14–23** shows how the line is tied to the line blocks.
7. Lay the remaining blocks to the line. These blocks must be plumbed with a spirit level.
8. At the end of each class session during which blocks are laid or when the mortar is thumbprint hard, strike the joints with a convex jointer.

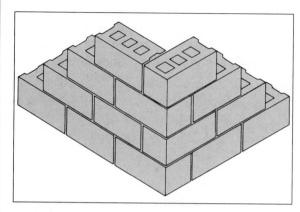

FIGURE 14–22 Four-block-high corner

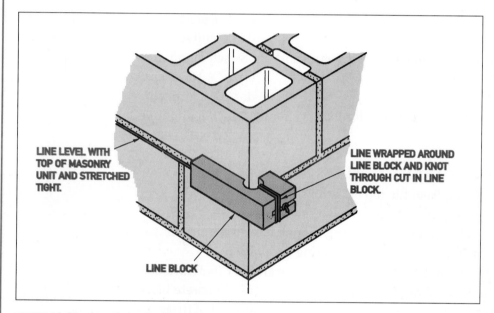

LINE LEVEL WITH TOP OF MASONRY UNIT AND STRETCHED TIGHT.

LINE WRAPPED AROUND LINE BLOCK AND KNOT THROUGH CUT IN LINE BLOCK.

LINE BLOCK

FIGURE 14–23 Line tied to line block

Applying Construction Across the Curriculum

Mathematics

The footings for the foundation in **Figure 14–10** are 16 inches wide and 8 inches deep in all places, except the wall between the main house and the garage, which is 24 inches wide and 16 inches deep. How many cubic yards of concrete will be required for these footings? Round your answer up to the nearest whole number.

Mathematics

If all of the foundation walls in **Figure 14–10** were to be 5 feet high, how many whole cubic yards of concrete would be required?

REVIEW

A. Multiple Choice. Select the best answer for each of the following questions.

1. What is the purpose of the footings for a house?
   a. To spread the weight over a larger area
   b. To provide a strong base for the foundation walls
   c. To prevent the foundation from settling
   d. All of these

2. Who is most apt to be concerned with designing footings?
   a. Building code official
   b. Mason
   c. Carpenter
   d. Excavator

3. Who is most apt to construct footings?
   a. Excavator
   b. Mason
   c. Carpenter
   d. Architect

4. What is the most common size for footings for an 8-inch foundation?
   a. 20″ × 16″ deep
   b. 16″ × 16″ deep
   c. 8″ × 16″ deep
   d. 16″ × 8″ deep

5. What is the most common size for footings for a 12-inch foundation?
   a. 30″ × 16″ deep
   b. 24″ × 8″ deep
   c. 24″ × 12″ deep
   d. 12″ × 24″ deep

6. How is the top of the footing leveled?
   a. By measuring from the top of the excavation
   b. By measuring from the bottom of the excavation
   c. Footings are not leveled; leveling is done later with mortar
   d. By driving grade stakes in the footing excavation

7. Which of the following materials is used to construct foundations?
   a. Pressure-treated wood
   b. Concrete
   c. Concrete blocks
   d. All of these

8. Which is the most likely order of work on a concrete block foundation?
   a. Each course is completed before the next is begun.
   b. The main portion of the wall is completed before the corners.
   c. The corners are completed before the rest of the wall.
   d. Each side of the foundation is completed before another is begun.

9. What is parging?
   a. A special cement used in foundations
   b. A portland cement plaster applied to foundations
   c. A waterproof membrane
   d. A mixture of tar and cement

10. How is the plumbness of blocks checked as a foundation is constructed?
    a. With a mason's line
    b. With a spirit level
    c. With a builder's level
    d. With a straightedge

B. Factors to Consider.   List four important things an architect considers in designing a basement that may be used as living space in the future.

CHAPTER 15

Floor Systems

OBJECTIVES

After completing this chapter, you should be able to:

▼ identify and describe the function of floor framing members; and

▼ explain the 4-inch module used in building construction.

KEY TERMS

subfloor

4-inch module

sill

girder

pier

pipe column

engineered lumber products

floor joist

joist header

stud

underlayment

The floor framing consists of all of the construction from the top of the foundation to the **subfloor** (the first layer of flooring). The girders and posts, which support the floor framing, are also considered part of the floor framing.

It is easy to take the floor in your home for granted. It's there and it keeps you from falling into the basement or walking on the ground, but like other parts of the building, floors serve very definite purposes. There are things that must be considered in designing and building a floor. In most light-frame buildings constructed today, the floor is one structural component in a whole system.

The platform-frame design used in most light-frame construction is the result of centuries of human experience in designing and constructing buildings. This system of floor construction has only been common for about 60 years. Platform-frame construction, which is described in this section, performs the following important functions:

▼ The floor is the first part of the frame to be completed, so it provides a good work surface (platform) for the remaining construction work.

- ▼ The floor frame supports the weight of the walls and partitions above. This system allows for easy adjustments to add strength where especially heavy loads will be expected, such as beneath a bathtub.
- ▼ If properly designed and constructed, the floor is rigid enough to prevent cracks in plaster or wallboard, shifting or sticking door openings, etc.
- ▼ The floor can be built tightly and insulated easily to prevent air movement and heat loss.
- ▼ This system of floor construction uses materials that are readily available from most building materials suppliers.

When the floor can be placed directly on the ground, the system described in Chapter 14 as slab-on-grade construction can be used, so a framed floor is not necessary. In this case the concrete slab serves all of the above functions, but the rest of the structure might use platform-frame construction methods. The slab floor is the platform.

Four-Inch Module

To understand the reasons for the size and spacing of many building materials, it is necessary to understand the **4-inch module**. Using 4 inches as a standard module, 4-foot wide building panels are exactly 12 modules wide, and 8-foot long panels are 24 modules long, **Figure 15–1.** Some building panels are made in 2-foot widths, so they are 6 modules wide. Likewise, the length is often varied in 2-foot increments to conform with the 4-inch module.

Most building panels, such as plywood, gypsum wallboard, and particleboard, require support in three or four spaces for each 4-foot width. This requirement is easily accommodated by the 4-inch module. If framing mem-

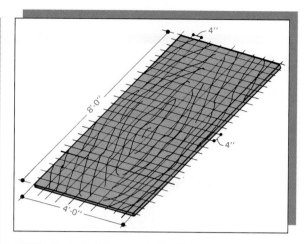

FIGURE 15–1 Building panels 4′ × 8′ are coordinated with the 4″ modules.

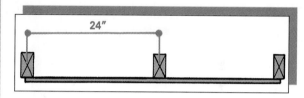

FIGURE 15–2 Four-foot panel on 24″ O.C. framing

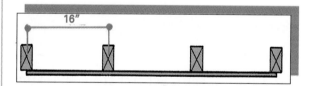

FIGURE 15–3 Four-foot panel on 16″ O.C. framing

bers are spaced on 24-inch centers, 4-foot panels are supported in three places, **Figure 15–2.** If framing members are spaced 16 inches on center (O.C.), the panels are supported in four places, **Figure 15–3.**

To avoid extra cutting and wasting of materials, buildings are generally designed around 24-inch major modules (six 4-inch modules), **Figure 15–4.** Although it is not always possible to design rooms around the 24-inch module, the overall dimensions of the building are easily held to this standard. With this system, if a 2-foot piece of wall material (1/

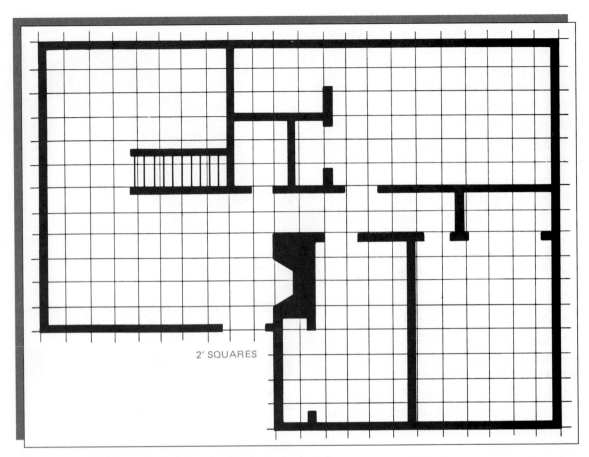

FIGURE 15–4   This floor plan is designed on two-foot major modules.

2 of a 4-foot panel) is left over at one end of the wall, it can be used on the opposite wall.

The Sill

The **sill** is the wood member directly on top of the concrete or masonry foundation. It provides a base for fastening the building frame to the foundation. The sill is usually $2'' \times 6''$ or $2'' \times 8''$ lumber. The method of anchoring depends somewhat on the kind of foundation. In concrete foundations, anchor bolts are inserted into the fresh concrete as it is placed. The carpenters then drill holes in the sill at the location of each anchor bolt. The same method can be used with concrete block foundations if the top courses of blocks are filled with concrete, **Figure 15–5**. Another method of anchoring the sill to masonry

foundations is to use solid blocks for the top course, then nail the sill in place with hardened nails.

Girders and Posts

Usually buildings are too wide for continuous pieces of lumber to span the full width. The floor framing is supported by one or more **girders** (beams) running the length of the building. The girder is supported at regular intervals by wood or metal posts or masonry columns called **piers**, **Figure 15–6**. Metal posts called **pipe columns** are used most often.

The girder may be one of several types. Steel beams are often used where strength is a critical factor. Carpenters construct built-up wood girders on the site. These consist of two or

FIGURE 15–5 Anchor bolts can be *grouted* into the top course of concrete blocks *(Courtesy of Richard T. Kreh, Sr.)*

FIGURE 15–6 The girder supports the floor joists.

FIGURE 15–7 Built-up wood girder

three pieces of 2″ × 8″ or 2″ × 10″ nailed together with staggered joints to form larger beams, **Figure 15–7**. Recently, many buildings have been constructed with beams made of laminated layers of wood. A common form of these **engineered lumber products** is called a micro-lam beam. Laminated beams are available in lengths up to 60 feet. They can be manufactured with less waste than sawed lumber and they are exceptionally strong.

Joists

The framing members that rest on the wood sill and support the first layer of flooring (subfloor) are called **floor joists**. In a typical house with one girder, the inner ends of the joists either overlap on top of the girder or butt against it. When the joists butt against the girder, they are supported by metal joist hangers or a ledger (a small strip of wood) nailed to the girder, **Figure 15–8**. The outer ends rest on the sill and are nailed to the **joist header**, sometimes called the *band*, **Figure 15–9**.

Traditionally, wood has been the standard material for house framing, including the joists. A form of engineered lumber, called a TJIs (short for Trus-Joist Institute, the organization

FIGURE 15–8 Metal joist hanger

FIGURE 15–10 TJIs are strong and can be manufactured in long lengths. *(Courtesy of Trus Joist MacMillan)*

FIGURE 15–9 The ends of the floor joists are nailed to the joist headers.

FIGURE 15–11 Metal joists are sometimes used instead of wood framing members. *(Courtesy of Bethlehem Steel Corporation)*

that invented them) is becoming increasingly popular for floor and roof framing, **Figure 15–10.** TJIs are made of a laminated web (center section) and a top and bottom reinforcement of laminated veneer. They provide the same

strength as solid wood joists or rafters with only about half the wood fiber. However, wood supplies are becoming more scarce. Engineered materials are readily available, so many architects specify metal framing members,

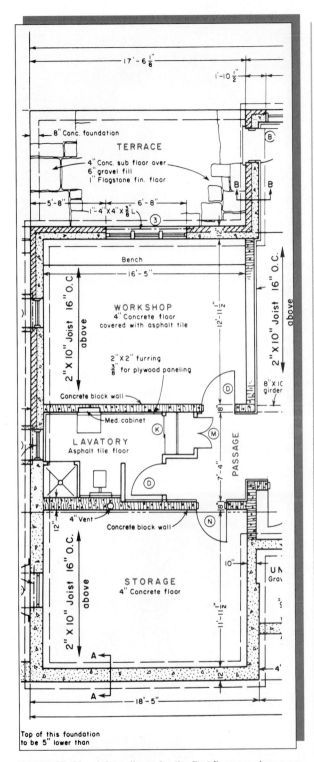

Top of this foundation
to be 5" lower than

FIGURE 15–12 Joist callouts for the first floor are shown on the foundation plan.

Figure 15–11 They are installed the same way as wood joists except for the method of fastening them. Metal joists are fastened to the foundation with powder-actuated stud drivers. These tools are powered by an explosive cartridge. They drive a nail-like fastener, called a **stud**, through the framing member and into the concrete or masonry foundation. The subfloor is later fastened in place by self-drilling, self-tapping screws and power screwdrivers.

The size and spacing of the floor joists are shown on the working drawings, **Figure 15–12.** The spacing is usually either 16 inches on center or 24 inches on center. Wood floor joists used in houses are usually 8, 10, or 12 inches deep. If the spacing is to be 16 inches on center, the carpenters mark the joist header every 16 inches from one end of the building. The first joist, called a *stringer joist*, is placed on the outside edge of the end wall of the foundation. Each successive joist is centered on the 16-inch marks. Where partitions are to be located, the joists can be doubled for added strength. Around openings for chimneys and stairs, the ends of the joists are nailed to a *double header*. The joists at each side are doubled with a *trimmer joist*, **Figure 15–13**

FIGURE 15–13 Double framing members are used around openings for stairs and chimneys. (Photo by Michael Dzaman)

FIGURE 15–14   Carpenters install subflooring over the floor joints. *(Courtesy of the American Plywood Association)*

FIGURE 15–15   Single-floor system uses thick plywood in a single layer. *(Courtesy of the American Plywood Association)*

Subflooring

The *subflooring* is the first layer of material applied over the floor joists, **Figure 15–14** Once the subfloor is in place, carpenters and other construction personnel have a smooth platform on which to work. Most subflooring used in residential construction is plywood. In conventional floor systems, 1/2-inch or 5/8-inch plywood is nailed to the floor joists. Then an **underlayment** (a second layer of plywood or particleboard), or wood flooring is laid over the subfloor.

Through their efforts to devise more efficient ways to use plywood, the American Plywood Association (APA) has developed a *single-floor system*. In this type of construction, a single layer of thicker plywood with tongue and groove edges is substituted for the subfloor underlayment system. This single-floor system reduces the labor involved in laying flooring. Because this system allows the joists to be spaced up to 48 inches on center, less labor and materials are needed for the framing, **Figure 15–15**

Plywood subflooring is installed with the long (usually 8-foot) edge spanning several joists. The joints are staggered, **Figure 15–16** If metal joists are used, the subflooring is installed with screws. If wood joists are used, the subflooring is nailed in place. In either case, glue can be applied to the joists first. This prevents floors from squeaking and nails from popping later.

FIGURE 15–16   The joints in plywood subflooring are staggered. (Photo by Michael Dzaman)

ACTIVITIES

Drawing a Joist Plan

A *joist plan* is a drawing showing the location of exterior walls and girders and the arrangement of joists as seen from directly above. Architects sometimes include a joist plan (**Figure 15–17A**), with the working drawings for a house, **Figure 15–17B** Sometimes the building contractor or carpenter's foreman sketches a joist plan to help plan the carpenters' work before they begin.

Equipment and Materials

Architect's scale
Plastic triangle
Pencils and paper

Procedure

Draw a joist plan for the building in **Figure 15–18** Use a scale of 1/4″ = 1′-0″.

Remember to double all headers and trimmers around openings and double the joists under walls.

Applying Construction Across the Curriculum

Science

A floor frame is a system of levers. Where is the fulcrum for number 7 in **Figure 15–19**? What class of levers is represented by number 7 in **Figure 15–19**?

Mathematics

How many 4 foot by 8 foot sheets of plywood are required to cover the floor in **Figure 15–4**? How many for the floor in **Figure 15–18**?

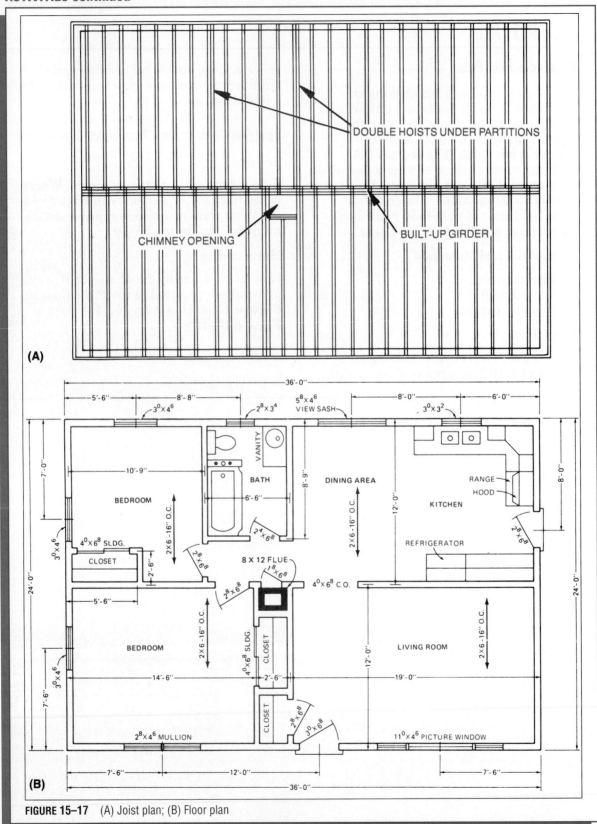

FIGURE 15-17 (A) Joist plan; (B) Floor plan

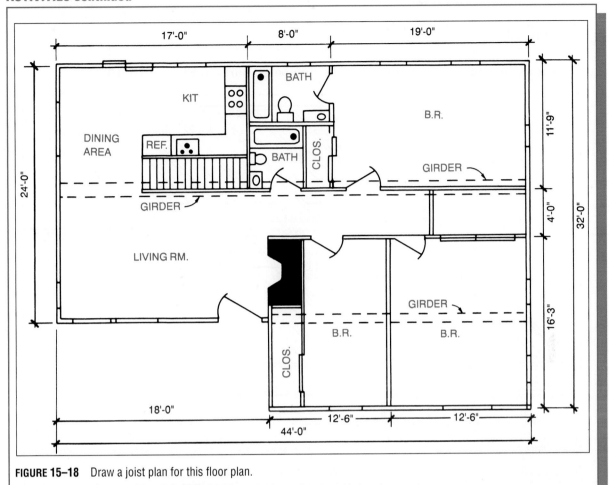

FIGURE 15–18 Draw a joist plan for this floor plan.

A. Identification. Identify each of the numbered parts in **Figure 15–19**

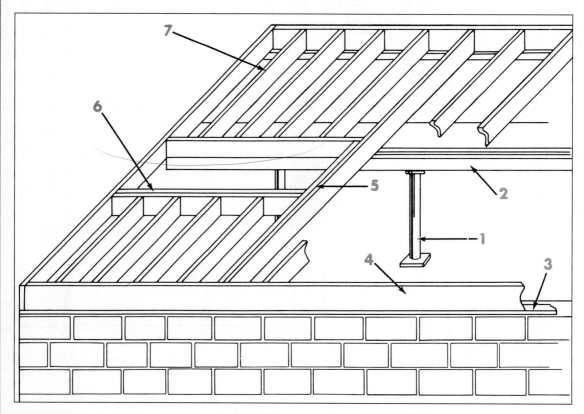

FIGURE 15–19

B. Matching. Match the use or description in Column II with the correct item in Column I.

<table>
<tr><td colspan="2">Column I</td><td colspan="2">Column II</td></tr>
<tr><td>1.</td><td>Box beam</td><td>a.</td><td>Steel post to support a girder</td></tr>
<tr><td>2.</td><td>Single-floor system</td><td>b.</td><td>Basic unit in building design</td></tr>
<tr><td>3.</td><td>Pipe column</td><td>c.</td><td>Available in great lengths</td></tr>
<tr><td>4.</td><td>Built-up beam</td><td>d.</td><td>Conserves material and makes lightweight, strong girders</td></tr>
<tr><td>5.</td><td>Four-inch module</td><td></td><td></td></tr>
<tr><td>6.</td><td>Subfloor</td><td>e.</td><td>Uses thick plywood</td></tr>
<tr><td>7.</td><td>Anchor bolts</td><td>f.</td><td>Solid wood girder built by a carpenter</td></tr>
<tr><td>8.</td><td>Steel beam</td><td>g.</td><td>First layer of flooring</td></tr>
<tr><td></td><td></td><td>h.</td><td>Fastens the sill in place</td></tr>
</table>

To build a house ...

Use measuring and layout tools to lay out the site (Chapter 13).

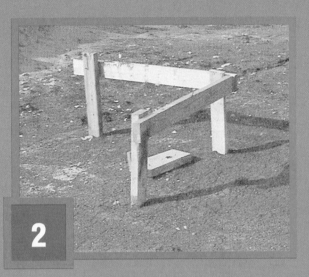

Construct batter boards to establish building lines (Chapter 13).

Excavate for footings (Chapter 13).

Build forms for concrete work and rough in any utilities that are to run under the concrete (Chapters 14 & 24).

5

Place concrete for foundations and slabs (Chapter 14).

6

Snap chalk lines or use a laser level to lay out the locations of walls (Chapter 16).

7

Cut wall plates to length (Chapter 16).

8

Assemble the walls laying flat on the slab or deck (Chapter 16).

9

Tip walls up and brace in plumb position (Chapter 16).

10

Anchor bolts and holddowns secure the wall frames in place (Chapter 15).

11

Set roof trusses or cut and assemble rafters (Chapter 17).

12

Apply wall sheathing (Chapter 16).

13

Apply roof deck and roof covering (Chapter 17).

14

Build cornices and install windows and doors (Chapter 18).

15

Rough in plumbing (Chapter 24)...

16

...and pressure test it for leaks (Chapter 24).

17

Install tubs and showers (Chapter 24).

18

Install electrical service (Chapter 26)…

19

…and rough in wiring (Chapter 26).

20

Insulate exterior surfaces (Chapter 25).

21

Build the fireplace (Chapter 16)...

22

...laying bricks to a line (Chapter 6)...

23

...and striking the joints as the mortar stiffens
(Chapter 6).

24

Apply wallboard (Chapter 27)...

25

...and tape the joints (Chapter 27).

26

Install trim (Chapter 27).

27

Set cabinets (Chapter 27).

28

Install finished plumbing and electrical fixtures (Chapters 24 & 26).

29

Paint inside and out (Chapter 28).

30

Install flooring and floor coverings (Chapter 27).

31

Landscape the yard (Chapter 29).

32

Inspect all work.

CHAPTER 16

Wall Framing Systems

OBJECTIVES

After completing this chapter, you should be able to:

▼ identify and describe the function of wall-framing members;

▼ explain the layout and construction of wall framing;

▼ cut and assemble members for a section of a wall; and

▼ list several ways in which walls can be made more energy efficient.

KEY TERMS

load-bearing wall

partition

plate

rough opening

header

trimmer

jack stud

sheathing

The walls in a house serve many purposes. Some walls support the weight of the structure above them. These are called **load-bearing walls**. Other walls, called **partitions**, do not support weight. They divide the house into separate rooms for privacy and convenient living. *Exterior walls* provide protection from the weather and make indoor living comfortable. All the walls of a house must create a pleasing appearance. The architect considers all of these functions in planning the design of a residence.

Walls can be constructed of masonry, concrete, or with wood or metal frames. In residential construction most walls are of the wood-frame type. Wood-frame walls are relatively inexpensive, easy to alter, and have enough strength for light construction. With wood building materials becoming scarcer, metal-framing members are becoming more popular. Construction with both materials is very similar.

Wall-Framing Members

The basic parts of a wall frame are the top and bottom plates and the studs, **Figure 16–1** The **plates** are the horizontal members at the top

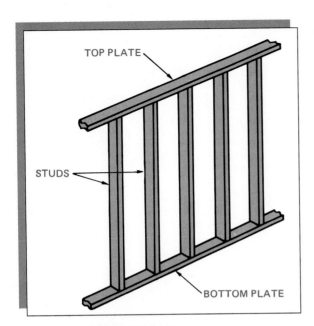

FIGURE 16–1 Studs and plates

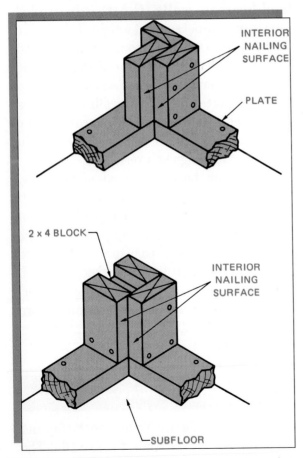

FIGURE 16–2 Two methods of building corners in exterior walls

and bottom of the wall. They hold the ends of the studs in place and provide a surface to fasten the wall in place on the floor framing. In most walls, the plates and studs are made of $2'' \times 4''$ lumber.

The wall frame also provides a surface on which to nail the inside and outside wall covering. Where two exterior walls meet at a corner, an extra stud must be added to provide this nailing surface. **Figure 16–2** shows two types of corner post construction.

Where an inside partition meets another wall, studs must be located to provide a nailing surface on both sides of the partition. Walls are usually built flat on the floor, then tilted up into place. For this reason, it is important that the carpenter who lays out the wall framing understands architectural drawings and works accurately. If extra studs for nailing surfaces are not properly located, the error will have to be corrected later at a great expense.

One or more of the studs must be cut off in the window and door openings. The weight ordinarily supported by those studs must be

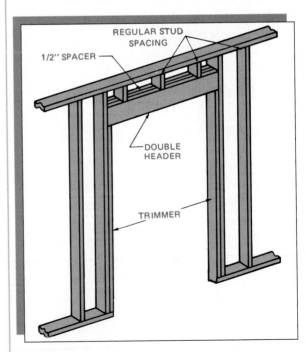

FIGURE 16–3 The header supports the weight over the opening.

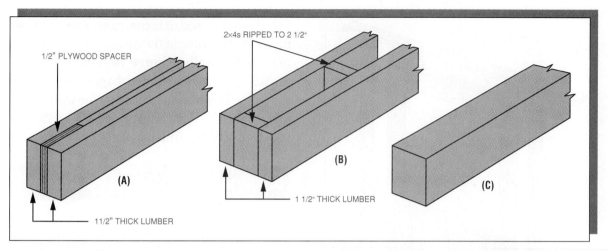

FIGURE 16–4 Three styles of wood headers: (A) 3-1/2″ thick header to fit 2 × 4 framing, (B) 5-1/2″ thick header to fit 2 × 6 framing, (C) Solid header

transferred to the studs at the sides of the **rough opening** (the framed opening into which the door or window is placed). The **header** distributes this load to the sides of the opening, **Figure 16–3**. In most parts of North America window and door headers are made by placing two pieces of 2″ dimensional lumber on edge. On the west coast it is common to use a single piece of 4″ or thicker timber placed on edge. When the header is built of 2″ lumber, spacers are placed between the two to build the header up to the thickness of the wall frame, **Figure 16–4**. The greater the width of the opening, the more rigid the header must be. A more rigid header may be achieved by increasing its depth (the width of the lumber used to build the header). Laminated engineered lumber products are also becoming popular for use as headers, **Figure 16–5**.

To support the extra weight the header transfers to the sides of the opening, a second stud is used. This stud, called a **trimmer**, is cut just long enough to fit under the ends of the header (see **Figure 16–3**). The trimmer supports the header and strengthens the full stud.

The framing for windows is the same as that for doors, except the rough opening does not extend all the way down to the floor for the

FIGURE 16–5 Micro-Lam manufactured header. *(Courtesy of Trus Joist MacMillan)*

windows. At the bottom of the rough opening for a window is a horizontal member called a *sill*. The sill extends from one stud to the other, **Figure 16–6**. Short lengths of framing, called **jack studs**, are installed below the rough sill. The jack studs provide a nailing surface for the interior and exterior wall covering.

Sheathing

The wall sheathing is not actually part of the wall framing. It is included here because it is usually installed by the carpenters as the framing is done. The **sheathing** is the first layer of wall covering on the exterior of the walls. Boards can be used, but the most common

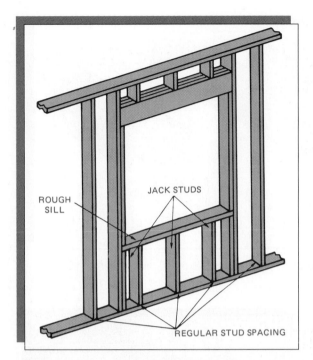

FIGURE 16–6 The bottom of the window opening is formed by the rough sill. Jack studs are installed under the sill.

sheathing materials are plywood, fiberboard, and rigid plastic foam boards. These materials are available in large sheets that allow the carpenters to enclose the structure more quickly. Plywood also makes the entire structure more rigid. If fiberboard or rigid foam board is used, it is common practice to use plywood sheathing at the corners of the building to brace the walls, **Figure 16–7**. When boards are used, they are installed diagonally to brace the walls. In addition to making the wall frame more rigid, the sheathing provides a surface on which to nail the finished siding.

Constructing Walls and Partitions

The first step in constructing the walls and partitions in a building is to determine their location and mark this on the subfloor. In large construction crews, the layout is done by the foreman or a senior carpenter. However, the carpenters who erect the walls usually lay them out. Wall and partition layout requires the ability to read plans and measure accurately. For many years, it has been customary to mark the locations of walls on the subfloor with a chalk line, **Figure 16–8**. Many builders now use a laser for this operation. The laser shown in **Figure 16–9** creates a 360° laser reference line. This reference line can be used to mark the wall location on the floor and ceiling at the same time.

FIGURE 16–7 Plywood is often used to brace the corners when insulating sheathing is used. *(Courtesy of Larry Jeffus)*

FIGURE 16–8   This carpenter is using a chalkline to lay out the positions of partitions.

FIGURE 16–9   A laser can be used to create a continuous reference plane. *(Courtesy of Spectra-Physics)*

With the location marked on the floor, two straight pieces of framing lumber are cut to length for the top and bottom plates. Next, the locations of the centers of all door and window openings are marked on the plates. From these marks the sides of the rough openings can be measured. The location of the centers of all studs are marked on the top and bottom plates. If the studs are to be 16 inches on center, a mark is made every 16 inches from the end of the plate. Tape measures and carpenter's folding rules usually have a mark every 16 inches. This helps the framing carpenter lay out 16-inch centers. The location of extra studs for corners at adjoining walls is marked.

If the walls are to be any height other than 8 feet, the carpenter must cut the materials for studs to the proper length. If the walls are to be 8 feet high, precut studs measuring $2'' \times 4'' \times 7'-9''$ are available. This eliminates the need for cutting them to length. When 2-inch nominal size framing material (1-1/2 inch actual size) is used, the two plates added to a 7'-9" stud makes an 8-foot wall.

The material is laid out on the subfloor. The plates and studs are nailed together. Two 16d common nails are face-nailed through the plates into the ends of the studs. Headers, trimmers, sills, and jack studs for openings are nailed into position.

The wall can be tilted up into place at this time, but many carpenters prefer to nail the wall sheathing on first. The large panels used for sheathing are easier to handle with the wall frame lying flat than after it is upright. The carpenter must make sure the wall is square before the sheathing is applied. The sheathing panels make the wall frame rigid, so it cannot be squared later. The squareness of the wall frame can be checked by measuring the diagonals, **Figure 16–10**

FIGURE 16–10   Measuring the diagonals to make sure the frame is square

Once the wall framing is completely assembled with the sheathing in place, it is tilted up and slid into position according to the chalk line on the subfloor, **Figure 16–11** The bottom plate is nailed to the floor framing with 16d common nails. The wall sections are plumbed and braced (**Figure 16–12**) to hold them plumb until the framing is completed. In many areas, especially in the western states, it is customary to erect the wall frame first, then apply the sheathing.

By the traditional method, carpenters use a spirit level to plumb the wall. Using a Laser-Level™, the top of the wall is aligned with the laser line on the ceiling joists. The corners are then nailed together.

After all of the framing is in place, the second member of the double top plate is nailed into position. The corners of the top member of the double plate are lapped over the bottom

FIGURE 16–12 Temporary braces hold the walls plumb.

members, **Figure 16–13** This helps tie the walls together and makes the structure stronger.

Special Wall-Framing Considerations

As the carpenters build the walls for a building, they must allow for features that will be added later. Openings sometimes have to be framed for heating, ventilating and air-conditioning (HVAC) ducts. This may require cutting studs and installing headers, **Figure 16–14** The design may call for 4-inch pipes to be installed by the plumbers. This requires thicker walls that are framed with $2'' \times 6''$ lumber instead of

FIGURE 16–11 When the wall section is assembled, it is tilted up and positioned on the floor.

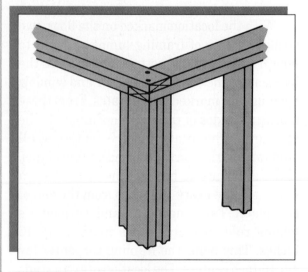

FIGURE 16–13 The upper membrane of the double top plate overlaps the lower member at the corners.

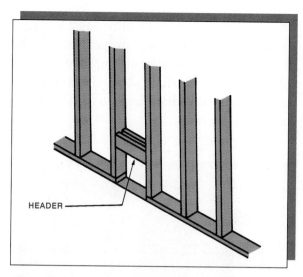

FIGURE 16–14 Framed opening for wall outlet for heating or air conditioning

$2'' \times 4''$ lumber. Sometimes it is necessary to install special framing on which cabinets will be mounted. It is often easier to install this as the wall is framed than to add it later.

In the past fifteen years the cost of energy to heat and cool homes has greatly increased. As a result, new construction systems have been developed to conserve energy. Most of these are intended to collect the sun's heat or to prevent the passage of heat through buildings' surfaces. Developments in wall framing have mostly been intended to reduce the loss of heat in winter or the gain of heat in the summer.

When wall framing is done with 2×4s spaced 16 inches on centers, a fairly large amount of solid wood is exposed to the surface of the wall. Wood conducts heat out of the building. Only the space between the solid wood framing can be filled with insulation. By using 2×6 studs spaced 24 inches on centers, the area of solid wood exposed to the surface of the wall is reduced by 20 percent, **Figure 16–15**. The amount of framing material is the same. Not only does this reduce the amount of wood exposed to the surface of the wall, it also

2 × 4's @ 16" O.C.

2 × 6's @ 24" O.C.

FIGURE 16–15 Using 2×6 studs spaced 24" center to center (24" O.C.) reduces the amount of wood exposed to the surface of the wall.

allows for 2 inches more of insulation.

Another method for reducing the area of exposed wood is by using 2-piece corner construction, **Figure 16–16** The third piece, which normally provides a nailing surface for the interior wallboard, is replaced by metal clips.

Modern construction often replaces plywood or board sheathing with rigid plastic foam boards, **Figure 16–17**. This *insulating sheathing* can add as much insulating value as that found within a conventional 2×4 wall frame. The insulating properties of the sheathing can be further improved by an aluminum foil surface. The aluminum foil surface reflects radiant heat so that it cannot penetrate the wall.

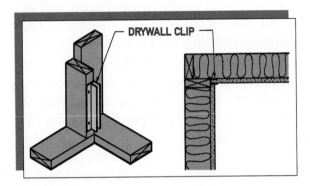

FIGURE 16–16 Two-piece corner construction

FIGURE 16–17 Rigid plastic foam provides extra thermal insulation in the walls.

ACTIVITIES

Drawing a Wall-Framing Elevation

Elevations are not normally drawn for the wall framing in a residence because the construction of walls is practically standardized. Nearly all walls for residential and light commercial construction are made as described in this chapter. However, to gain experience in laying out the members in wall framing, drawing an elevation is a useful technique.

Equipment and Materials

Paper and pencils
Architect's scale
Drafting triangle

Procedure

Draw an elevation of a wall frame using a scale of 1/2″ = 1'-0″ and **Figure 16–18** Dimension the location of all members.

Building a Wall Frame

The framing carpenter must keep in mind how the wall will fit the frame. Carpenters must follow architectural drawings and measure accurately. Carpentry mistakes add unnecessary costs to a project.

Equipment

56 linear feet of 1″ × 2″ lumber
3 linear feet of 1″ × 3″ lumber
Supply of 6d common nails
Tape measure or folding rule
Saw
Hammer
Rafter square

Procedure

1. Use a scale of 6″ = 1'-0″.
2. Cut all members to length.
3. Select two pieces to be used as plates. Mark the location of all studs 16 inches O.C.
4. Face nail all studs in place.
5. Construct the header over the rough opening.
6. Install the trimmers at the sides of the rough opening. These should be as long as the opening is high, **Figure 16–19**
7. Install the rough sill.
8. Install the bottom part of the trimmers and the jack studs.

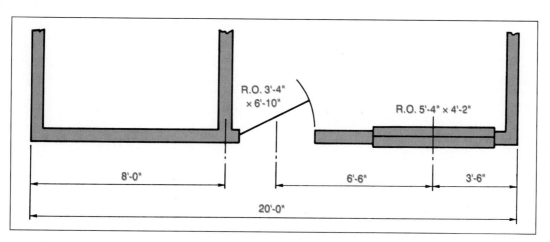

FIGURE 16–18

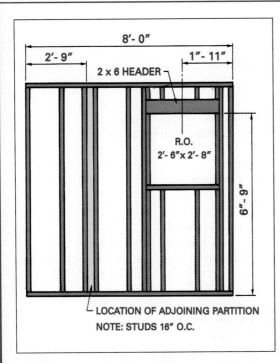

8'- 0"

2'- 9"   1"- 11"

2 x 6 HEADER

R.O.
2'- 6"x 2'- 8"

6"- 9"

LOCATION OF ADJOINING PARTITION
NOTE: STUDS 16" O.C.

FIGURE 16–19 Wall framing elevation for "Building a Wall Frame" activity

Applying Construction Across the Curriculum

Mathematics

Estimate the number of $2'' \times 4''$ studs required for the exterior walls in **Figure 15–18** The studs are placed 16 inches on centers. Allow one extra stud for every corner and two extra studs for every opening or where an interior partition intersects the wall. *Hint:* Determine the linear feet of bottom wall plate needed for exterior walls, then calculate the number of studs at 16" O.C. Finally, add the number of studs required for corners and interior partitions.

Mathematics

How many $2'' \times 6''$ studs would be required to build the exterior walls in **Figure 15–18?**

A. Identification. Identify the parts of the wall frame in **Figure 16–20.**

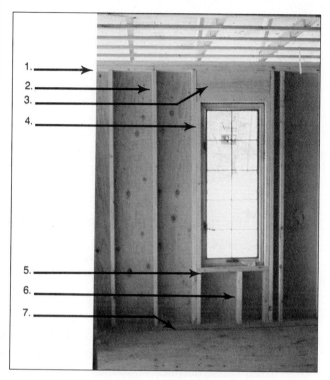

FIGURE 16–20

B. Matching. Match the item in Column II with its use or description in Column I.

Column I

1. Specially constructed to provide a nailing surface for interior wall covering
2. Wall that supports the weight of the structure above
3. Divides space but supports no weight
4. Transfers the weight from above to the sides of a rough opening
5. Is nailed to the floor framing
6. Is overlapped at the corners
7. Vertical members installed under a rough sill
8. Makes the entire wall frame rigid
9. A framing system which allows for more insulation and less exposed wood
10. Replaces the third piece in a corner

Column II

a. Sheathing
b. Load-bearing wall
c. Bottom plate
d. Partition
e. Double top plate
f. Jack studs
g. Header
h. Corner post
i. Clips
j. 2 × 6s at 24″ on center

CHAPTER 17

Roof Systems

OBJECTIVES

After completing this chapter, you should be able to:

▼ identify the most common types of roofs used on residences;

▼ explain the function of roof-framing members;

▼ apply asphalt shingles; and

▼ compare conventional rafters with trussed rafters.

KEY TERMS

gable roof	pitch
gambrel roof	hip rafter
hip roof	hip jack rafter
mansard roof	bird's mouth
shed roof	collar beam
common rafter	roof truss
ceiling joist	top chord
ridge board	bottom chord
span	lookout
run	fascia header
rise	sheathing
measuring line	underlayment
plumb cut	square
tail	gusset

The roof of a house protects the structure and its occupants from rain and snow. The style of the roof should be in keeping with the architectural design. The roof must also be able to support a heavy load, especially where several feet of snow may fall on the roof. Carpenters who construct roofs must be familiar with various roof designs and their construction.

Types of Roofs

There are five types of roofs that are commonly used in residential construction, **Figure 17–1**. Variations of these may be used to create certain architectural styles.

Gable Roof. The gable roof is one of the most common types used on houses. The gable roof consists of two sloping sides which meet at the ridge. The triangle formed at the ends of the house between the top plates of the wall and the roof is called the *gable*.

Gambrel Roof. The gambrel roof is similar to the gable roof. The sides of this roof slope very steeply from the walls to a point about halfway up the roof. They then have a more gradual slope.

Hip Roof. The hip roof slopes on all four sides. The hip roof has no exposed wall above

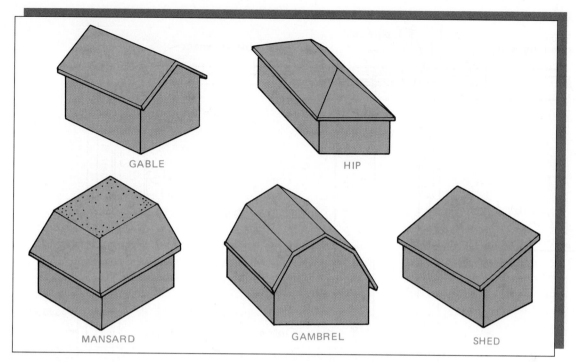

FIGURE 17–1 Common roof types

the top plates. This results in all four sides of the house being equally protected from the weather.

Mansard Roof.   The mansard roof is similar to the hip roof, except the lower half of the roof has a very steep slope and the top half is more gradual. This roof style is widely used in commercial construction, such as on stores.

Shed Roof.   The shed roof is a simple sloped roof with no ridge. The shed roof is not as common as other types for residential construction. It is used on some modern houses and additions to houses.

Conventional Rafter Framing

The roof-framing members that extend from the wall plates to the ridge are called **common rafters**. On a shed roof the common rafters span the entire structure.

Ceiling Joists

Before the carpenters can begin setting rafters, **ceiling joists** must be installed,

Figure 17–2. These are similar to floor joists. The ceiling joists perform two functions. They provide a surface on which to attach the ceiling and they prevent the rafters from pushing the walls outward. The size and direction of the ceiling joists are shown on the plan for the floor directly below in the same way floor joists are indicated, **Figure 17–3.** The ceiling joists are either toenailed to the double top plate or fastened with special metal anchors.

FIGURE 17–2 The ceiling joists rest on the top plate.

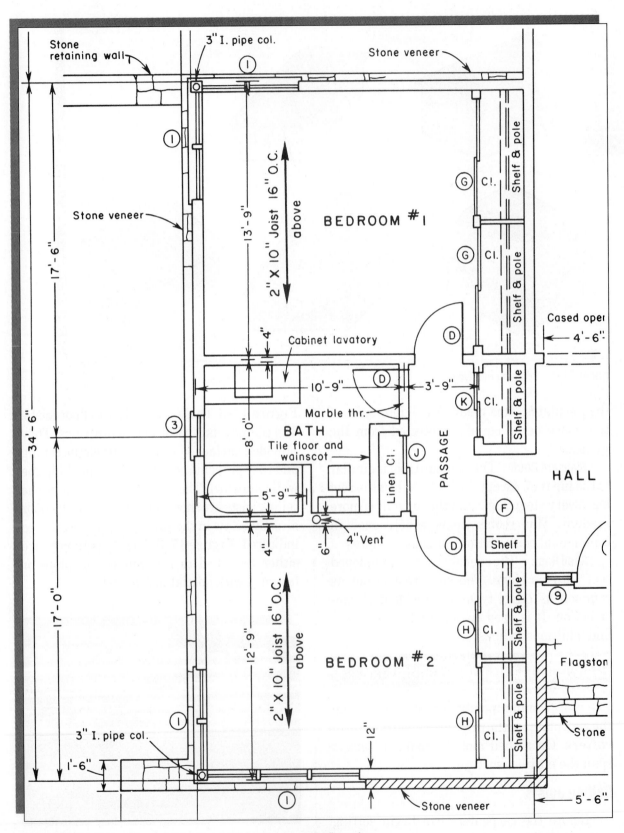

FIGURE 17–3 The positions of the ceiling joists are shown on the floor plan.

Houston Astrodome

On April 9, 1965, the Houston weather forecast called for rain, possibly thunderstorms, with strong winds. That was the day the Astrodome—a Texan's refusal to be shoved around by Mother Nature, as one reporter commented—was unveiled. The world's first all-weather, multipurpose stadium is the model by which every sports arena since has been measured.

The cost of the project, including off-site improvements, land acquisition, engineering fees, site roads, and parking lot paving, was $31.6 million.

The stadium structure covers nine and a half acres of land. With the rest of Astrodomain and parking areas, the complex covers 260 acres.

The outside diameter of the Astrodome structure is 710 feet. The stadium's floor is approximately 25 feet below normal ground level. This allows minimum vertical travel by fans.

The permanent roof of the dome, which has a clear span of 642 feet (twice that of any previous structure), has a maximum height of 208 feet above the playing field. This would allow the construction of an 18-story building under the roof.

The roof design consists of a steel frame with trussed beams arching upward to meet at the center and braced in a diamond pattern. This steel skeleton supports a roof containing 4,596 cast acrylic sheet skylights (7'2" by 3'4"). The roof can withstand hurricane winds of 135 mph and gusts of 165 mph.

The playing field is illuminated to an average brightness of 165 footcandles by 800 floodlights. The total electric power used exceeds 18,000 KVA, more than that needed by a city of 9,000 population.

The stadium is completely air-conditioned by electronic filters and activated charcoal odor removers. The air is then distributed through the system. Smoke and hot air are drawn out through the top of the roof. ■

Roof-Framing Terms

The main parts in a roof above the ceiling joists are the rafters and the ridge board. The **ridge board**, **Figure 17–4**, runs the length of the roof between the rafters of the two sides. The ridge board is the nailing surface for the top end of the rafters.

When the rafters are ready to be cut, the first one is carefully laid out and used as a pattern. A few terms must be understood for any discussion of how rafters are laid out, **Figure 17–5**.

▼ **Span** is the total width to be covered by the rafters. This is usually the distance between the outside walls of the house.
▼ **Run** is the width covered by one rafter. If the roof has the same slope on both sides, the run is one half the span.
▼ **Rise** is the height from the top of the wall plates to the top of the roof.
▼ **Measuring line** is an imaginary line along the center of the rafter. This is where the length of the rafter is measured.
▼ **Plumb cuts** are the cuts made at the top and bottom of each rafter. These cuts are plumb (vertical) when the rafter is in place.
▼ **Tail** is the portion of the rafter that extends from the wall outward to create the overhang at the eaves.
▼ **Pitch** is the steepness of the roof. This is usually expressed in terms of the number of inches of rise per foot of run. For example, if the height of the roof changes 5 inches for every 12 inches horizontally, the pitch is referred to as 5 in 12.

Rafter Tables

Carpenters use rafter tables to determine the length of the rafters. These tables are

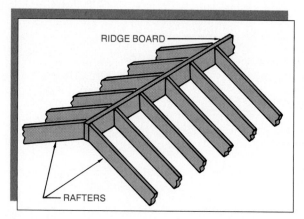

FIGURE 17–4   The ridge board runs the length of the roof.

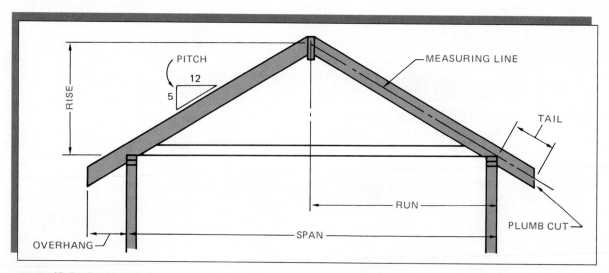

FIGURE 17–5   Roof framing terms

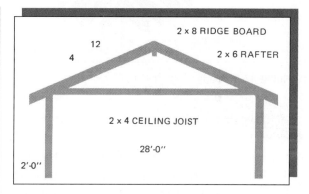

	2|3	2|2	2|1	2|0	1|9	1|8	1|7	1|6	1|5	1|4	1|3	1|2
LENGTH		COMMON	RAFTERS	PER FOOT	RUN	21 63	20 81	20	19 21	18 44	17 69	16 97
"	DIFF	HIP OR	VALLEY	"	"	24 74	24 02	23 32	22 65	22	21 38	20 78
"		IN LENGTH	OF JACKS	16 INCHES	CENTERS	28 7/8	27 3/4	26 11/16	25 5/8	24 9/16	23 9/16	22 5/8
"	SIDE		2 FEET			43 1/4	41 5/8	40	38 7/16	36 7/8	35 3/8	33 15/16
		CUT	JACKS	2 JACKS	USE	6 11/18	6 15/16	7 3/16	7 1/2	7 13/16	8 1/8	8 1/2
"	"	"	HIP OR	VALLEY		8 1/4	8 1/2	8 3/4	9 1/16	9 3/8	9 5/8	9 7/8

Bottom edge graduations: 2\|2 2\|1 2\|0 1\|9 1\|8 1\|7 1\|6 1\|5 1\|4 1\|3 1\|2 1\|1

1|2	1|1	1|0	9	8	7	6	5	4	3	2	1
16 97	16 28	15 62	15	14 42	13 89	13 42	13	12 65	12 37	12 16	
20 78	20 22	19 70	19 21	18 76	18 36	18	17 69	17 44	17 23	17 09	
22 5/8	21 11/16	20 13/16	20	19 1/4	18 1/2	17 7/8	17 5/16	16 7/8	16 1/2	16 1/4	
33 15/16	32 9/16	31 1/4	30	28 7/8	27 3/4	28 13/16	26	25 5/16	24 3/4	24 5/16	
8 1/2	8 7/8	9 1/4	9 5/8	10	10 3/8	10 3/4	11 1/16	11 3/8	11 5/8	11 13/16	
9 7/8	0 1/16	10 3/8	10 5/8		10 7/8	11 5/16	11 1/2	11 11/16	11 13/16	11 15/16	

Bottom edge graduations: 1\|0 9 8 7 6 5 4 3 2 1

FIGURE 17–6 Rafter table on the face of the square. The top line is the length of common rafters per foot of run.

available in handbooks and are usually printed on framing squares, **Figure 17–6**. To find the length of a common rafter:

1. Find the number of inches of rise per foot of run at the top of the table. These numbers are the regular graduations on the square.
2. Under this number, find the length of the rafter per foot of run. A space between the numbers indicates a decimal point.
3. Multiply the length of the common rafter per foot of run (the number found in step 2) by the number of feet of run.
4. Add the length of the tail and subtract one half the thickness of the ridge board. This is the length of the common rafter as measured along the measuring line.

NOTE: If the overhang is given on the working drawings, it can be added to the run of the rafter instead of adding the length of the tail.

EXAMPLE:
Find the length of a common rafter for the roof in **Figure 17–7**. (Refer to the rafter table in **Figure 17–6**.)

1. Rise per foot of run = 4"
2. Length of common rafter per foot of run = 12.65"

FIGURE 17–7

3. Run of one rafter including overhang = 16'-0"
4. 16 × 12.65" = 202.40" (round off to 202-1/2")
5. Subtract 1/2 the thickness of the ridge board. 202-1/2 – 3/4" = 201-3/4"

Hip Rafters

Hip rafters run from the corner of the building to the ridge at a 45-degree angle, **Figure 17–8**. The second line of the table on most framing squares is used to calculate the length of hip rafters. (Refer to the rafter table in **Figure 17–6**.) This table is based on the unit-run-and-rise method for finding the length of common rafters explained earlier in this chapter.

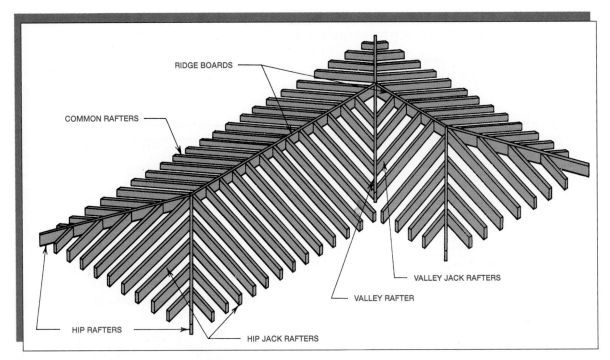

FIGURE 17–8   Parts of a hip-and-valley roof frame.

To calculate the length of a hip rafter, you must know the run of the common rafters in the roof and the unit rise of the roof, **Figure 17–9**. The length of the hip rafters is then found by using the table for the length of hip and valley rafters in the same way the length of common rafters was found earlier.

Step 1.   Find the unit rise (number of inches of rise per foot of run) at the top of the table.

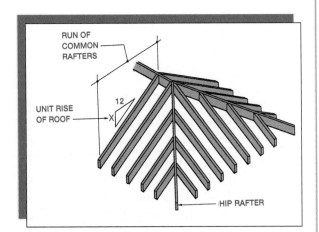

FIGURE 17–9   To use the table for the length of hip rafters, use the run of the common rafters in the roof.

These numbers are the regular graduations on the square.

Step 2.   Under this number find the length of the hip rafter per foot of run of the common rafters.

Step 3.   Multiply the length of the hip rafter per foot of common-rafter run (the number found in step 2) by the number of feet of run of the common rafters (1/2 the width of the building).

Step 4.   Subtract the ridge allowance. Because the hip rafter meets the ridge board at a 45-degree angle, the ridge allowance is one half the 45-degree thickness of the ridge board, **Figure 17–10** The 45-degree thickness is the length of a 45-degree line across the thickness of the ridge board. The 45-degree thickness of a 1-1/2-inch (2-inch nominal) ridge board is 2-1/8 inches. Therefore, the ridge allowance for a hip rafter on a 1-1/2-inch ridge board is 1-1/16 inches.

NOTE: If the hip rafter includes an overhang, add the overhang of the common rafters to the length used for common rafters.

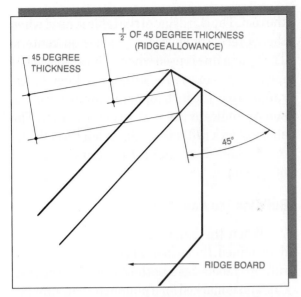

FIGURE 17–10 Use one-half of the 45° thickness of the ridge board as a ridge allowance for hip rafters.

EXAMPLE:

Find the length of a hip rafter for the roof shown in **Figure 17–11**

1. Rise per foot of run = 4"

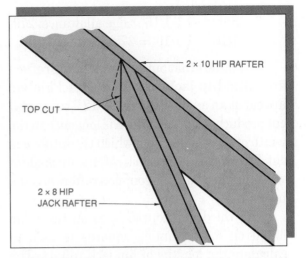

FIGURE 17–12 Because the top of the hip jack is cut on an angle, the hip rafter must be deeper.

2. Length of hip rafter per foot of common-rafter run = 17.44"
3. Common rafter run, including overhang = 16'-0"
4. $16 \times 17.44" = 279.04"$ (round off to 279-1/16")

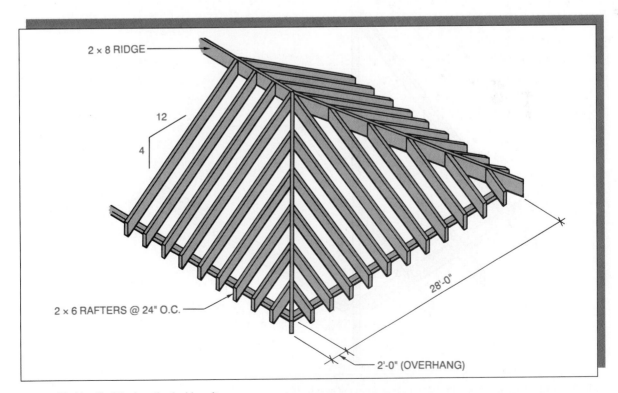

FIGURE 17–11 Find the length of a hip rafter.

5. Subtract 1-1/16" ridge allowance: 279-1/16" – 1-1/16" = 278"

The rafters that butt against the hip rafters are called **hip jack rafters**. Hip jack rafters are cut at an angle, **Figure 17–12** This angled cut produces a surface which is longer than the width of the lumber from which the rafter was cut. Therefore, the hip jack rafters are made of wider lumber than the common rafters and the hip rafters.

The third and fourth lines on the table found on most framing squares is used to calculate the lengths of hip jack rafters. The length of each jack rafter in a roof varies from the one next to it by the same amount, **Figure 17–13** The amount of this variance depends on the spacing of the rafters used and the pitch of the roof. The third line of the table is used when the rafters are spaced 16 inches on centers. The fourth line is used when they are spaced 24 inches on centers. As with other lines of the rafter table, the inch numerals at the top of the square indicate the unit rise of the roof. The shortest jack rafter is whatever length is shown on the table. Each jack rafter after that is that much longer.

Building the Roof Frame

When the length of the rafters has been determined, the carpenter cuts one rafter. This rafter is labeled as a pattern to lay out the rest. One end is marked for a plumb cut. The number 12 on the blade of the square is placed at the edge of the rafter. The number on the tongue,

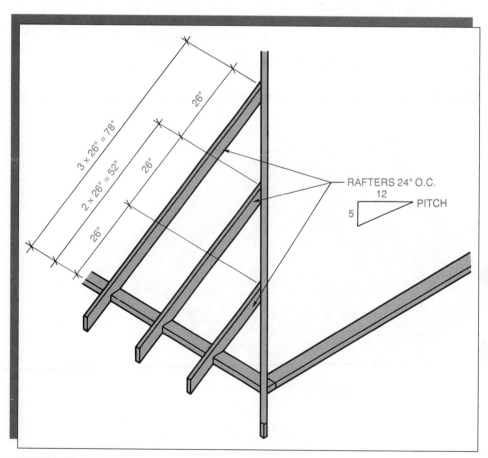

FIGURE 17–13   Each jack rafter in a string varies from the next by the same amount.

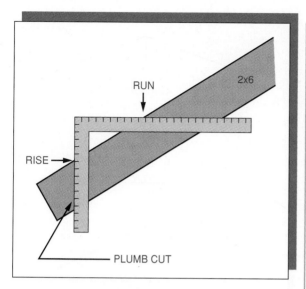

FIGURE 17–14 The rafter square is used to lay out a plumb cut.

which corresponds to the rise per foot of run, is lined up with the edge of the rafter, **Figure 17–14.** Stair gauges are helpful if several rafters are to be laid out. The end of the rafter is cut at this line. The opposite end is cut in the same manner and is parallel to the first end. Finally a notch, called a **bird's mouth**, is cut where the rafter rests on the wall plate.

The ridge board is temporarily held in place by braces. The rafters are then nailed to the ridge board and the wall plates.

It is common practice to install collar beams on every other or every third pair of rafters. **Collar beams** are horizontal members that brace the rafters about halfway up the roof, **Figure 17–15**

Trussed Rafters

Through engineering advances, a system has been devised that speeds roof framing, reduces material needs, and produces stronger roofs. This is the use of trussed rafters. Trussed rafters, commonly called **roof trusses**, are units assembled in a shop. These units are then transported to the construction site and set on the walls, **Figure 17–16**

This is the system of roof construction used most often today. The top members of a truss, corresponding to rafters, are the **top chords**. The **bottom chord** is on the bottom of the truss and corresponds to the ceiling joists in conventional rafter framing. Depending on the

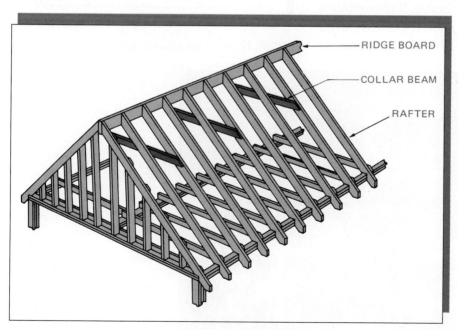

FIGURE 17–15 Collar beams are often added to every second or third pair of rafters for extra strength.

FIGURE 17–16 Hoisting roof trusses into place *(Courtesy of Northern Homes)*

design of the truss, there are several braces, called *webs*, between the top and bottom chords. **Figure 17–17** shows several roof truss designs.

Gable Framing

Whether conventional framing or roof trusses are used, the triangular ends of a gable roof must be filled in with framing. If the architect specifies a trussed roof, gable trusses are set on the end walls, **Figure 17–18** In conventional

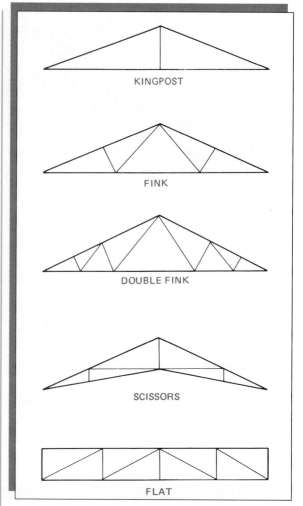

KINGPOST

FINK

DOUBLE FINK

SCISSORS

FLAT

FIGURE 17–17 Some common types of roof trusses

FIGURE 17–18 The gable truss has studs to which the sheathing and siding are fastened.

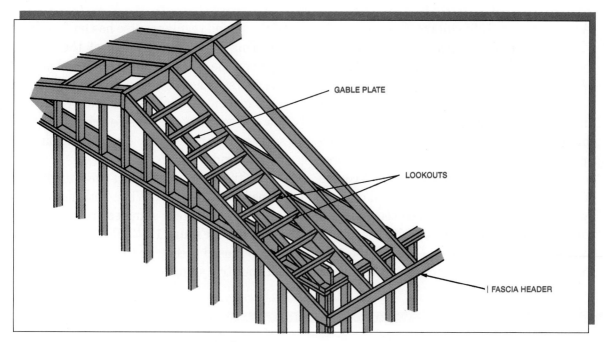

FIGURE 17–19 Framing for an overhang at the gable end

framing, gable end studs are placed directly above the regular wall studs. Gable end studs are toenailed to the top wall plate and notched to fit against the end rafter.

Rake Framing

When the roof overhangs at the ends, special framing is required. This involves the use of *gable plates* and **lookouts**. The overhanging rafter is nailed to the ends of the lookouts. The lookouts rest on top of the gable plates, **Figure 17–19**

The final member of the roof frame is the **fascia header**. This is a piece of lumber the same size as the rafters nailed to the ends of the rafters. The fascia header will support trim to be added later.

Roof Covering

Most types of roof coverings require that the roof frame be covered with boards or plywood first. This is called the *roof sheathing* or *roof decking*. Roof **sheathing** is applied over the rafters in the same way that subflooring is applied over the floor joists.

The roof covering is applied by *roofers* (people who specialize in this field of work) or carpenters. Roofers and carpenters who do roofing must be comfortable working in high places. Most roofing contractors work on both residential and commercial constructions.

The first step in covering the sheathing is to apply roofing **underlayment**. *Roofing underlayment* serves two purposes. It prevents chemical reactions between the resins in the wood and the roof covering material. It also provides added weather protection. Asphalt-saturated felt is the most common roofing underlayment material.

The most common roof coverings for residential construction are asphalt-rolled roofing, asphalt shingles, and wood shingles and shakes. Asphalt strip shingles are the most common. Asphalt shingles and rolled roofing are available in several weights. The weight is specified according to the weight of material required to cover one square.

(A **square** is 100 square feet of roof.) Underlayment felt is generally a 15-pound weight. Strip shingles are generally 225- to 240-pound asphalt.

Shingles are applied starting at the eaves and working toward the ridge. Each row of shingles across the roof is called a *course*. The first course of shingles is a solid starter course or a strip of 240-pound rolled roofing. A course of regular roofing shingles is applied directly over the starter course. The next course is laid with the bottom edge lined up with the tops of the tabs of the first course. (A *tab* is the part that is exposed to the weather.) The first shingle of the second regular course has half a tab cut off. This insures that none of the cutouts between the tabs will fall over an end joint in the course beneath. **Figure 17–20** shows the arrangement of shingles on a roof. After every few courses the roofers strike a chalk line to be sure that the courses are straight and parallel with the ridge.

At the ridge, the tops of the shingles are lapped over the ridge. When both sides of the roof are completely shingled, ridge shingles are applied, **Figure 17–21** Ridge shingles can be purchased precut, or they can be made by cutting strip shingles apart at each tab.

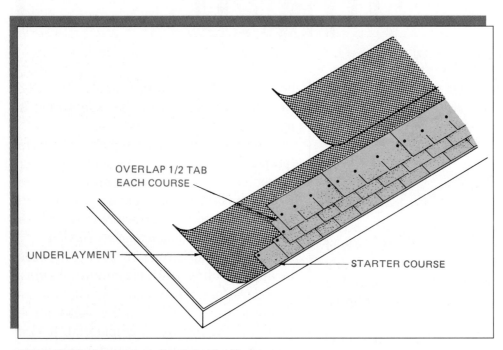

FIGURE 17–20 Arrangement of asphalt strip shingles

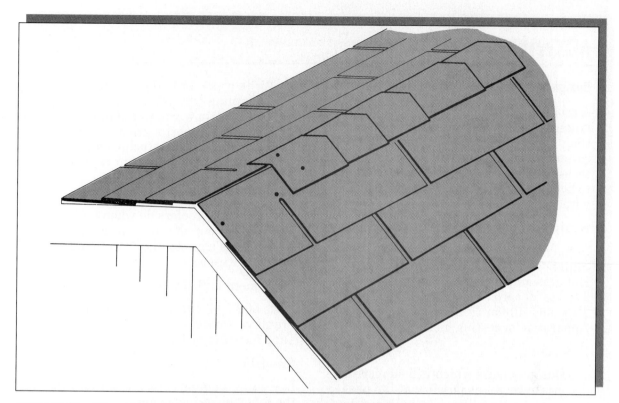

FIGURE 17–21 Ridge shingles can be made by cutting strip shingles at each tab.

ACTIVITIES

Design and Test a Roof Truss

A roof framed with efficiently designed trusses uses less material and is stronger than a roof framed with conventional rafters. In this activity you will design a truss, using as little material as possible to achieve the greatest strength possible with those materials.

Equipment and Materials

Supply of 1/8" × 3/8" balsa wood
Supply of 1/8" × 1/4" balsa wood
Hot glue gun or liquid wood glue
Thin cardboard of 1/16" balsa wood sheet
Tape measure or rule
Apparatus for testing model trusses

Procedure

1. Design a truss which will support as much weight as possible at the peak, meeting the following requirements:

   16-inch span
   Pitch of 3 in 12 to 5 in 12
   To be constructed at a scale of 1/16" = 1"

   You may use one of the truss designs in **Figure 17–17** or a design of your own, but you should design your truss to be as strong as possible, using as little material as possible.

   You may wish to add **gussets** to the joints in your truss. A *gusset* is a flat piece of reinforcement added to a joint to increase the rigidity and strength of the joint, **Figure 17–22**.

   Your truss will be judged on the following:

   Load supported before breaking
   Most efficient use of materials
   Workmanship

2. Using a model truss testing apparatus supplied by your teacher or one of your own design, apply a load of 1 pound to your truss. Increase the load gradually, noting the load with each increase, until the truss breaks.

3. Write a one- or two-page report explaining why your truss behaved as it did and how you would modify the design to support a greater load or use less materials. Include a dimensioned drawing with your report.

Laying Out Rafters

The first thing roofers do when building a roof is to cut out one rafter as a pattern. With this as a guide, they can then lay out the entire roof.

Equipment and Materials

Piece of 2" × 4" lumber at least 4 feet long
Framing square
Tape measure or folding rule

Procedure

Lay out a common rafter for a roof with the following dimensions:

Span of roof—6'-0"
Pitch—3 in 12
Ridge board—2" × 6"
Overhang—0'-6"

Building a Roof Truss

Equipment and Materials

8 pieces of plywood, 1/4" × 10" × 16"
2 pieces of 2" × 4" lumber, 8 feet long

FIGURE 17–22   Gusset

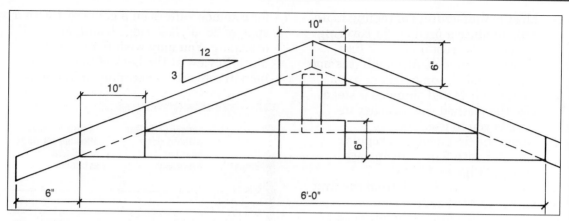

FIGURE 17–23

(rafter from the first activity can be substituted for one piece)
Supply of 2d box or common nails
Hammer
Framing square
Saw
Tape measure or folding rule

Procedure

1. Cut the top chords, bottom chord, and web to length according to **Figure 17–23**. The ends of the bottom chord can be laid out with the framing square, **Figure 17–24**.
2. Mark the length of the tail on each top chord. The tail le ngth can be found by finding the length of a common rafter for 1/2 foot of run.
3. Position all members on a flat surface. Trace the shape of the gussets (the plywood reinforcements at joints).

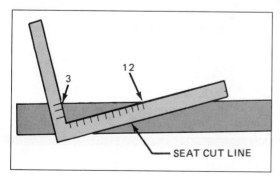

FIGURE 17–24   Laying out the bottom chord

4. Being careful to keep the parts aligned, nail the gussets in place. Drive four nails into each member of the truss at each gusset.

Shingling

Roofers work with a variety of building materials and hand tools. Because they work in high places and in unusual positions, they must be safety conscious.

Equipment and Materials

Sheet of 1/2 to 3/4-inch plywood
Bundle of asphalt strip shingles
15-pound saturated felt
Supply of 3d smooth shank drywall nails*
Hammer
Utility knife (to cut shingles)
Square
Chalk line

Procedure

1. Nail scrap pieces of lumber to the back of the plywood. By doing this, the nails used for shingling will not penetrate through the plywood into the surface below.
2. Cover the plywood with underlayment.

*Asphalt shingles are normally applied with galvanized roofing nails. However, smooth shank drywall nails are suggested for this activity to make removal and salvage easier.

3. Most manufacturers of roofing materials produce a product for covering the edge of a roof under the first course of shingles. Although it is not recommended, some roofers use a course of inverted shingles instead. Use the method your teacher instructs you to use.

4. Apply the first course of regular shingles with four nails in each shingle, **Figure 17–25**.

5. Cut away half of a tab from the first shingle for the second course.

6. Continue to shingle the plywood according to **Figure 17–25**. Strike a chalk line along the tops of the cutouts of every third course.

Applying Construction Across the Curriculum

Mathematics

Use the common rafter table on a framing square to calculate the information necessary to complete the following table for common rafters on a gable roof with a span of 26'-0", 3/4" ridge board, and 1'-0" overhang. You may wish to refer to the Math Review at the back of the book for help in converting decimals to fractions.

Slope	Length of Rafter without ridge board or overhang	Length of Rafter allowing for ridge board and overhang
3 in 12	_____	_____
6 in 12	_____	_____
12 in 12	_____	_____

Mathematics

How many bundles of asphalt shingles are required for a common gable roof 30 feet long, with rafters which are 14'-4" long? Asphalt shingles cover one-third square per bundle.

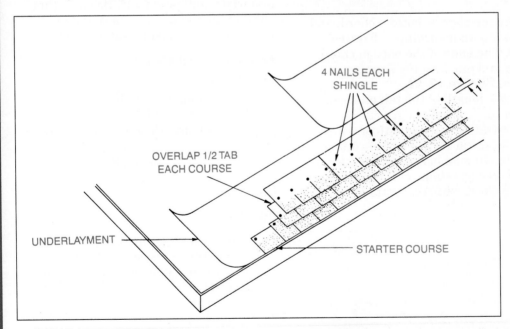

FIGURE 17–25

REVIEW

A. Identification. What type of roof is indicated in each drawing in **Figure 17–26**?

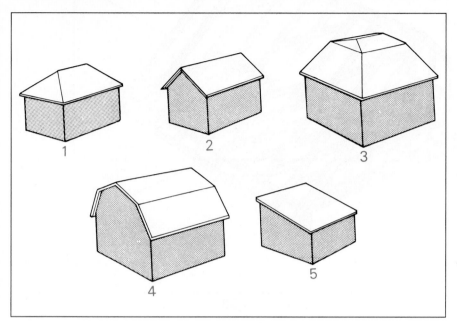

FIGURE 17–26

B. Identification. Show where these items are indicated in **Figures 17–27** and **17–28**.

Top chord	Gable stud
Common rafter	Bottom chord
Ridge board	Span
Web member	Tail
Gusset	Fascia header

C. Fill-Ins. Fill in each blank with the proper term.

1. A _____ cut is made at the end of the rafter so that the end will be vertical when in place.
2. The height from the top of the wall plates to the top of the roof is the _____ .
3. The total width covered by a pair of rafters is the _____ .
4. The width covered by a single rafter is the _____ .
5. The imaginary line along which a rafter is measured is the _____ .
6. The portion of a rafter that extends outside the wall plates is the _____ .
7. The steepness of a roof is the _____ .
8. A rafter that extends all the way from the wall plate to the ridge board is called a _____ .

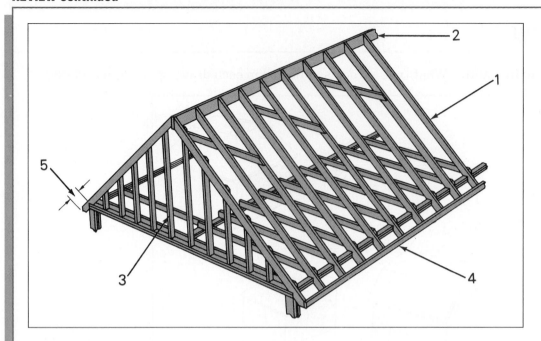

FIGURE 17–27

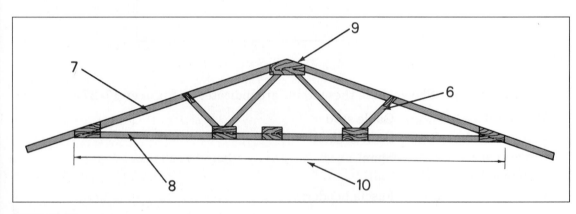

FIGURE 17–28

CHAPTER 18

Enclosing the Structure

OBJECTIVES

After completing this chapter, you should be able to:

▼ explain the use of trim to enclose structural elements;

▼ describe the various types of siding or masonry veneer, doors and windows, and exterior trim; and

▼ compare a variety of materials used to finish residential building exteriors.

KEY TERMS

rake cornice	gain
frieze	sash
soffit	double-hung window
fascia	sliding window
closed cornice	casement window
open cornice	awning window
boxed cornice	exposure
flush door	masonry veneer
panel door	bond
stile	lath
rail	scratch coat
casing	brown coat
prehung door	finish coat

Cornice

The building frame, wall sheathing, and roof deck and covering provide strength and cover the large areas of the structure. Where these structural elements meet or where the sheathing and decking have not completely closed small openings in the building frame there is a need to enclose the building more tightly. Also, because many framing and sheathing materials do not lend themselves to smooth cuts and tight joints, there is a need to use trim and finish materials to tighten up the building. The trim that encloses the edges of the roof, corners of walls, and joints between the walls and the roof structure is called cornice.

The *eaves cornice* is the construction where the rafters meet or overhang the walls. Where the cornice follows the slope of the rafters, such as at the ends of a gable roof, it is called a **rake cornice**.

The cornice is usually made of wood or a combination of wood and aluminum. Wood cornices are made by carpenters. On many jobs this is done by finish carpenters. Finish carpenters install trim and millwork after the framing carpenters have completed the basic

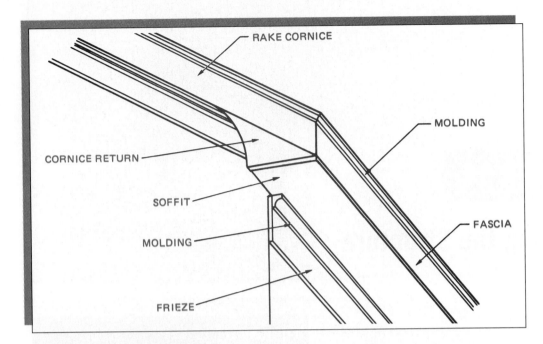

FIGURE 18-1 Parts of a boxed cornice

frame of the structure. Finish carpenters work to close tolerances. They must do very neat work. When aluminum trim is used in the cornice, it is installed either by aluminum siding and trim installers or by carpenters.

The main parts of a cornice are the **frieze**, **soffit**, and **fascia**, **Figure 18-1**. The *frieze* is a horizontal piece against the wall of the building. The *soffit* is the covering on the underside of the rafters on a boxed cornice. The *fascia* is

nailed to the ends of the rafters at the eaves or to the side of the rake (end) rafters. Molding can also be used to create a pleasing appearance.

There are basically three types of cornices used in residential construction. A **closed cornice** is used where there is no overhang. An **open cornice** is used with overhanging eaves and exposed rafters. A **boxed cornice** completely encloses the rafters, **Figure 18-2**.

Boxed cornices can be made with the soffit fastened directly to the underside of the rafters, **Figure 18-4**. This type is called a *sloping cornice*. Another type of boxed cornice is the *level cornice*, such as the one in **Figure 18-2**. On a level

FIGURE 18-2 A boxed cornice encloses the ends of the rafters. (Photo by Michael Dzaman)

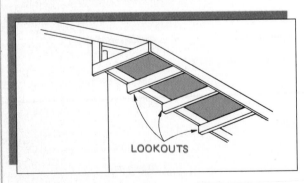

FIGURE 18-3 The soffit is nailed to lookouts which extend from the wall to the fascia.

FIGURE 18–4 Sloping cornice (Photo by Michael Dzaman)

FIGURE 18–5 The soffit must be ventilated to allow air to circulate between the rafters.

cornice, *lookouts* are installed from the ends of the rafters to the wall of the building. The soffit is nailed to these lookouts, **Figure 18–3**.

The soffit has openings to allow air to flow into the space between the rafters, **Figure 18–5**. This keeps ice from building up at the

eaves. The free flow of air also helps ventilate the attic.

The working drawings almost always include detail drawings of the cornice, **Figure 18–6**. Carpenters and siding and trim installers

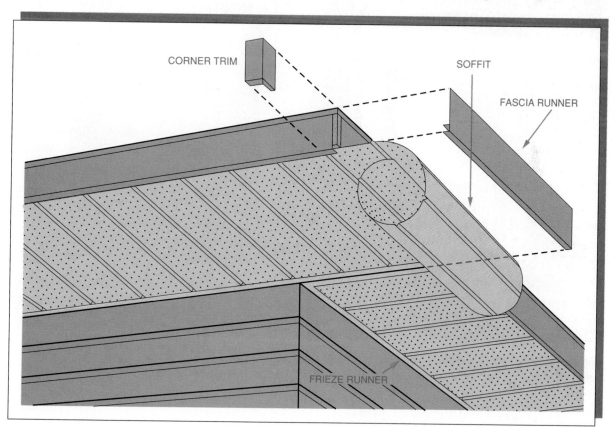

CORNER TRIM

SOFFIT

FASCIA RUNNER

FRIEZE RUNNER

FIGURE 18–6 Typical cornice detail drawing (*Courtesy of Home Planners, Inc.*)

FIGURE 18–7 Several styles of entrances: (A) solid door with side lights, (B) door with glazing, (C) solid door with side and transom glazing, and (D) double glazed doors with transom light. *(Courtesy of Morgan Products Ltd.)*

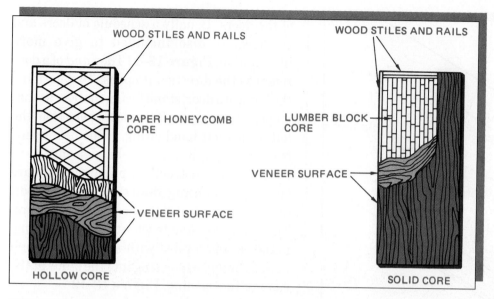

FIGURE 18–8 Construction of flush doors

consult these detail drawings before beginning work on the cornice.

Exterior Doors

Exterior doors give protection and privacy. They are also an important part of any architectural design. They may be of the swing type or sliding type. The main entrance is usually a decorative swing-type door. Where the door opens into an entrance hall with no other windows, the door assembly often includes *side lites* (small windows beside the door) and a glazed door. A *glazed door* has a window. Several styles of exterior doors are shown in **Figure 18–7**.

Flush doors have a smooth facing that is glued to a solid or hollow core, **Figure 18–8**. **Panel doors** have two or more panels in a framework of **stiles** and **rails**, **Figure 18–9**. *Stiles* are the vertical members. *Rails* are the horizontal members.

Doors are made of wood, aluminum, and steel. Wood has long been the standard material for doors. Metal doors are also becoming popular.

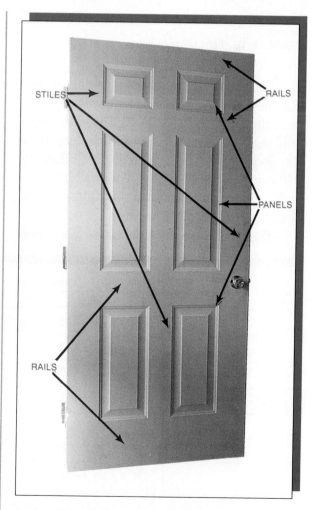

FIGURE 18–9 Construction of a panel door

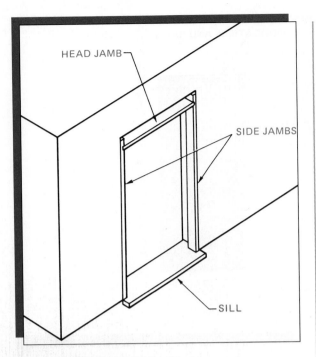

FIGURE 18–10 Parts of a door frame

Doors are hung on hinges in a door frame. The door frame is made up of four main parts and the trim, **Figure 18–10** The side jambs and head jamb are cut to the proper length for the door and the necessary width to fit flush on both sides of the wall. The sill is made of either oak or metal to resist wear. The interior and exterior trim is called **casing**.

The floor plans show the size and location of the doors. Usually a schedule of doors and windows is also included to give more information, **Figure 18–11** The *hand* of a door refers to the direction it opens. To determine the hand of a door, stand facing the door so that it opens toward you. If the door opens from the left, it is a left-hand swing. If the door opens from the right, it is a right-hand swing.

In some construction, prehung doors are specified. **Prehung doors** include the door, door frame, and all of the necessary hardware. These units are simply set in the rough opening, plumbed and leveled with a spirit level, and nailed through the casing into the studs at the sides of the opening, **Figure 18–12**

When the doors are not prehung, the carpenter installs the individual parts and fits the door. Hanging a door requires a skilled carpenter and careful work. The door jamb and sill are first assembled and fitted into the opening. Next, the door is trimmed with a plane to fit the opening. Then **gains** are cut for the hinges. A *hinge gain* is a recess made in both the door edge and the jamb that accepts the hinge. The hinges are installed and the door is hung in the jambs. If the jambs do not include a door stop rabbet, stop molding is nailed to the jambs. Finally, the lock set is installed.

MARK	QUANTITY	SIZE	TYPE	MATERIAL	REMARKS
		DOOR SCHEDULE			
A	1	1¾" x 3'-0" x 6'-8"	SEE ELEV	PINE	PAINT
B	8	1⅜" x 2'-6" x 6'-8"	FLUSH	BIRCH	HOLLOW CORE
C	3	1⅜" x 2'-6" x 6'-8"	LOUVERED	PINE	
D	1	1¾" x 2'-6" x 6'-8"	10 LIGHTS	PINE	
E	3	1⅜" x 5'-0" x 6'-8"	FOLDING	PINE	LOUVERED
F	1	1⅜" x 2'-0" x 6'-8"	FLUSH	BIRCH	HOLLOW CORE
G	1	1¾" x 2'-6" x 6'-8"	LIGHTS	PINE	SEE REAR ELEVATION

FIGURE 18–11 Door schedules are included on working drawings to give complete information about all of the doors.

FIGURE 18–12 Setting a prehung door in the rough opening

Windows

Windows are made up of the sash and the frame. The **sash** is the glass and the frame that holds it. The *frame* of a window holds the sash in the wall.

Windows are named according to the way they open, **Figure 18–13** The simplest kind of window has a *fixed* sash that does not open. **Double-hung windows** have two sashes that slide up and down. A **sliding window** is similar to a double-hung window turned on its side. Sliding sashes slide from side to side. **Casement windows** are hinged on the side

and swing out like doors. **Awning windows** are hinged at the top so the bottoms swing outward.

The size of windows is usually referred to by sash size. Other sizes sometimes referred to are the glass size and the rough-opening size, **Figure 18–14** The floor plan shows the location and sash size of the windows. The type of window is also indicated on the floor plan by a symbol. More information about the windows is given in the specifications and on the window schedule.

Windows are installed by carpenters in much the same way as doors. They slide the complete window unit, without the interior casing, into the rough opening, **Figure 18–15** *Shims* are placed under the window sill to level the window. The unit must be perfectly level and square to operate properly. When the window is accurately positioned, finishing nails or casing nails are driven through the exterior casing and into the studs. These nails are set below the surface of the wood and covered with plastic wood or putty.

Exterior Wall Coverings

The purpose of exterior wall covering is to make the structure weathertight and to create an attractive appearance. The kind of exterior wall covering on a house plays a very important part in the overall architectural style. Exterior wall coverings used on most residences can be classified as either siding or masonry veneer.

Wood Siding

Wood siding can be either vertical or horizontal. Horizontal siding creates a more traditional appearance. Vertical board-and-batten siding creates a contemporary appearance. Whether the siding is vertical or horizontal, it must be applied so that water can run down the wall without running behind the

FIGURE 18–13 Different styles of windows include: (A) double hung, (B) sliding, (C) casement, and (D) awning. *(Courtesy of Andersen Corp., Bayport, MN 55003)*

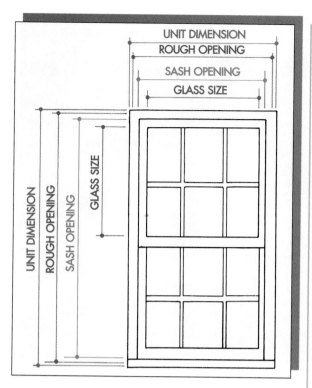

FIGURE 18–14   The size of a window can be measured in several ways. The size usually listed is the sash opening.

FIGURE 18–15   The carpenter first positions the window unit in the opening, then nails through the casing or a nailing flange and into the sheathing and studs. (*Courtesy of Andersen Corp., Bayport, MN 55003*)

siding. Each kind of wood siding shown in **Figure 18–16** will shed water as it runs from top to bottom.

Horizontal wood siding is installed starting at the bottom of the wall. The first piece is leveled, then nailed to the wall frame with aluminum or galvanized nails. Regular steel nails are not used because they would rust and stain the painted siding. The second course of siding is applied so that a specified amount of the first course is left uncovered. This is called the **exposure** of the siding. Wood siding is made in widths from 6 inches to 16 inches, depending on the intended exposure. After every few courses, a chalk line is snapped to guide the siding installers.

For appearance and weather protection, it is important that the siding be accurately measured and cut. Large joints between the ends of two pieces would allow water to seep through. At the corners the siding is either butted snugly against corner boards, **Figure 18–17**, or covered with metal corners.

Vertical siding is plumbed with a spirit level as it is applied. Tongue-and-groove siding is nailed with finishing nails driven at an angle through the tongue. Board-and-batten siding is nailed near the edges where it will be covered by the battens. The battens are face-nailed with aluminum or galvanized nails.

Vinyl Siding

Many homes are covered with vinyl siding. This eliminates the need for painting and reduces maintenance. Vinyl siding is made in a wide range of styles to duplicate the

FIGURE 18–16   Wood siding *(Courtesy Wendy Troeger)*

appearance of most kinds of wood siding, **Figure 18–18** Vinyl siding is made from colored plastic, so repainting is never necessary. The trim for this siding is made from vinyl of a matching color.

FIGURE 18–17   Horizontal wood siding is butted against the edge of the corner board.

The first step in applying horizontal vinyl siding is to nail a starting strip to the bottom of the wall. Channels, called *J channels*, are nailed around windows and doors. Corners are trimmed with corner posts. The siding is cut to length. Where more length is needed, the pieces are overlapped slightly. The bottom of the first piece is slid up onto the starting strip, then the top edge is nailed. The top edge of the siding is shaped to hold the bottom of the next piece. For added insulation and support, *backer boards* can be slipped into the back of the siding. **Figure 18–19** shows a typical vinyl siding installation.

Sheet Siding

Plywood and hardboard are made with special patterns for use as exterior wall coverings. These are usually a vertical pattern, **Figure 18–20**. They make attractive, weathertight coverings and reduce the amount of labor involved. Most sheet sidings are simply face-nailed to the building.

FIGURE 18–18 Vinyl siding looks like wood, but does not require painting. *(Courtesy of Reynolds Metals)*

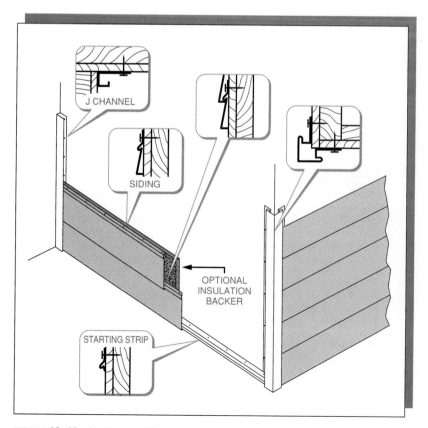

J CHANNEL

SIDING

OPTIONAL
INSULATION
BACKER

STARTING STRIP

FIGURE 18–19 Horizontal siding installation

Masonry Veneer

Masonry veneer covers a structure with masonry materials, such as brick or stone. Most houses that appear to be made of bricks are actually brick veneer. Masonry veneer is attractive and needs almost no maintenance.

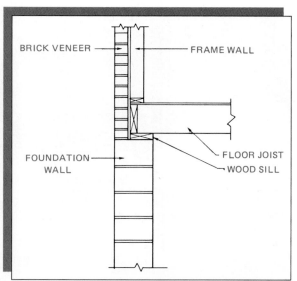

FIGURE 18–21 For masonry veneer construction, the foundation walls must be thick enough to provide a base for the veneer.

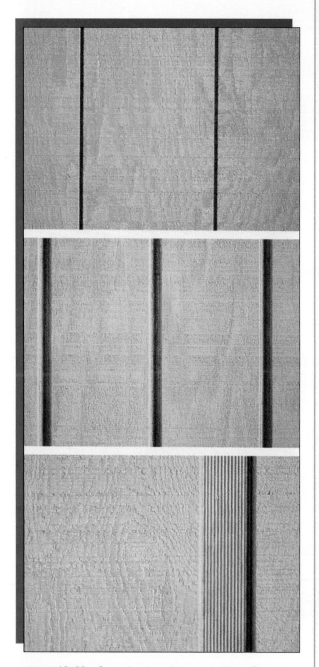

FIGURE 18–20 Several styles of plywood siding *(Courtesy of the American Plywood Association)*

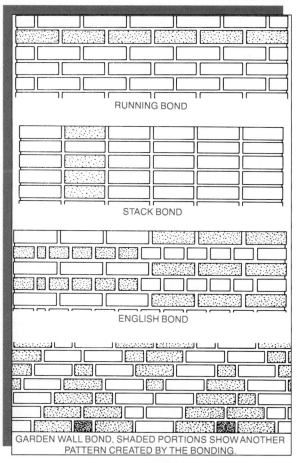

FIGURE 18–22 A variety of effects can be created by using different bond patterns.

FIGURE 18–23 Wire lath ready for stucco

FIGURE 18–24 Spraying on the stucco plaster

Unlike other kinds of wall covering, masonry veneer requires that the foundation be designed for this kind of wall covering. The weight of the masonry veneer is more than most soil can support, so it must rest on the foundation, **Figure 18–21**

Masonry veneer is built by masons after the building frame is completed. The masonry units are laid in mortar, just as masonry units are for foundation walls. Architects may specify special arrangements of bricks in veneer construction to create decorative effects, **Figure 18–22**. The arrangement of bricks to create a pattern is called a **bond**. Masons must know how to lay bricks in a variety of bond patterns.

To prevent the masonry veneer from pulling away from the building frame, wall ties are used. Wall ties are metal devices that are nailed to the wall sheathing and embedded in the mortar joints of the veneer.

Stucco

Stucco is plaster made with portland cement. The stucco is applied in two or three coats over wire lath. The wall sheathing is covered with waterproof building paper. Next the **lath** (usually wire netting) is stapled to the wall, **Figure 18–23**. Finally, the stucco is troweled on. First, a rough **scratch coat** is put on. Then a **brown coat** is put on to build up to the approximate thickness. Finally a **finish coat** is used, **Figure 18–24**. Outside corners and edges are formed with galvanized metal beads nailed to the wall before the scratch coat is applied.

Peter George Davies

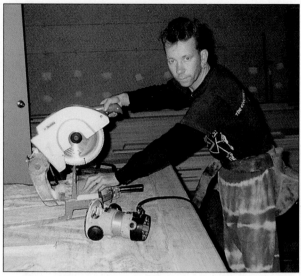

Occupation:

Finish Carpenter

How long:

15 Years

Typical day on the job:

Peter has recently relocated from California to New York state. He is currently working with a builder on all kinds of framing while he establishes his reputation in the new area. Most of his experience has been with specialized framing systems, like the ones used in the construction of recording studios and set construction for movies. No nails are used in sound studio framing. All framing is pressed together with carefully-fitted joints, so this a very specialized form of carpentry. Set construction also has very special requirements. Movie sets are framed with 1×3s instead of 2×4s, and they must be built in such a way that they can be dismantled quickly, yet be safe for actors and crew.

Most of Peter's work now consists of installing doors and trim in custom-built homes. The day begins with organizing the crew of carpenters and laborers who will work with him, getting the power generator running when there is no house electricity available, and planning the day's work. Work crews range in size from three to thirty people.

Education or training:

There is quite a bit of math required in the calculation of stair construction and angled rooms, and in bidding jobs. Peter learned his trade in England. He attended a regular British school until he was 12 years old, then he entered an apprenticeship for carpentry. During his four-year apprenticeship he worked five days a week for low wages and went to school one day a week.

Previous jobs in construction:

Peter was part owner of a company that specialized in recording studio construction. He also worked on several well-known movie sets, including *Terminator II, Last Action Hero,* and *The Last Boy Scout.*

Future opportunities:

Peter is working for a builder now to establish himself in a new area so that he can again specialize in true finish carpentry on his own, later.

Working conditions:

Most framing work is done between April and October, so that the winter months are spent inside a completed building shell. An unheated building in the winter is sometimes colder than working outdoors.

Best aspects of the job:

Peter enjoys the satisfaction of making customers happy with his work. Also, he has the self-satisfaction of a job well-done.

Disadvantages or drawbacks of the occupation:

This is very seasonal work. Also, if you work for yourself, there are not many fringe benefits.

ACTIVITIES

Constructing a Boxed Cornice

Finish carpenters install wood cornices. Their work must be accurate if the completed project is to look neat and attractive. Finish carpenters must be able to read architectural drawings. These drawings show how the exposed trim is installed.

Equipment and Materials

Roof and wall frame module
Lumber: 2" × 4" × 2'
       1" × 2" × 4'
       1" × 4" × 4'
       1" × 8" × 4'
3/4-inch cove molding, 4 feet long
Square
Tape measure or folding rule
Saw
Hammer
Supply of 6d finishing nails

Procedure

1. Cut and install 2" × 4" lookouts on each rafter.
2. Cut all cornice materials to length. Install them with 6d finishing nails according to **Figure 18–25**. Remember that the cornice is part of the exposed trim on the house. Neatness is the mark of a fine carpenter.

Hanging a Door

Hanging a custom fitted door requires a great amount of skill. The carpenter must measure, cut, and trim carefully to ensure that the door will swing smoothly, close tightly, and be attractive. Most standard size doors are prehung at the factory, then installed in the opening at the site. However, finish carpenters and cabinetmakers still have to fit and hang special sized doors. Also, hanging a custom fitted door is an excellent way to practice tools skills.

Equipment and Materials

Lumber: 1" × 4" × 6'-6"
       1" × 8" × 4'
One pair 2-inch butt hinges
Wood chisel or router
Jack plane
Square
Hammer
Saw
Supply of 6d common nails
Screwdriver
Tape measure or folding rule

Procedure

1. Construct the simulated door frame shown in **Figure 18–26**.

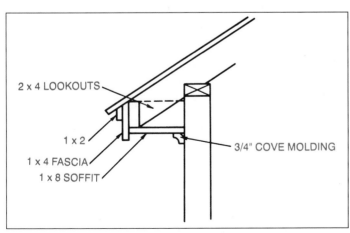

FIGURE 18–25 Cornice detail for "Constructing a Boxed Cornice" activity

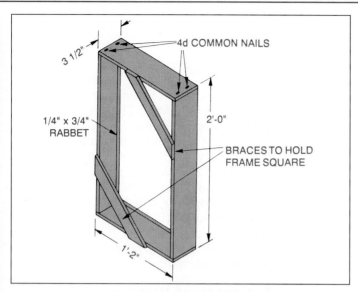

FIGURE 18–26 Simulated door frame for "Hanging a Door" activity

2. Glue two 4-foot pieces of 1" × 8" together to make a panel slightly larger than the opening. This will be the door.
3. Trim the door to fit the opening. Saw off large amounts of excess. Then plane the door for an accurate fit.
4. Mark the location of the hinges on the door edge. The hinges should be three inches from the top and bottom of the door.
5. Using a chisel or electric router, cut the hinge gains in the door edge, **Figure 18–27**.

CAUTION: If a chisel is used, review its use in Chapter 10. If an electric router is used, review its use in Chapter 12.

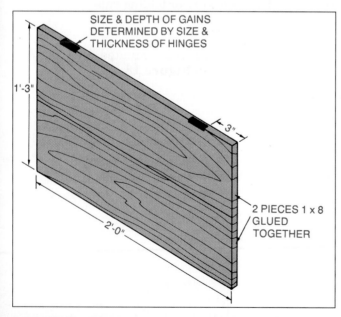

FIGURE 18–27 Gains are cut in the edge of the door and inside the door jambs to receive the hinges.

ACTIVITIES continued

6. Install the butt hinges on the door edge.
7. Place the door in the opening. Mark the location of the hinges on the jamb.
8. Cut the hinge gains in the jamb.
9. Install the door in the door frame. Plane any areas that bind.

Installing Vinyl Siding

Equipment and Materials

Sheet of 1/2- to 3/8-inch plywood with 2-foot square opening
Vinyl siding to cover plywood
12-foot J channel for siding
8-foot starting strip for siding
Tape measure or folding rule
Framing square
Aviation snips
Utility knife
Chalk line
Hammer
Supply of siding nails

Procedure

1. Nail J channel along both ends and around the opening in the plywood.
2. Nail the starting strip along the bottom edge of the plywood.
3. Apply siding to the plywood.
4. Siding can be cut with aviation snips or tin snips.
5. To form a beaded edge at the top of a cut piece of siding, slip the bead from the top edge over the cut edge, **Figure 18–28**.

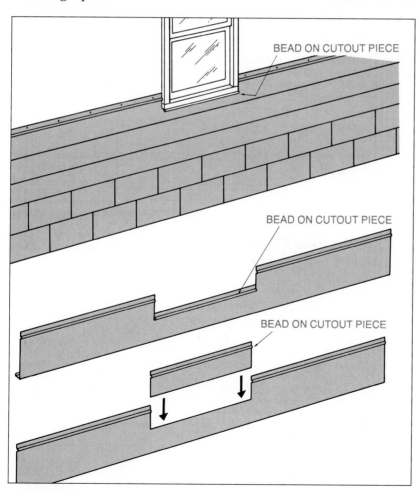

BEAD ON CUTOUT PIECE

BEAD ON CUTOUT PIECE

BEAD ON CUTOUT PIECE

FIGURE 18–28 To cover the cut edge, slip the cutout piece over it.

Applying Construction Across the Curriculum

Mathematics

Estimate the number of bricks required to cover the exterior walls of the house in **Figure 15–17** to a height of 9 feet. Allow 675 bricks per 100 square feet of wall area.

REVIEW

A. **Questions.**   Give a brief answer for each question.

1. List three types of cornice construction.
2. Name the occupation that ensures wood trim is applied to a house.
3. Name the occupation that applies masonry veneer.
4. Name the occupation that installs vinyl cornices.
5. What is a glazed door?
6. What is a prehung door?
7. What is a sash?
8. Name two important functions of siding.
9. What is the difference between a foundation for a masonry veneer building and the foundation for a building to have siding?
10. What is the function of wall ties?

B. Identification

1. Label the lettered parts of the cornice in **Figure 18–29**.
2. Identify the kinds of windows in **Figure 18–30**.

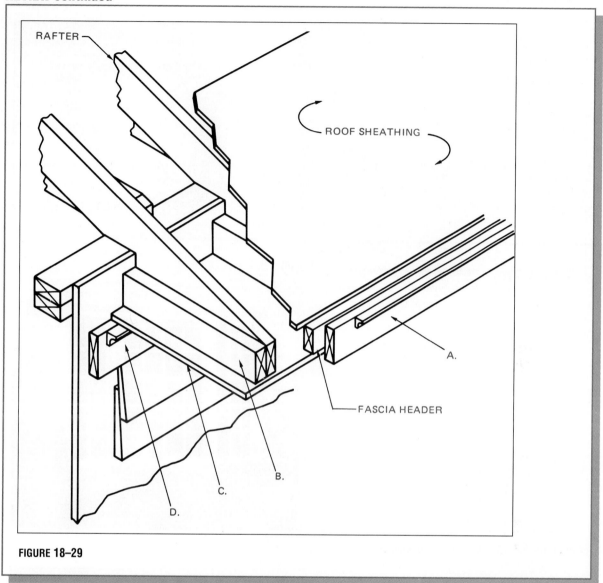

RAFTER

ROOF SHEATHING

A.

FASCIA HEADER

B.

C.

D.

FIGURE 18–29

FIGURE 18–30 *(Courtesy of Marvin Windows, Warroad, MN 56763)*

Section Five

Heavy Construction Systems

Heavy construction differs from light construction in the materials used and the intended use of the end product. The units in this section draw upon the resources and inputs covered in earlier sections to discuss construction processes. Heavy construction includes industrial and civil projects, as well as large buildings.

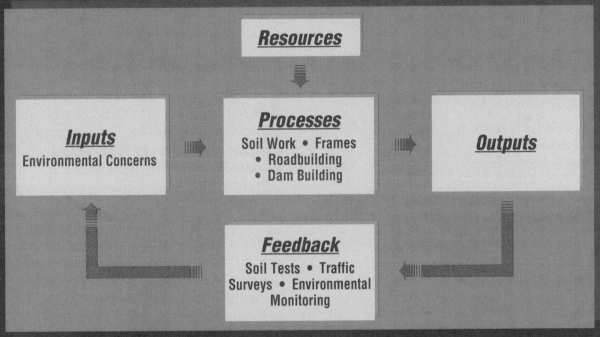

Resources

Inputs
Environmental Concerns

Processes
Soil Work • Frames
• Roadbuilding
• Dam Building

Outputs

Feedback
Soil Tests • Traffic Surveys • Environmental Monitoring

**Chapter 19
Soils and Foundations for Heavy Construction**
explores the support systems for heavy construction. This area of soils and foundations is where the greatest difference exists between light and heavy construction.

**Chapter 20
Floor Systems for Heavy Construction**
presents an overview of the structural systems used in the floors of large buildings. These are mainly based on the use of structural steel, reinforced concrete, and prestressed concrete.

**Chapter 21
Structural Frames and Walls**
examines the structural systems used for the walls of large buildings. These are based on structural steel frames, curtain-wall panels, and reinforced concrete.

**Chapter 22
Highway Construction**
outlines the steps in planning and constructing a highway interchange. The major elements are surveying and planning, site work, soil and foundations, superstructure, and bridge construction.

**Chapter 23
Industrial Construction**
presents a case study of a medium-sized power plant project. This unit involves environmental impacts, a dam, a canal, and hydroelectric turbines.

CHAPTER 19

Soils and Foundations for Heavy Construction

OBJECTIVES

After completing this chapter, you should be able to:

▼ describe the important engineering characteristics of soil;

▼ explain the principles of piers, pilings, steel-grillage, and mat foundations; and

▼ outline the steps in designing and constructing foundations.

KEY TERMS

standard sieve analysis

percolation test

bedrock

load

compaction

soils exploration

Proctor curve

sand-cone method

pier

grade beam

superstructure

pile

steel grillage

mat foundation

cofferdam

Large commercial and industrial structures need more carefully engineered foundations than do residences. A reinforced concrete building or a steel-frame structure places much more weight on its foundation. Equally important are the characteristics of the ground which must support the foundation.

Early in the design stages, the building planners obtain soil surveys, **Figure 19–1** Usually these are available from the Soil Conservation Service, U.S. Department of Agriculture. The soil survey tells the designers some important facts about soil conditions on that site. In order to understand the design of heavy-construction foundations, some knowledge of soil characteristics is necessary.

Soil Types

Civil engineers and soil scientists can predict how a particular type of soil will behave under load. Therefore, an important part of any soil testing is the classification of soil types. Most natural soils are made up of several types with varying grain sizes. Soil experts recognize sand, gravel, loam, silt, and clay by their color and texture.

More precise classification is done by a

FIGURE 19–1 Soil surveys are prepared by soil scientists. *(Courtesy of the U.S.D.A. Soil Conservation Service)*

standard sieve analysis. In this test the soil is passed through a series of screens, **Figure 19–2**. The percentage of the soil that passes through each sieve size is recorded, **Figure 19–3**. The sieve sizes are numbered according to the number of openings per inch in the screen. Soil types are named according to their grain size, **Figure 19–4**.

Soil Characteristics

Shrinking and Swelling

Some soils swell considerably when they are wet and shrink when they are dry. These are soils with a high content of clay. In extreme cases, this shrinkage and swelling may be enough to crack concrete or masonry walls, **Figure 19–5**. Soil engineers calculate shrinkage and swelling by first measuring a sample of wet soil, drying it in an oven, and then measuring the dried soil.

Drainage

Some soils allow water to drain off more readily than others. Poor drainage can cause

FIGURE 19–2 Soil sieves

several problems. Soil that is saturated does not support heavy construction machinery and excavations may become flooded, **Figure 19–6**. In extreme cases, there may be a flood hazard after construction.

The drainage characteristics of soil can be determined by pouring a measured amount of water into a hole and measuring the time it

Sieve Size	% Passing	Contract Spec. Sel. Gran. Fill % Passing
2"	100.0	100
1 1/2	97.8	
1	89.4	
3/4	85.4	
1/2	80.4	
3/8	72.2	
1/4	61.9	30 to 100
#4	53.8	
8	36.2	
16	22.1	
30	13.3	
40	11.1	0 to 50
50	10.0	
100	9.1	
200 (washed)	8.4	0 to 15

Dk Bn Gravel & Sand, tr. Silt

Complies with Contract Gradatton Spec. for "Select Granular Fill"

FIGURE 19–3 Soil sieve analysis report

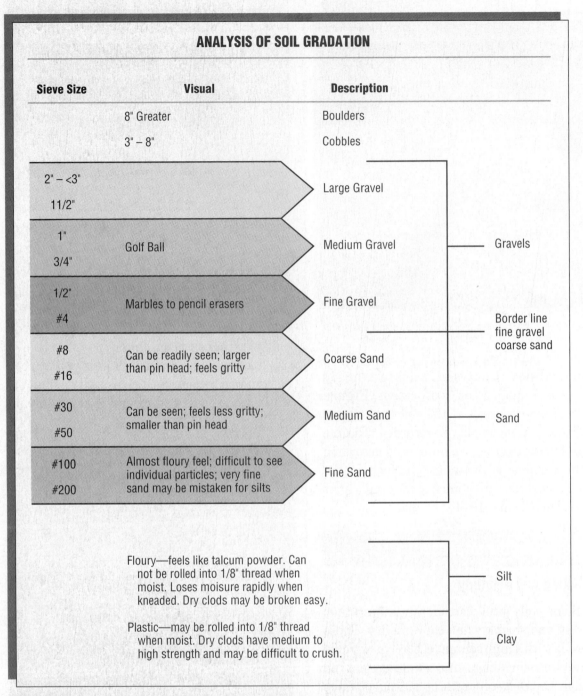

FIGURE 19-4 Soils are named according to their grain size.

takes for the soil to absorb it. This is called a **percolation test**.

Depth to Bedrock

The surface of the earth is made up of a hard rock crust called **bedrock**. The bedrock is covered with a thin layer of soil, **Figure 19-7**. This layer of soil varies from a few inches to hundreds of feet deep. Some kinds of soil will support the weight of buildings. However, skyscrapers and other types of heavy construction are supported by bedrock when

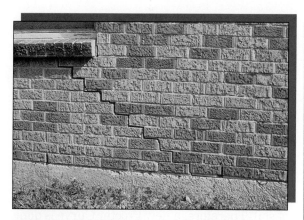

FIGURE 19–5   A soil scientist examines damage to a foundation wall. It cracked because of shrinking and swelling of the soil on which it is built. *(Courtesy of the U.S.D.A. Soil Conservation Service)*

FIGURE 19–6   This construction site is in a poor drainage area.

possible. The engineers designing the structure must know how deep the soil is so they can design the foundation.

Several methods are used to find the depth of the bedrock. One of the simplest is the *sounding rod method*. Sections of steel rod are driven into the ground with a sledge hammer. As each section is driven into the surface, another section is screwed onto it. When the rod strikes bedrock, it makes a ringing sound, **Figure 19–8**. Several soundings are made to make sure the rod has not hit a boulder.

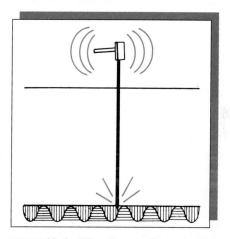

FIGURE 19–8   When the sounding rod strikes bedrock, it "rings up."

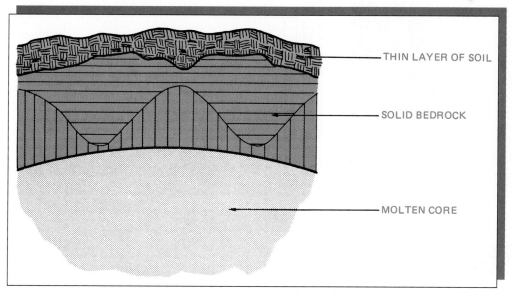

THIN LAYER OF SOIL

SOLID BEDROCK

MOLTEN CORE

FIGURE 19–7   Layers in the earth's surface

Compaction

The degree to which soil is compacted plays a very important part in how heavy a load the soil can support. If a building or a highway is built on top of soil that is not fully compacted, the results can be disastrous. Any soil that is not fully compacted before construction can be expected to compact (settle) over a period of time. If the soil compacts more under one part of the building than another, the building will settle unevenly and may actually come apart. To prevent problems with **compaction** after the building is finished, the civil engineer or an independent soils laboratory does **soils exploration** both before and during construction.

Before construction of a major project, test borings are taken with a soil auger. These borings tell the engineers what soil types will be found at important points.

During construction, tests are done on the soil near the surface. There are two types of tests that are used together to ensure that the soil is fully compacted.

A *Standard Proctor Soil Density* test determines what the density of fully compacted soil should be. The **Proctor curve**, as it is commonly called, will be different for each type of soil. Therefore, a new Proctor Curve is done for each new source of soil to be used on a construction project. In this test, the soil is tightly packed into a mold of a known size. The soil is pounded 25 times using a 5.5-pound rammer dropped from a height of 12 inches, **Figure 19–9**. The soil is then weighed. A standard mold contains 1/30 of a cubic foot. Therefore, the weight of a compacted sample of

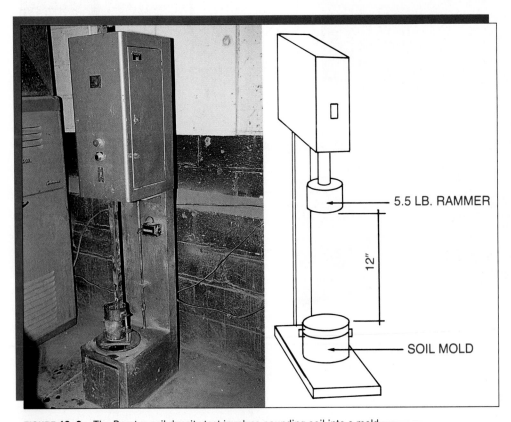

FIGURE 19–9 The Proctor soil density test involves pounding soil into a mold.

soil is multiplied by 30 to find the weight per cubic foot.

By comparing the weight per cubic foot of soil at the construction site with the weight per cubic foot obtained in the Proctor curve, it is possible to know how well the construction soil has been compacted. For example, if the Proctor-test soil weighed 138 pounds per cubic foot, and the site test showed 124 pounds per cubic foot, the site soil is 90 percent compacted (124/138 = .90).

One of two methods is used to find the density (weight per cubic foot) of the soil at the site. The newest and easiest method is to use an electronic instrument called a surface moisture density gauge. In a matter of seconds, this instrument measures the density and displays it on a digital readout. However, the **sand-cone method** has been the standard for many years and is still widely used, **Figure**

FIGURE 19–10 Cone and graduated jar for the sand-cone method of measuring soil density

19–10 Using the sand-cone method, a small hole is dug in the compacted site soil. This hole is then filled with sand of a known density. The sand is measured out from a calibrated container so the amount of sand needed to fill the hole is known. Then, by comparing the volume of sand required to fill the hole with the weight of the soil taken out, the density of the soil taken out can be calculated.

Foundations

Structural engineers design the load-carrying parts of structures, including the foundation. Throughout the design process, the structural engineers calculate the weight of the parts of the structure. Each structural part is designed to support the **load** (weight) placed on it. The entire weight of the structure is the *foundation load*. The foundation load plus the weight of the foundation itself is supported by the soil beneath it.

In residential construction, the load on the soil is relatively small, so simple spread footings are adequate. In heavy construction, much greater loads are carried by the soil, so the foundation design is often more complex.

The structural engineer works closely with the architect during the design stages to keep track of building loads and to design an adequate structure. The structural engineer also consults with soil engineers to design the foundation. Generally, soil survey maps are available to give some information. Soil tests are done at the site, however, to get more accurate information. Depending on conditions, the engineer designs one of four kinds of foundations: pier, pile, steel-grillage, or mat foundation.

Piers

When the soil is not capable of supporting the structure for several feet below the surface,

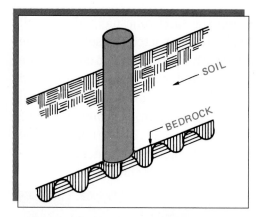

FIGURE 19–11 Piers carry their load to a stable base.

a pier foundation is used. A **pier** is a column, usually concrete, which carries the load down to a suitable depth, **Figure 19–11** The most common type is a round, reinforced concrete pier. When a greater bearing capacity pier is needed, a *bell-bottom pier* is used, **Figure 19–12**. The flared bottom spreads the load over a greater area.

Piers are generally cast into place in the ground. A hole is bored with a soil auger, then the reinforcement is set in this hole. However, the sides of the boring may cave in. To prevent this, steel or wooden forms extend at least

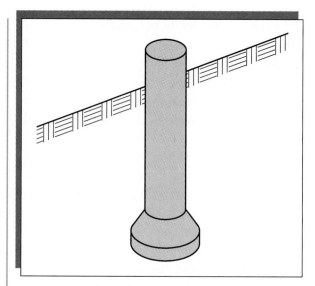

FIGURE 19–12 Where bedrock is too deep to reach, bell piers can be used.

below loose soil. Steel reinforcement rods are placed inside the form and tied with wire to hold them in place. Concrete is placed inside the casing (form), which may or may not be left on the pier.

Piers are spaced several feet apart. **Grade beams** are placed on top of them, **Figure 19–13**. Grade beams may be made of timber, steel, or reinforced concrete. The **superstructure**

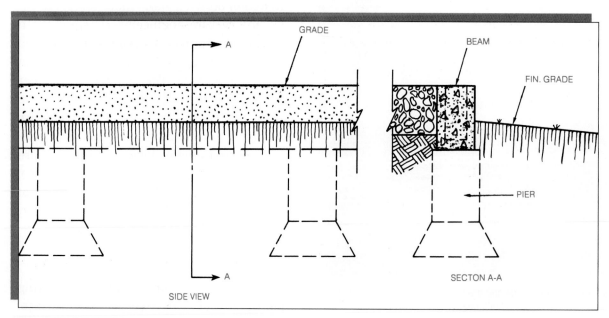

FIGURE 19–13 Grade beam on piers

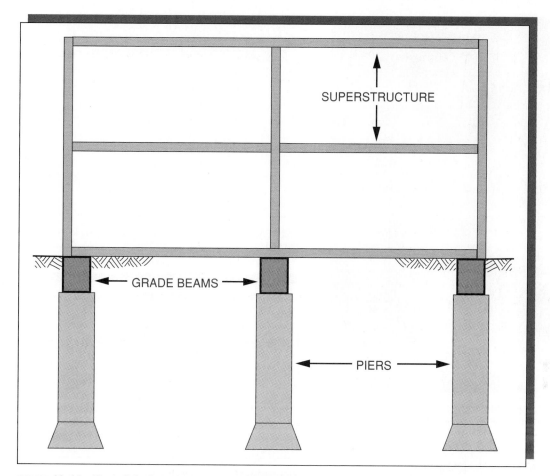

FIGURE 19–14 The building's superstructure rests on the grade beams.

(all of the above-ground structure) rests on the grade beams, **Figure 19–14**

Pile Foundations

Piles are similar to piers, except they are usually smaller in diameter and often longer than piers. Piles are made of concrete, steel, or wood. While piers are cast of concrete in a hole in the ground, piles are driven into undisturbed soil, **Figure 19–15** Piles can be driven very deep to bedrock, **Figure 19–16** When the bedrock is too deep, tapered piles are used. *Tapered piles* support their load by friction against the soil. As the tapered pile is driven, it compacts the soil along its sides. This compacted soil bears the load of the pile, **Figure 19–17.**

Piles are used in clusters. When the piles are driven, they are cut off at the proper height. A concrete cap is cast around the top of each cluster of piles, **Figure 19–18** In some construction, entire floor areas are supported by pile caps. Piers or grade beams rest on the pile caps.

Steel Grillage

Where columns place a very heavy load on the foundation, steel grillage may be used. **Steel grillage** is a kind of spread footing. In lighter construction the column load is spread by the use of a steel plate. However, when very heavy loads exist, the footing must spread more. In steel grillage, short steel beams form the footing, **Figure 19–19.** Steel-grillage

FIGURE 19–15 A pile driver drops a huge weight on the pile to drive it into the ground. *(Courtesy of Northwest Engineering)*

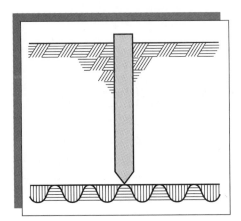

FIGURE 19–16 Straight piles rest on bedrock.

foundations are constructed on top of concrete footings or pile caps by ironworkers.

Mat Foundations

Lighter-weight buildings on low-bearing capacity soil can be built on mat foundations. A **mat foundation** spreads over a very large

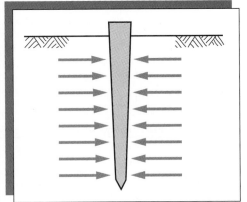

FIGURE 19–17 Tapered piles are supported by friction with the soil.

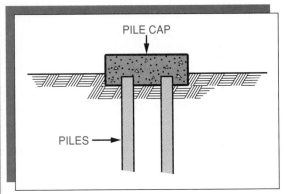

FIGURE 19–18 Concrete pile cap

area so that it "floats" on the soil. A good example of a mat foundation is a building on which the basement floor is the footing, **Figure 19–20**. As the soil moves, the entire structure moves with it.

Cofferdams

Often it is necessary to hold back earth or water at the edge of an excavation. When deep excavations are near existing structures, the sides of the excavation are braced by a **cofferdam**. Construction in waterways and very wet areas is possible by using a cofferdam to keep the water out, **Figure 19–21**. Cofferdams are usually made of steel sheet piles that are driven into the ground and interlocked at their edges.

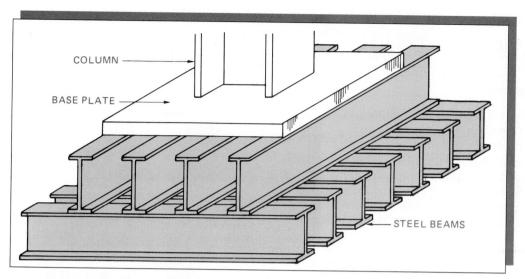

FIGURE 19–19   Steel-grillage foundation

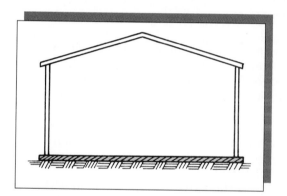

FIGURE 19–20   A mat foundation spreads the load over a very large area.

FIGURE 19–21   This sheet-steel cofferdam keeps water out of the construction area. *(Courtesy of George B. Gary and Sons)*

ACTIVITIES

Percolation Test

Before a foundation can be designed, the soil on the construction site must be studied. Soil engineers measure shrinking and swelling. They test the soil's drainage and load-bearing capacity. The depth of the bedrock is also determined. All this information helps the structural engineers design an adequate foundation.

Equipment and Materials

Dirt shovel
5-gallon bucket
Folding rule or yardstick

Procedure

1. Dig a hole 12 inches in diameter by 30 inches deep.
2. Pour five gallons of water into the hole and measure the time required for the water level to drop one inch. This is the percolation time for that area.

Sand-Cone Method

The sand-cone method has long been the "old stand-by" for measuring soil density. This test is run many times during the earthwork stages of a construction project.

Equipment and Materials

Sand cone, jar, and base plate
Dry sand
Garden trowel or large spoon
Compacted soil to be tested
Balance scale to weigh 0–20 pounds
Calculator

Procedure

1. Assemble the jar and cone and fill with sand. Weigh the filled apparatus. Record the weight.
2. Position base plate where test is to be done.
3. Scoop out a hole about 6 inches deep through the hole in the base plate.

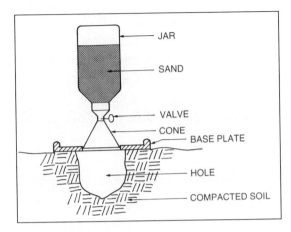

FIGURE 19–22

Save the soil you remove in a clean container. (If you use a very lightweight container, such as a cut off plastic jug, do not allow for the weight of the container.)

4. With the cone attached to the jar of sand and the valve closed, position the cone on the base plate, **Figure 19–22.**
5. Open the valve and fill the hole with sand. Any sand that remains on the plate should be carefully returned to the jar.
6. Weigh the jar and the remaining sand. Subtract this weight from the weight of the filled jar before conducting the test. The difference is the weight of the sand in the hole.
7. Divide the known weight of 1 cubic foot of sand by the weight of the sand needed to fill the hole. The result is the volume of the hole.

EXAMPLE:
If the sand is known to weigh 135 pounds per cubic foot and you used 13.25 pounds of sand to fill the hole:

$13.25 \div 135 = .09814$ cubic feet

This can be rounded off to 3 decimal places or .098 cubic feet.

8. Weigh the soil originally dug from the hole. Divide the weight of the original soil by the volume of the hole (volume of the sand needed to fill the hole). The result is the density of the soil being tested in pounds per cubic foot.

   EXAMPLE:
   If the soil removed from the hole weighs 11.5 pounds and the volume of the hole is .098 cubic feet:

   $11.5 \div .098 = 117.34$ pounds per cubic foot

If your teacher knows what the maximum soil density is by a Proctor curve, compare your test with that figure. If your soil is within 90 percent of the Proctor curve, it is safe for many building projects. If not, it should be compacted.

Kinds of Foundations

Locate several large structures in your community and find out what kind of foundation they have. Locate at least three different kinds of footings. Sketch a cross section of each and label the important parts.

Applying Construction Across the Curriculum

Science

Collect soil samples (enough to fill a coffee can) from several different places throughout your area. Do not use the top few inches of soil because it may not be representative of what is below. Classify your soil samples according to **Figure 19–4**.

Obtain a soil survey of your area from your county soil and water conservation department or the cooperative extensive office. Compare the information on the soil survey with the samples you analyzed. How accurate were you? Can you explain the differences?

REVIEW

A. Multiple Choice. Select the best answer for each of the following questions.

1. Which of the following information is given on a soil survey map?
   a. Property boundaries
   b. Areas that may flood
   c. Location of buildings
   d. Best crops to plant in an area
2. Which type of soil swells and shrinks the most?
   a. Sand
   b. Loam
   c. Gravel
   d. Clay
3. What causes soil to swell?
   a. Wetting
   b. Drying
   c. Heating
   d. Cooling
4. What is measured by a percolation test?
   a. Bearing capacity
   b. Drainage
   c. Swelling and shrinking
   d. Depth to bedrock

5. Why do engineers need to know the depth to bedrock?
   a. To predict earthquakes
   b. To determine drainage characteristics
   c. To figure the depth of water mains
   d. To figure the depth of the foundation

6. Which of the following has the highest bearing capacity?
   a. Clay
   b. Bedrock
   c. Wet sand
   d. Loam

7. What professional determines the load on a foundation?
   a. Structural engineer
   b. Architect
   c. Mechanical engineer
   d. Soil engineer

8. What type of foundation uses grade beams?
   a. Piles
   b. Steel grillage
   c. Piers
   d. Mat

9. What type of foundation relies on friction against the soil?
   a. Piles
   b. Steel grillage
   c. Piers
   d. Mat

10. What type of foundation is best where earthquakes may occur?
    a. Piles
    b. Steel grillage
    c. Piers
    d. Mat

B. Constructing a Foundation. Arrange the following in the proper order for constructing the foundation in **Figure 19–23.**

1. Carpenters build forms for pile caps.
2. Piles are driven.
3. Steel casings for piers are set in place.
4. Construction surveyor lays out building lines.
5. Soil survey maps are ordered.
6. Cement masons place concrete for piers.
7. Cement masons place concrete for pile caps.
8. Rodsetters place reinforcement for piers.
9. Tops of piles are cut off.
10. Excavation is done.

FIGURE 19–23

CHAPTER 20

Floor Systems for Heavy Construction

OBJECTIVES

After completing this chapter, you should be able to:

▼ describe common floor systems; and
▼ list the major advantages and disadvantages of common floor systems.

KEY TERMS

deck

joist

open-web steel joist

site casting

precast concrete

ribbed-steel decking

cellular-steel decking

slab on grade

Two Functions: Support and Enclosure

Any floor system must perform two basic functions. It must provide the necessary support for any loads placed on it, and it must provide a flat surface or **deck**. The support for a floor comes from the beams, girders, columns, posts, and walls of the building. The deck is made of lumber, plywood, metal, or concrete which is supported by the structural members. The floors for all but the lowest level of the building also form the ceiling for the space below.

As with many other major parts of a structure, the floor system is designed by the architect and structural engineer, who work together. The architect designs the general shape, layout, and appearance of the floor system, **Figure 20–1**. The structural engineer calculates the load on the floor system parts and designs a system that will support this load.

Supporting Members

Wood and Lightweight Steel Joists

The system of floor framing used in most residential construction is also common in light

FIGURE 20–1 The designer of this building specified a large, open space. (Courtesy of Portland Cement Association)

FIGURE 20–2 Open-web steel bar joists are a type of truss. (Courtesy of Cargill, Wilson & Acree, Inc.)

Open-Web Steel Joists

Open-web steel joists, sometimes called bar joists, are a type of truss, like roof trusses, **Figure 20–2**. These joists consist of a top chord, bottom chord, and diagonal web members, **Figure 20–3**. Open-web joists are supported by bearing walls and girders, as are wood joists. Utilities are run through the open webs of these joists, **Figure 20–4**. They can span fairly wide areas and can support a heavier load than wood or light steel solid-web joists. Open-web joists are easy to handle. They can be installed without heavy equipment. These joists are either riveted or bolted in place by ironworkers, or welded in place by welders.

commercial construction. In this system **joists**, which are closely spaced, rest on bearing walls or girders. The joists may be dimensional lumber or lightweight steel if the floor will not be required to support heavy loads. Commercial and industrial structures often involve heavier loads, due to more people using the building or heavy equipment. Therefore, commercial and industrial buildings often require the use of stronger floor framing materials.

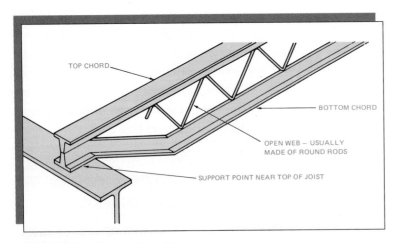

FIGURE 20–3 Parts of a bar joist

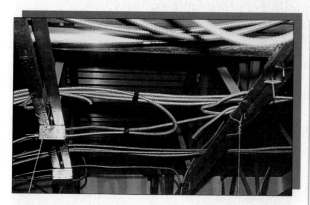

FIGURE 20–4 Utilities, such as wires and pipes, can be run through the open webs.

Structural Steel Floor Frames

Steel S beams and wide-flange beams are used in floor framing, **Figure 20–5**. S beams (standard beams) and wide-flange (WF) beams are the most common I-shaped steel beams. Although structural steel floor framing is most common with steel-frame buildings, it is sometimes used with other types of construction. With the proper decking, structural steel-framing members can be placed much farther apart than joists. Structural steel can also support much greater loads than any of the types of joists discussed. Structural steel does, however, have some disadvantages. It is heavier

FIGURE 20–6 Steel beams are protected against the possible heat of a fire when they are encased in concrete. *(Courtesy of Bethlehem Steel Corp.)*

than wood or light metal joists. A crane is needed to handle it. Also, bare metal softens when it gets hot enough. It often must be protected from overheating in case of fire, **Figure 20–6**.

Although several workers are needed to erect structural steel frames, its wide spacing makes it fast to assemble. A typical crew for erecting a structural steel floor frame consists of ironworkers, laborers, and an operating engineer to work the crane.

Concrete Floor Supports

Concrete is used in a variety of ways to support heavy loads. The use of concrete involves building forms and long waiting periods for curing, but it is capable of supporting very heavy loads.

Concrete members can be either site cast or precast. **Site casting** refers to placing concrete in forms at the construction site, **Figure 20–7**. This requires that forms be built and reinforcement be set. Then the concrete is mixed, placed, and allowed to cure. Although this is time consuming, there is no limit to the size of members that can be cast.

A faster kind of concrete construction can be done through the use of **precast concrete**.

FIGURE 20–5 The floor above this manufacturing area is framed with structural steel. *(Courtesy of Bethlehem Steel Corp.)*

FIGURE 20–7 These cement masons are casting a concrete beam in place. *(Courtesy of Symons Corporation)*

Concrete is placed in reusable forms at a precast yard, **Figure 20–8**. When the concrete has cured, the member is removed from the form and stored. Some companies specialize in making precast members.

Precast concrete is delivered to the construction site ready to use. The method of

FIGURE 20–8 These concrete sections are being precast because it would be difficult to build the curved forms on site. *Courtesy of Rotecoury-Beltcrete)*

FIGURE 20–9 This worker is directing the placement of a precast concrete beam. *(Courtesy of Portland Cement Association)*

assembly is similar to that for structural steel. A crew of cement masons, laborers, and an operating engineer position the concrete members, **Figure 20–9**. They fasten the concrete members with steel fittings which were cast into the members at the precast yard.

Decking

Plywood or Timber Deck

As in light construction, wood is a popular material for floor systems in heavy construction.

FIGURE 20–10 Wood nailers are bolted to the structural steel to permit nailing the wood joists.

Most wood-frame floors have decks of plywood topped with a more durable wear surface. Wood, either plywood or timbers, is also used on other types of floor framing. When wood is to be laid over metal framing, it is either bolted to the frame or a wood nailing strip is bolted to the framing, **Figure 20-10**. Although wood chars during fires, heavy timbers burn slowly and offer good fire protection.

Steel Deck

Sheet steel provides a rapid means of installing a deck. Steel decking is either ribbed or cellular, **Figure 20-11**. **Ribbed-steel decking** has ribs several inches wide along its length to make it rigid. **Cellular-steel decking** is similar to ribbed steel, except both sides are enclosed, forming open cells in the center, **Figure 20-12**.

Steel decks are lightweight, so the structural members of the building can also be

FIGURE 20-11 Steel decking provides a base for concrete topping.

made lighter. They are stronger than lightweight wood decks, but not as strong as some concrete decks.

The steel decking is installed by ironworkers. The pieces of decking are fastened in place with screws or welded in place. The

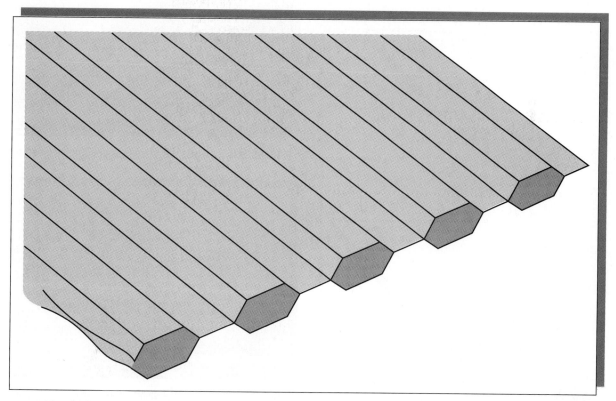

FIGURE 20-12 Cellular steel decking

concrete wear surface is placed and finished by cement masons.

Reinforced Concrete Slabs

The simplest floor to construct is a **slab on grade**. This is a concrete slab that is reinforced with welded wire mesh and placed directly on the ground. Concrete slabs are also used to build the floor decking in large structures that are several stories high.

Some very strong floors are made of flat slabs supported by columns or beams, **Figure 20–13**. This type of floor system provides a flat surface for the ceiling of the space below.

In order to produce a deck capable of greater spans between supports, vertical ribs are included. A popular type of precast floor uses single-T or double-T units, **Figures 20–14** and **20–15**. For floors with light loads, these tees are capable of spanning as much as 100 feet

FIGURE 20–13   A flat-plate concrete floor leaves a smooth ceiling beneath. *(Courtesy of Portland Cement Association)*

FIGURE 20–15   This floor system uses precast concrete Ts.

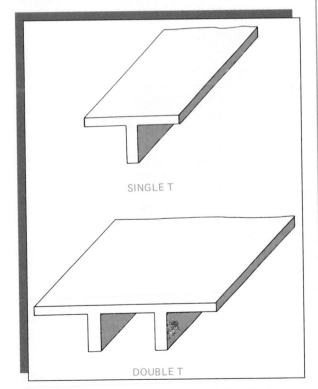

SINGLE T

DOUBLE T

FIGURE 20–14   Precast single- and double-T units allow long spans.

FIGURE 20–16   This waffle-slab floor is nearly ready for the concrete. *(Courtesy of Portland Cement Association)*

FIGURE 20–17 A completed waffle-slab concrete floor *(Courtesy of Portland Cement Association)*

between supports. The underside of such a floor is either covered with a suspended ceiling or left exposed for an interesting effect. Another interesting floor construction is the waffle-slab system. In this system, reinforcing ribs are created as the floor is cast, **Figures 20–16** and **20–17**.

ACTIVITIES

Floor Systems

Visit at least two large commercial buildings. Study the construction of the floor system in each. If the floor construction is not visible because of the finished ceiling, the building superintendent may be able to describe its construction to you. Usually there are some areas which have exposed ceilings, such as in loading and receiving areas, indoor parking areas, and mechanical rooms. Sketch a cross section of the floor construction in two buildings of different design. Be sure to include both supporting members and decks. Label the important parts.

Identification

List the trades involved with the construction of the two floors sketched in the first activity. Arrange this list in the order of the work done by the various trades.

Applying Construction Across the Curriculum

Social Studies and Communications

Choose a commercial or industrial building in your area with historical importance. Visit the building and talk to the owner or superintendent of the building or someone from your local historical society and find out as much as you can about the structural design of the building. Write a report, including sketches, drawings, and photographs to explain what type of construction was used.

REVIEW

B. Matching. Match the item in Column II with its correct use or description in Column I.

Column I

1. Used mainly in residential and light commercial construction
2. Precast concrete deck unit capable of long spans
3. Reinforced concrete floor placed directly on the ground
4. Consists of top and bottom chords with diagonal bracing between them
5. Heavy framing material which often must be protected from overheating in case of fire
6. Concrete deck cast over removable pans
7. No limit to size
8. Decking which is later covered with a thin layer of concrete
9. Slow-burning deck material
10. Structural concrete which allows for rapid construction

Column II

a. Open-web joists
b. Wood joists
c. Structural steel
d. Precast concrete beams
e. Single T
f. Waffle slab
g. Cellular steel
h. Site-cast concrete
i. Timbers
j. Slab on grade

CHAPTER 21

Structural Frames and Walls

OBJECTIVES

After completing this chapter, you should be able to:

▼ describe the difference between bearing-wall construction and skeleton-frame construction;

▼ explain how vertical and lateral loads are supported in buildings; and

▼ explain the principles of basic construction methods for walls.

KEY TERMS

bearing-wall construction

skeleton-frame construction

dowel

tilt-up construction

reinforced concrete

curtain wall

veneer

dead load

live load

lateral load

wind bracing

shear wall

Bearing-Wall Construction Versus Skeleton-Frame Construction

The types of construction used in buildings can be classified as either bearing-wall or skeleton-frame construction. The basic difference between the two types is in the way the building loads are supported. In **bearing-wall construction**, the walls are structural parts of the building. That means they support the structure, **Figure 21–1** In **skeleton-frame construction**, a structural frame supports the structure, **Figure 21–2** In skeleton-frame construction, the walls only serve to enclose a space.

Bearing Walls

Masonry

Masonry units are widely used in bearing-wall construction up to a few stories high, **Figure 21–3**. For very tall structures, masonry bearing walls are seldom used. To support the load of a skyscraper, masonry walls would have to be abnormally thick at their base. Masonry bearing walls are constructed by masons, just as are masonry foundation walls for the foundations of residences.

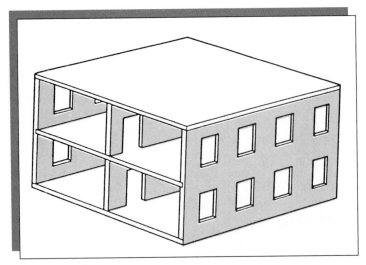

FIGURE 21–1 With bearing-wall construction, the walls support the structure.

FIGURE 21–2 Skeleton-frame construction *(Courtesy of Bethlehem Steel Corp.)*

FIGURE 21–3 Masonry bearing-wall construction

Precast Concrete

Precast concrete panels are very common material for bearing-wall construction, **Figure 21–4.** Reinforced concrete provides excellent compressive strength, which is necessary in bearing-wall construction. The precast method allows the walls to be assembled quickly after they are delivered to the site. The precast panels are made with steel fittings for lifting with a crane. They are lifted into position, then fastened by means of steel plates cast into

FIGURE 21–4 Precast concrete panels used for bearing-wall construction *(Courtesy of Portland Cement Association)*

them when they are made, or with steel rods called **dowels** that fit into adjoining members.

A typical construction crew for setting precast concrete members consists of the following:

▼ Masons fasten the members into position.
▼ Laborers handle materials and assist other trades.
▼ Riggers handle cables and pulleys and make slings to lift the members into place.
▼ Operating engineer operates the crane.
▼ Signaler directs the crane operator (operating engineer) by hand signals or two-way radio.

Tilt-Up Walls

To eliminate the need for transportation, heavy precast panels can be cast at the construction site. Forms are made for the edges of the panels and for the window and door openings. Reinforcing steel is placed in the forms and tied or welded, then the concrete is placed. When the concrete has cured, the panels are tilted up into place, **Figure 21–5**. Other than being cast at the construction site, **tilt-up construction** is very similar to precast wall construction.

Skeleton-Frame Construction

Reinforced Concrete

Because of its great compressive strength, **reinforced concrete** is widely used for the structural frames of buildings, **Figure 21–6**. However, it is not practical to lift it to the tops of very tall buildings. The weight of reinforced concrete places a great load on the lower members of the structure. For these reasons, reinforced concrete frames are not often used above a few stories.

Concrete skeleton frames can be site cast, precast, or a combination of the two. It is practical to cast vertical columns and piers on the construction site. Forms are erected by form carpenters, the steel reinforcement bars are set, and the concrete is placed. When the concrete is cured and the forms are removed, the finished column or pier is in place and ready for use. Horizontal beams and girders present more of a problem. Until the concrete cures and the member can support its own weight, more support is needed. Wet concrete is very heavy, but has little strength. Con-

FIGURE 21–5 Tilting up a concrete wall panel *(Courtesy of Portland Cement Association)*

FIGURE 21–6 Reinforced concrete building frame *(Courtesy of Portland Cement Association)*

Royce Crowell

Occupation:

President, Crowell and Son Plastering

How long:

30 Years

Typical day on the job:

Royce arrives on the job at about 6:30 AM and leaves about 6:00 PM,but he points out that being the owner of a business is a 24-hour a day job. He spends the first few hours in the morning estimating and preparing bids for new business. That is the time of day when his mind is freshest. He spends much of the rest of the day supervising work crews, ordering materials, and doing the bookkeeping. Crowell and Son is a union shop, so Royce also spends some time training apprentices.

Education or training:

Royce started in the business by working with his cousin, who was a contractor. His cousin helped him get into the union apprentice program, where he learned most of his trade. He learned the business aspects of his job by starting his own small business.

Previous jobs in construction:

After working as a plasterer for his cousin, Royce spent 18 years as a foreman, before starting his own business.

Future opportunities:

He is pleased with the job he has and is not anxious to do anything else.

Working conditions:

Most of his work Crowell and Son does is exterior stucco. They are located in California, where there is not much unpleasant weather.

Best aspects of the job:

Estimating and bidding. Its like a challenging competition with other contractors. There is also a very positive feeling when you see the productive results of your work.

Disadvantages or drawbacks of the occupation:

You have to be very particular about who you work for, so you don't have too much trouble collecting your fees.

ventional forms are not strong enough to hold it up alone. for this reason horizontal members are often precast, then set in place.

Structural Steel

Structural steel provides good strength with much less weight than concrete. This makes it a common material for frames of tall structures. Steel frames can also be put up quickly, thereby saving valuable construction time.

To protect steel building frames from the weakening effects of fire, the steel members are sometimes encased in fire-resistant material. By covering the steel with a few inches of concrete or gypsum plaster, it can retain most of its strength up to four hours even in very serious fires.

The individual pieces of steel are lifted into position with a crane. They are either bolted together with high-strength steel bolts or welded. Until the middle of this century, nearly all connections in structural steel were made with rivets. More recently, welding and bolts have taken the place of rivets. It is common for both methods to be used on one construction job. Often subassemblies, made up of several pieces of steel, are welded together. These subassemblies are then bolted to the building frame.

Welding has become an important trade in the construction industry. Welding is done by welders who specialize in this work and by ironworkers who work on building frames. One of the most outstanding characteristics of ironworkers' jobs is that they work in high places under dangerous conditions, **Figure 21–7**.

Curtain Walls

Curtain walls do not support any of the structural load. They only enclose a space. They are made of panels that are attached to

FIGURE 21–7 Iron workers work in high places similar to these sky-scrapers. *(Courtesy of Cargill, Wilson & Acree, Inc.)*

FIGURE 21–8 The exterior walls of this building are of curtain-wall construction.

the frame. These panels fill in the spaces in the frame, **Figure 21–8**. Curtain walls can be made of masonry, precast concrete, glass, steel, or a combination of materials. Commonly, curtain-wall panels are a sandwich of insulation between a weather-resistant exterior surface and an interior finish, **Figure 21–9**. Where windows are needed, the panels are made of glass. Curtain-wall panels are shipped to the site, ready for installation.

Veneer Walls

A **veneer** is a thin layer of facing material attached to a strong base. In construction, veneer walls are made of a durable face applied

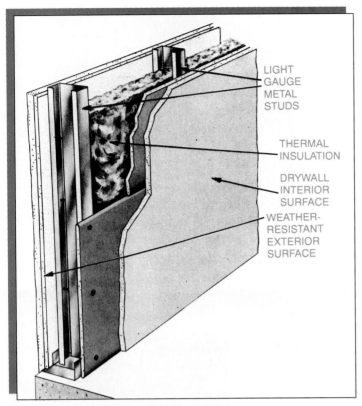

FIGURE 21–9 This curtain-wall panel has light-metal studs and fiberglass insulation in place. *(Courtesy of United States Gypsum)*

over the structural part of the wall. Veneers are usually brick, decorative concrete blocks, stone, or metals. Masonry veneers are constructed in front of the wall and fastened to it with metal wall ties.

Siding

Any of the types of siding listed for residential construction in Chapter 18 can be used on nonresidential structures. Industrial buildings are often sided with ribbed metal siding, **Figure 21–10** This kind of siding is durable and can be applied quickly. It is either screwed or nailed to the building frame.

Structural Functions of Walls and Frames

The obvious purposes of a wall are to protect the interior from the weather, give privacy, and to divide a space into smaller rooms. However, walls and the framework which supports them also serve some important structural functions. While the architect is mainly concerned with the functions mentioned above, the structural engineer is concerned with other things.

FIGURE 21–10 Industrial building with ribbed steel siding

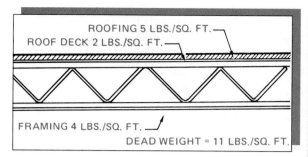

FIGURE 21-11 The dead weight is the total weight of the building materials.

Vertical Loads

There are two types of vertical loads that must be supported by building members. The **dead load** is the weight of all the parts of the building or structure. For example, if the roofing material, roof deck, and roof frame together weigh 11 pounds per square foot, the dead load on the roof frame is 11 pounds per square foot, **Figure 21-11**

Variable loads, such as people, furniture, and rain and snow are called **live loads**. If the roof in the above example is also supporting 20 pounds of snow per square foot, then the live load is 20 pounds per square foot. The total load on this roof is 31 pounds per square foot. The roof frame and all the structure below must be engineered to support that load. In a 50-foot by 100-foot building, if the dead load of the roof is 11 pounds per square foot and the live load is 20 pounds per square foot, the walls or vertical framing members must support 155,000 pounds of roof load.

In shorter buildings the vertical loads are the only loads considered in the design of the structure. The vertical columns, posts, or studs support the vertical loads.

Lateral Loads

In taller buildings the wind applies considerable force to the sides of the building. In some areas, earthquakes also apply force

FIGURE 21-12 Diagonal bracing helps the frame resist wind loads

sideways to the building. These horizontal forces are called **lateral loads**.

Lateral loads are resisted by wind bracing or shear walls. **Wind bracing** refers to diagonal members of the frame which form triangular shapes, **Figure 21-12** This prevents the building frame from flexing in that direction. Wind bracing can be used on exterior wall frames or it can be incorporated in interior framing. The advantage in using wind bracing in interior framing is that this does not conflict with the placement of windows.

When a building does not have a frame that can be braced with diagonal members, another method is used. **Shear walls** are substituted for diagonal braces. These shear

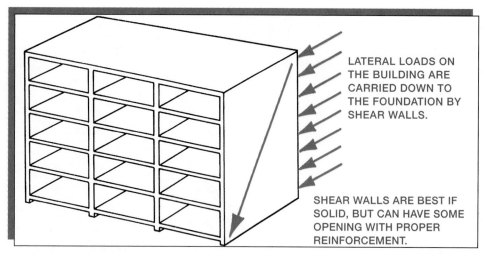

LATERAL LOADS ON THE BUILDING ARE CARRIED DOWN TO THE FOUNDATION BY SHEAR WALLS.

SHEAR WALLS ARE BEST IF SOLID, BUT CAN HAVE SOME OPENING WITH PROPER REINFORCEMENT.

FIGURE 21–13   Shear walls

walls are made of rigid material, such as plywood panels or precast concrete, which serves the same purpose as wind bracing, **Figure 21–13** Because exterior walls are often weakened by window openings, shear walls are often inside the structure where they serve as partitions.

An interesting example of the use of shear walls is in the Marina Twin Towers in Chicago, **Figure 21–14** These tall buildings have vertical columns near the all-glass outside walls for vertical support. The center of the buildings is a 32-foot diameter cylinder. This core provides lateral support for the towers.

FIGURE 21–14   The Marina Towers use cylindrical shear walls.
(Courtesy of Bertrand Goldberg Associates)

ACTIVITIES

Model Structure

Construct a structural model of one of the building types described.

Equipment and Materials

Architect's scale
Paper straws
Cardboard
Model cement

Procedure

Construct a structural model of a structural steel skeleton frame for an 8-story building. Each floor is to be 40 feet by 60 feet. The rise should be 10 feet between floors. Use a scale of 1/4" = 1'-0". Be sure to include wind bracing. Write a report describing your building, including what loads are supported by each of the structural elements. Include illustrations in your report.

OR

Construct a structural model of a precast concrete bearing-wall building for an 8-story building. Each floor is to be 40 feet by 60 feet. The rise should be 10 feet between floors. Use a scale of 1/4" = 1'-0".

Be sure to include shear walls. Write a report describing your building, including a discussion of how lateral earthquake loads are resisted. Include illustrations in your report.

Applying Construction Across the Curriculum

Mathematics

If the roof of a large commercial building measures 62 feet by 140 feet and is made up of components with the following weights, calculate the dead load per square foot and the total dead load imposed on the walls by the weight of the roof:

Material/component	Lbs. per square foot
Steel Girders	2.5
Open-web bar joists:	12
Steel Decking	3.5
Asphalt underlayment	1.25
Neoprene Roofing	3.3

REVIEW

B. Matching. Match the item in Column II with its correct use or description in Column I.

Column I

1. Diagonal members
2. Weight of building material
3. Weight of furniture
4. Handles cables and hooks for lifting
5. Most common method for tall structures
6. Force of wind and earthquakes
7. Walls to resist horizontal force
8. Panels cast flat on the site
9. Uses walls for vertical support
10. Operates heavy equipment
11. Panels attach to building frame
12. Very strong, but too heavy for tall buildings

Column II

a. Rigger
b. Tilt-up construction
c. Shear wall
d. Wind bracing
e. Bearing-wall construction
f. Operating engineer
g. Dead load
h. Live load
i. Lateral load
j. Curtain wall
k. Concrete frame
l. Structural steel frame

CHAPTER 22

Highway Construction

OBJECTIVES

After completing this chapter, you should be able to:

▼ point out the similarities between building construction and highway construction;

▼ point out the differences between building construction and highway construction;

▼ read a civil engineer's scale;

▼ list the major steps in the construction of a highway; and

▼ list the major steps in the construction of a simple bridge.

KEY TERMS

traffic survey

conceptual design

eminent domain

extent of work

cut

fill

lift

abutment

pylon

prestressed concrete

asphalt concrete

Highway construction is a major part of the civil construction field. Highway construction projects range from small resurfacing or repair projects, **Figure 22–1**, to major new interstate highway projects, **Figure 22–2**. This unit explains the construction of a major highway interchange. This type of project involves most of the activity found in any highway project.

Planning a Highway

Architects design buildings to satisfy clients' needs. Civil engineers design highways to satisfy needs, too. In the case of a highway interchange, the need is to allow traffic to safely flow from one highway to another. This need is specifically identified by a **traffic survey**. A traffic counter measures the number of vehicles passing a given point each hour or day, **Figure 22–3**.

Traffic engineers look at the results of the traffic survey, the speed of the traffic, and number of traffic lanes on each highway. These items, along with other information on the needs of the motorist will be used to develop a **conceptual design**. This basic design is the

317

FIGURE 22-1 Spreading asphalt for a resurfacing project *(Courtesy of* Texas Highways*)*

FIGURE 22-2 Major new highway construction *(Courtesy of* Texas Highways*)*

FIGURE 22-3 Traffic volume is counted with a traffic counter. *(Courtesy of Streeter Richardson Division, Mangood Corporation)*

FIGURE 22-4 Surveying the route of a new highway *(Courtesy of The Lietz Company, copyright ©1986)*

beginning idea of how to solve a traffic problem, not always the final solution.

When the basic conceptual design is approved, route surveyors measure the land for the highway accurately, **Figure 22-4**. Surveyors record detailed information in a field notebook. When all of the information is collected, civil drafters and the surveyors draw plans of the project. These plans, or maps, show where the actual roadway will be, the contour of the site, and the boundaries of the right of way, **Figure 22-5**.

Highways are usually owned by some level of government—federal, state, or local. Any land within the right of way that is not already owned by that government must be bought. Exercising its right of **eminent domain** (the right of the government to buy land for public

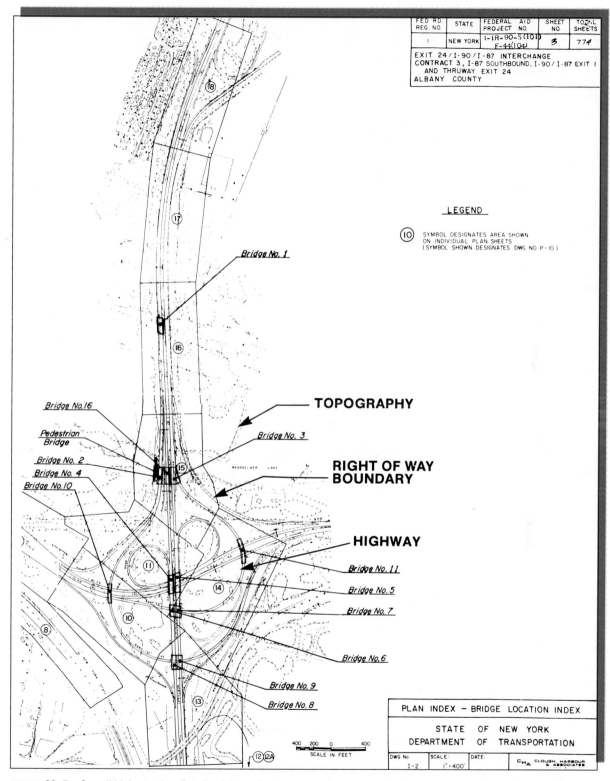

FED RD. REG NO.	STATE	FEDERAL AID PROJECT NO.	SHEET NO.	TOTAL SHEETS
I	NEW YORK	I-IR-90-5(101) F-44(104)	3	774

EXIT 24 / I-90 / I-87 INTERCHANGE
CONTRACT 3, I-87 SOUTHBOUND, I-90 / I-87 EXIT I
AND THRUWAY EXIT 24
ALBANY COUNTY

LEGEND

(10)   SYMBOL DESIGNATES AREA SHOWN
       ON INDIVIDUAL PLAN SHEETS
       (SYMBOL SHOWN DESIGNATES DWG NO P-10)

Bridge No. 1

TOPOGRAPHY

Bridge No. 3

Bridge No.16

Pedestrian
Bridge

Bridge No. 2

Bridge No. 4

Bridge No.10

**RIGHT OF WAY
BOUNDARY**

HIGHWAY

Bridge No. 11

Bridge No. 5

Bridge No. 7

Bridge No.6

Bridge No. 9

Bridge No. 8

400 200 0   400
SCALE IN FEET

PLAN INDEX — BRIDGE LOCATION INDEX

STATE   OF   NEW   YORK
DEPARTMENT   OF   TRANSPORTATION

DWG No. I-2	SCALE: 1"=400'	DATE:	CHA CLOUGH, HARBOUR & ASSOCIATES

FIGURE 22–5 Overall highway plan. Details of the various parts are shown separately at a larger scale. *(Courtesy of NY State Department of Transportation)*

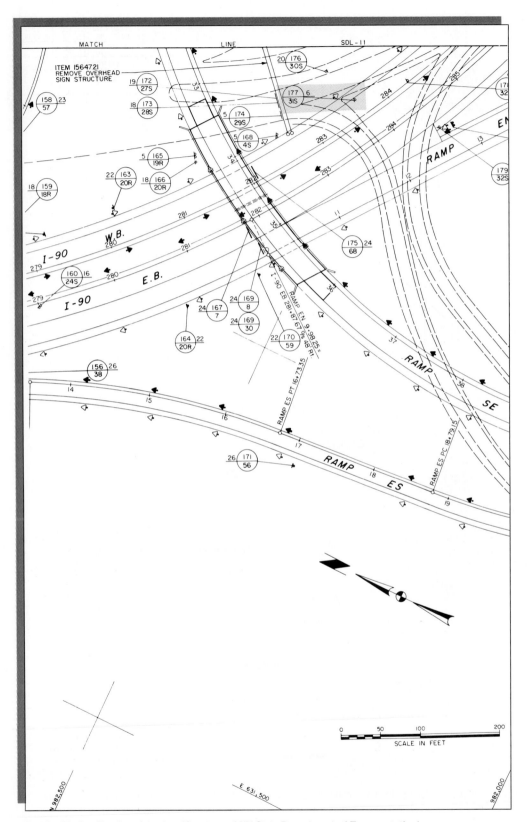

FIGURE 22–6 Sign location plan *(Courtesy of NY State Department of Transportation)*

use and pay a fair price for that land), the government, therefore, buys the necessary land. Without this right, it would be possible for the owner of a small lot to force a change in the route of a major highway.

Civil Drawings

The final planning step is to design the actual roadway, bridges, and safety devices. Just as a building design is communicated with drawings and specifications, so is a highway design. The specifications spell out the type and quality of materials to be used, methods for compacting soil, and the procedures for inspection and testing.

The drawings for a complete project may be well over one hundred pages long. The overall route map, drawn by the surveyors, is only one of the required drawings. Where complex on and off ramps are needed, they are drawn separately. Section views are drawn to show more detail about the roadway and embankments. Complete elevations, sections, and details are drawn for bridges. Sign drawings are a good example of the details necessary for a highway project. **Figure 22–6** shows part of a sign location drawing. One of the sign schedules used on this project is shown in **Figure 22–7**. A key at the bottom of the location drawing, **Figure 22–6**, explains where each sign is shown on the schedule. For example, the highlighted symbol on the drawing matches the highlighted sign on the schedule.

Civil drawings are usually dimensioned in feet and decimal parts of a foot. Drawings for civil engineering and construction are drawn and read with a civil engineer's scale, **Figure 22–8**. The triangular scale contains six scales, one on each edge. These scales are graduated in multiples of ten. A small number near the zero end of the scale indicates the ratio of the scale. The number 10 indicates a 1:1 ratio, **Figure 22–9**. The number 20 indicates a 1:2 or

SIGNS TO BE REMOVED					
ITEM NO	LOCATION NO	TEXT NO	TEXT	APPROX. SIZE OF SIGN	TYPE OF MOUNTING
647.16	172	27S	(90) EAST Albany STRAIGHT AHEAD	130.0 S.F.	O.H.
647.16	173	28S	EXIT 1N (87) Northway Albany Airport Montreal 3/10 MILE	180.0 S.F.	O.H.
647.16	174	29S	EXIT 1S TO U.S. 20 Western Ave	110.0 S.F.	O.H.
647.15	176	30S	DO NOT ENTER	21.0 S.F.	GR. MTD. (Back To Back)
647.02	177,196	31S	MERGING TRAFFIC	16.0 S.F.	GR. MTD.
647.01	178,179,184, 294	32S	GROOVED PAVEMENT	9.0 S.F.	GR. MTD.
647.02	181,183	33S	25 MPH	15.3 S.F.	GR. MTD.
647.01	186	34S	NO PARKING ANY TIME	1.5 S.F.	POLE MTD.
647.02	201,202,291	35S	XX MPH	18.0 S.F.	GR. MTD.

FIGURE 22–7 Sign schedule. These signs must all be removed. *(Courtesy of NY State Department of Transportation)*

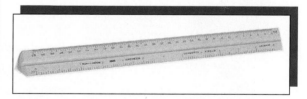

FIGURE 22–8 Civil engineer's scale *(Courtesy of Koh-I-Noor Rapidograph, Inc.)*

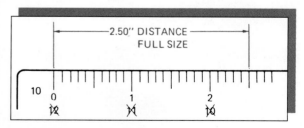

FIGURE 22–9 Civil engineer's scale—1:1

Bart's Transbay Tube

The San Francisco Bay Area Rapid Transit District (BART) is a 71.5 mile rapid-transit rail system. Since opening in 1972, BART has carried over 925 million people nearly 6 billion miles. BART is made up of 19 subway and tunnel miles, 23 ariel (elevated railway) miles, 25 surface miles, and nearly 4 miles in the transbay tube. BART is a modern rapid-transit system with many engineering and civil-construction records. The tube which carries BART trains under the San Francisco Bay has been recognized the world over as one of history's most outstanding civil engineering achievements.

The concept of an under-the-bay tube has been around for many years. In 1920 Major General George T. Goethals, the builder of the Panama Canal, proposed building such a tube under the San Francisco Bay. In 1947, a joint Army-Navy commission recommended the construction of an underwater tube to relieve automobile congestion on the San Francisco–Oakland Bay Bridge. In 1959, six years before construction began, engineering studies were started and the design of today's tube began.

Parsons Brinckerhoff-Tudor-Bechtel, BART's general engineering consultants, were charged with design and construction

management of the total project. The plan was to build the tube in sections. There were to be 57 in all, each averaging 330 feet in length, longer than a football field. These were fabricated on dry-land shipways, at the Bethlehem Shipyards in South San Francisco. From there they were launched, towed into the bay, and sunk in their proper position.

The tube sections look like huge binoculars in cross section. They are 24 feet high and 48 feet wide. Trackways in each bore carry trains in each direction. They are separated by an enclosed central corridor for pedestrian access, ventilation, and utilities.

Construction began in the mid-1960s as a joint venture of four large contractors—Peter Kiewit Sons' Co.; Raymond International, Inc.; Tidewater Construction Corp.; and Healy-Tibbitts Construction Co. This group of companies operated under the name Trans-Bay Constructors. Their low bid was $90 million. This included ventilation structures at either end, 2.8 miles of aerial and subway approaches in Oakland and San Francisco, trackage, final finish work, and electrification. The full cost of the project was $180 million in 1970 dollars.

The contract called for a demanding two-and-a-half year schedule for completion of the basic structure. This meant maintaining a pace of building and placing two tube sections per month. Subcontracts were let and soon an army of welders set to work fabricating the steel skin of the sections at the Bethlehem Shipyards.

First came the tube shell, made from 3/8-inch steel plate and reinforced with steel T-beams set six feet apart. The inside of the completed shell was then laced with steel reinforcing bars for concrete. After a section was completed and watertight bulkheads placed at each end, it was launched from the

shipways and towed to a nearby dock. There, about 70,000 square feet (4,200 cubic yards) of concrete was placed to form the 2.3 foot thick interior walls and track bed.

The first of the 57 sections was launched in February of 1967. Barely buoyant after the addition of the concrete, it was towed gingerly out to its assigned position. There it was weighted with 500 tons of gravel ballast placed in bins on top of the section. It was then slowly lowered into place. Final weight of each section is approximately 10,000 tons.

Meanwhile, excavation of the trench was progressing. For this job, the contractors had assembled a small navy of specialized vessels and clamshell dredges to cut a ditch in the bay floor 70 to 100 feet deep, sloping to a 60-foot-wide bottom. In all, the contractor removed about 5.7 million cubic yards of material. This would be a considerable earth-moving job even on land, much less 135 feet beneath the water's surface.

At the same time, surveyors worked around the clock with construction crews to keep the trench precisely aligned through two horizontal and six vertical direction changes. Using lasers from shore positions, engineers were able to pinpoint the exact position required for the dredge barges.

To permit leveling of the tube to exacting specifications, the engineers specified that a two-foot layer of gravel bedding be placed along the entire length of the trench. This required some special ingenuity. To place and level the gravel, the contractor specially designed a large "screened barge" 85 feet wide, 240 feet long, and floating 44 feet high on pontoons. Installed on top was a traveling bridge. This carried the machinery for funneling gravel to the floor, and for moving a box-like leveling device called a "screed."

Once the trench was ready, another specially designed rig had to be built to lower the heavy tube sections into place. It was made up of two barges, connected by

means of overhead "bridges," separated just enough to nestle a floating tube section between them.

Once in place, each new section was snugged tightly against the previous one by means of four 50-ton, hydraulically operated railroad-type couplers. The procedure was to lower the new section into line about two feet away from a tube section already in place, engaging the couplers. Then the hydraulic rams were activated to draw the new section tightly against the old section. Once this linkup was completed, a barge-mounted crane packed gravel and stone against the sides of the section to lock it in place. An added five-foot layer of sand and gravel provides a top protective blanket.

Once the sections were joined and sealed by a neoprene rubber gasket around the rim, water trapped between the end bulkheads was bled off. Water pressure then exerted enough force to keep the seal tight. Later the bulkheads were removed from inside the structure. Permanent steel connections were welded into place. Concrete was added to complete the joint construction.

The last of the 57 sections of the tube was launched and placed in April of 1969. Track laying, electrification, and installation of train control equipment and ventilation were completed by early 1973. On August 10, 1973, the first powered, automatically controlled train made the first round trip through the tube. ■

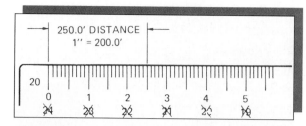

FIGURE 22–10 Civil engineer's scale—2:2

1/2 scale, **Figure 22–10**. The number 30 indicates a 1:3 or 1/3 scale, and so on. **Figure 22–11** shows several of the scale options that can be obtained using a civil engineer's scale.

Because highways cover so much area, it is sometimes necessary to use scales with one inch representing several hundred feet. Area is, in fact, one of the greatest differences between highway construction and building construction. The supervisor of a highway project may have a fairly long drive from one end of the project to the other.

Preparing the Site

With the design work completed, it is time to begin actual work on the site. As with building construction, the first step in site preparation is staking the boundary of the site. Two sets of stakes are put in place. The first marks the actual boundary of the property included in

FIGURE 22–12 Stakes are driven several feet inside the boundary to mark the limit to which clearing will be done.

CIVIL ENGINEER'S SCALE					
Divisions	**Ratio**	**Scale Used With This Division**			
10	1:1	1" = 1"	1" = 1'	1" = 10'	1" = 100'
20	1:2	1" = 2"	1" = 2'	1" = 20'	1" = 200'
30	1:3	1" = 3"	1" = 3'	1" = 30'	1" = 300
40	1:4	1" = 4"	1" = 4'	1" = 40'	1" = 400'
50	1:5	1" = 5"	1" = 5'	1" = 50'	1" = 500
60	1:6	1" = 6"	1" = 6'	1" = 60'	1" = 600

FIGURE 22–11 Scale options with a triangular civil engineer's scale

the right of way. A second set of stakes is positioned a few feet inside the property lines, **Figure 22–12**. The second line of stakes marks the **extent of work**. This is the line beyond which no work is to be done.

More stakes are driven to mark the actual roadway. However, highways often need extensive earthwork, so the roadway must be surveyed often throughout the construction process.

It is sometimes necessary to remove old pavement before the new road can be built, **Figure 22–13**. The pavement is broken into small enough pieces so it can be loaded into trucks and hauled away.

The people who operate the rippers, loaders, and other heavy road-building equipment are

FIGURE 22–13 Old pavement is broken up and loaded onto trucks. *(Courtesy of Deere and Co.)*

FIGURE 22–14 Bulldozer clearing rubble stone from the right of way. *(Courtesy of Caterpillar Tractor)*

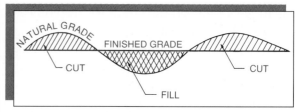

FIGURE 22–15 Cut is where earth must be removed. Fill is where earth must be added.

called *operating engineers*. This skilled trade can be learned in some technical schools or through apprentice programs. Operating engineers receive a fairly high hourly wage when they are on the job. However, when the weather is bad, they may not work. The constant pounding of the machinery is a disadvantage of this job for some people. Others consider this a small price to pay for the enjoyment of operating powerful machinery.

When the highway is being built through wooded land, the site must be cleared. Large trees are cut down with chain saws. Smaller trees, stumps, and brush are cleared with bulldozers, **Figure 22–14.**

Some excavation and rough grading may be necessary. The soil found near the surface can contain organic matter (rotted vegetation). Soils that have large amounts of organic matter cannot be compacted adequately, so they must be removed. Large boulders have a tendency to work their way to the surface. This is a disaster under a highway, so they must be removed.

In some cases, it may be necessary to remove or add earth to get the desired grade, **Figure 22–15.** Soil that is removed in order to create the desired grade is called **cut**. Engineers usually use the cut as **fill** nearby (soil which is

added to a low spot to create the desired grade). Earth (soil and rock) is a useful material for building a road bed, but it can also be a nuisance if a highway builder has to pay for a place to get rid of it. Highway designers do extensive calculations to ensure that they do not need to dispose of extra earth or purchase it for use as a material. They determine the number of cubic yards to be removed from a cut area, then match that amount of material with the fill needed nearby. After debris, such as stumps and boulders, has been removed from the site, scrapers are used for most of the cut-and-fill work, **Figure 22–16.**

Embankments and Foundation Work

Once the unwanted material has been removed from the area, most of the soil moved will become part of the finished highway. In

Pamela Neill

Occupation:
Operating Engineer—Roller Operator

How long:
1 1/2 years

Typical day on the job:
The day usually starts at about 6:30 or
7:00 AM. Early in a job, the normal week
is four ten-hour days, with a three-day
weekend. Near the end of the job, there is
often a rush to get the job done and she
might have to work six or even seven
days a week. Each morning she checks
the fuel, oil levels, and condition of the
machine she will be operating. Then,
most of the day is spent on the vibrating
roller, compacting soil for road and bridge
construction. She must pay attention to
the condition of the compacted soil and
call a water truck if more water is
needed.

Education or training:
Pamela started as an apprentice operating engineer at a paving company. First
she operated a roller, learning from more
experienced operators as new challenges
came up. Then she learned to operate
other equipment. She has driven water
trucks and worked as tailman on a pipe
crew.

Previous jobs in construction:
Pamela worked as the bookkeeper for a
computer company that did work for a
construction company. That experience

gave her the idea that she might like a
career in construction.

Future opportunities:
Pam hopes to get additional training so
that she can become a back hoe operator,
a blade operator (bull dozer or scraper),
or possibly, a crane operator.

Working conditions:
She has worked in 114°F heat and 4°F
cold, but there is not usually much work
in the coldest weather.

Best aspects of the job:
Winters are free to do other things and
the pay is very good.

Disadvantages or drawbacks of the occupation:
Some work is done in very uncomfortable
weather, and sometimes the weather
prevents work.

FIGURE 22–16 Scrapers are used to move earth from one area to another. *(Courtesy of Caterpillar Tractor)*

FIGURE 22–18 A sheepsfoot roller is used to compact soil. *(Courtesy of* Texas Highways*)*

highway construction the foundation, which is soil, uses more material than the superstructure, **Figure 22–17**. That is why it is important to think of soil as a construction material.

When the roadway is elevated, such as at the approach to a bridge, an embankment is created. Embankments are built up in layers, called **lifts**. The lifts are no more than two feet. Each lift is compacted to a specified percent of the maximum soil density before the next lift is placed. (Soil density was explained in Chapter 19.)

There are two types of rollers that are used for compacting soil in embankments. A vibrating smooth roller has the same compacting ability as a nonvibrating roller weighing several times more. A sheepsfoot roller, **Figure 22–18**, is often used for soil compaction.

The embankment may be built mostly of subgrade materials. These are gravels containing cobbles up to six inches in diameter. The top lift is select gravel, with much finer aggregate.

Bridges

Bridges are needed to span streams or for one roadway to pass over another. Major highway construction projects often include bridges. Although one government engineer may be responsible for both the roadway and the bridge, they may be built by different companies.

The simplest type of bridge is made up of **abutments** and **pylons** with a deck supported on beams, **Figure 22–19**. In this type of bridge the abutments are the foundation and the roadway is the superstructure. Bridges that span greater distances need either a more complex superstructure or added foundation supports, **Figure 22–20**.

The abutments and pylons for the bridge are made before the embankment is placed, **Figure 22–21**. Building a small bridge abutment is not too different from building a reinforced concrete building foundation. The abutment is built on piles or another type of heavy foundation. Steel reinforcing is tied in place and concrete forms are built. Concrete is placed in the forms and vibrated to eliminate

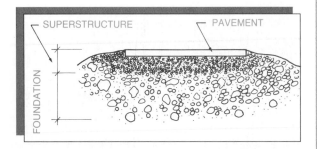

FIGURE 22–17 Roadway foundations and superstructure

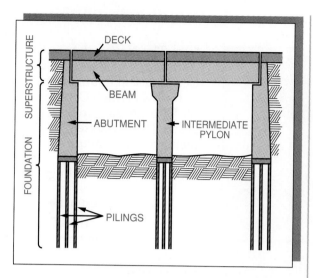

FIGURE 22–19   Major bridge parts

any voids (air pockets). Anchor bolts or steel dowels are left protruding from the top of the abutment. The anchor bolts or dowels will provide a means of attaching the superstructure.

After the embankment is in place, but before the pavement is placed, the bridge superstructure is built. The superstructure of a bridge, consisting of beams supported by abutments and pylons, begins with the

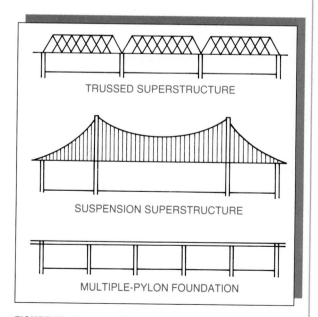

FIGURE 22–20   Long Bridges require more superstructure or foundation.

FIGURE 22–21   Central pylon for a multi-level interchange *(Courtesy of* Texas Highways*)*

beams, **Figure 22–22**. The beams, which span from one support to the next, are usually either steel or **prestressed concrete**. Steel beams are welded together in a shop and trucked to the site.

Prestressed concrete beams are also made in a shop and trucked to the site. These beams are cast with steel cables inside them. The cables are tightened (stressed) before the concrete cures. The cured concrete bonds to the cable, holding it in stress. The effect of the cable pulling against the concrete beam is similar to the effect of holding a row of books together tightly enough to lift them as a single unit, **Figure 22–23**. The stressed steel cable helps to make the concrete beam more rigid.

Light steel decking is placed over the beams,

Susan Levy

Occupation:

Highway Specialty Sub-
contractor

How long:

14 Years

Typical day on the job:

Susan is a working owner, which means that she is very involved in the day-to-day activities of her business. She typically arrives at her office at 6:00 AM to begin preparing daily instructions for her staff and to dispatch the work crews. Usually, work crews must leave the office at 7:00 AM, so they can be on the job by 8:00 AM. She personally supervises the unloading of equipment at the job site and lays out the day's work. Susan's company, Femi-nine Contracting Corp., does saw cutting, milling, and grooving operations to prepare highways for repaving. Contracts often require that the crews and equipment be completely off the site by 4:30 PM, when rush-hour traffic starts. Back in the office, Susan spends about 45 minutes opening mail and returning phone calls.

Education or training:

Susan took high school business courses. She then studied business administration for two years in college. She has also completed several home-study courses from The National Association of Women in Construction Foundation in their Certified Construction Associate program. These courses included such things as contract negotiating, estimating, and bidding.

Previous jobs in construction:

Susan worked in the office of a heavy equipment leasing company for twelve years.

Future opportunities:

Susan looks forward to the continued growth of her business. She started with one piece of equipment and worked only on jobs close to home. She now has over $1 million worth of equipment and does work in a three-state area.

Working conditions:

Much of the day is spent on the job site, in a variety of weather conditions; but also a big part of the work is in the office.

Best aspects of the job:

The job is a mixture of inside, office work and outside, hands-on work.

Disadvantages or drawbacks of the occupation:

This is a seasonal business. From April through October, she might work as much as sixteen hours in a day. Then in the winter, there might not be any work at all.

FIGURE 22–22 Steel beams for bridges *(Courtesy of* Texas Highways*)*

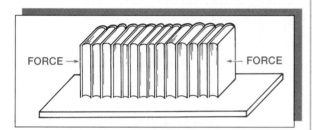

FORCE → ← FORCE

FIGURE 22–23 The principle of prestressed concrete

FIGURE 22–24 Steel decking in place and ready for the concrete topping

Figure 22–24. The steel decking will be the form for the bottom surface of a concrete surface. Reinforcing steel is tied in place and the concrete is placed. The construction of a trussed bridge or a suspension bridge is similar except for the added superstructure.

Paving and Finishing

With the earthwork complete and bridge decks in place, the roadway is ready for paving. Both portland-cement concrete and asphalt are widely used for pavement. Concrete makes a very hard and wear-resistant pavement. However, portland-cement concrete needs much more curing time and is less flexible than asphalt concrete. Therefore, asphalt concrete is used more often than portland-cement concrete in highway construction.

In **asphalt concrete** the cement that binds the aggregate together is asphalt (a black petroleum-based material). For paving, the

FIGURE 22–25 A paving machine receiving asphalt from a dump truck *(Courtesy of* Texas Highways*)*

asphalt concrete is mixed, loaded into dump trucks, and delivered to the construction site. The asphalt concrete, which is commonly referred to simply as asphalt, is dumped into a paving machine, **Figure 22–25**. The paving machine spreads the asphalt, screeds it, and compresses it. A smooth, nonvibrating roller finishes the paving operation.

There may be considerable work to be completed before the highway is ready for final inspection. Guard rails are installed to prevent out-of-control vehicles from leaving the roadway. Paint stripes mark traffic lanes. Signs are installed and the roadside is seeded.

ACTIVITIES

Roadway Construction

Draw a cross section of a roadway with 8-inch pavement, 32 feet wide and 10 feet above the surrounding grade. Make your drawing to a scale of 1" = 20'. Include the dimensions of the major parts of the roadway. Then, label each major part with a description of how that part is built.

Bridge Construction

Building bridges or towers with model-making materials is a popular activity in technology education. Bridge building contests are fun and challenge the designer's skills at analyzing forces, selecting appropriate materials, and workmanship. In this activity you will work in small teams, using the materials and the design your team chooses to create a model bridge. Your bridge must span 18 inches and it is to have a road surface capable of allowing two vehicles 1-1/2 inches wide to pass. Your challenge is to use materials as efficiently as possible and build the strongest bridge possible. Each bridge entered in the contest will be loaded with weights until it breaks. Each team must make a presentation about their bridge to the other teams. The presentation is to include:

▼ Why did you choose the materials you used?
▼ How did you join the structural elements and why did you choose that method?
▼ What other designs did your team consider?
▼ How could you redesign your bridge if you were starting over?

Materials

Your team can use any reasonable materials available to you, but the total weight of the bridge is not to be more than 1-1/2 pounds.

Applying Construction Across the Curriculum

Social Studies

Find an example in your area or community where the right of eminent domain was exercised to build something. Describe the project and why the right of eminent domain was exercised. What should have been the impact on society if that right had not been exercised?

Mathematics

Visit a current or recent construction activity in your area that has involved moving earth (cut and fill). Estimate the dimensions of the cutting or filling that was done. It might not be practical to actually measure the area affected and the depth of the cut or fill, but you should guess as nearly as you can. How much earth did you estimate had to be moved? (Earth is usually measured in cubic yards.) Was all of the cut used as fill? Was it necessary to bring in additional fill, or was there enough material in the cut to provide all of the necessary fill?

Science

Examine a bridge and draw a sketch showing all of the structural members in that bridge. Label your sketch, indicating which members are in tension and which are in compression. Identify the components that act as levers and the class of lever for each.

The environmental impact statement for a large civil or industrial project is similar to the one described and illustrated in Chapter 13, but contains more information. The review and approval process is often much more complex for a major project because it has the potential to affect many more people.

REVIEW

A. Multiple Choice. Select the best answer for each of the following questions.

1. What profession is most directly involved in designing the construction details of a roadway?
   a. Traffic planner
   b. Route surveyor
   c. Civil engineer
   d. Architect

2. Which of the following is considered in the design of a highway interchange?
   a. Traffic volume
   b. Traffic speed
   c. Neighborhood development
   d. All of these

3. At what point do surveyors generally get involved in a highway project?
   a. Before the land is purchased
   b. After the plans are drawn
   c. When the right of way is being cleared
   d. When the pavement is placed

4. What is the term for the power of the government to take over land for public use?
   a. Power of public service
   b. Utility acquisition
   c. Right of eminent domain
   d. Government authority

5. Which of the following would *not* be included in highway specifications?
   a. Methods to be used for soil compaction
   b. Amount of payment to the contractor
   c. Schedule of inspections
   d. Soil types to be used for roadway foundations

6. Which of the following is a typical scale measured with a civil engineer's scale?
   a. 1" = 100'
   b. 3" = 1"
   c. 1/4" = 1'
   d. None of these

7. What machine is used to compact the earth under a roadway?
   a. Bulldozer
   b. Smooth nonvibrating roller
   c. Vibrating roller
   d. Ripper

8. What machine is used to compact freshly placed asphalt concrete pavement?
   a. Bulldozer
   b. Smooth nonvibrating roller
   c. Vibrating roller
   d. Sheepsfoot roller

9. Which of the following is *not* a material widely used in highway construction?
   a. Portland-cement concrete
   b. Soil
   c. Asphalt concrete
   d. All of these are used

10. What is the structural part of a single-span bridge that serves as its foundation?
    a. Spread footing
    b. Concrete slab
    c. Bedrock
    d. Abutment

B. **Bridge Nomenclature.** Which of the following terms is indicated by each of the numbered items in **Figure 22–26**?

abutment
piling
beam
steel decking
concrete fill
intermediate pylon
truss

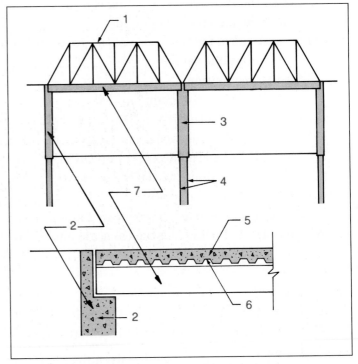

FIGURE 22–26 *(Courtesy of Bay Area Rapid Transit District)*

CHAPTER 23

Industrial Construction

OBJECTIVES

After completing this chapter, you should be able to:

▼ point out the similarities between building construction and industrial construction;

▼ point out the differences between building construction and industrial construction; and

▼ list the major elements in a typical industrial construction project.

KEY TERMS

environmental impact statement

endangered species

seismic activity

overburden

detonate

concrete pump

geodesic dome

head pressure

Any form of on-site construction intended to provide facilities for processing or controlling materials, energy, power, or other resources is called industrial construction. Many industrial projects involve a great amount of civil construction. Dams, for example, involve mainly civil construction techniques. They also usually include hydro-electric power generation plants. Many industrial projects also include large amounts of building construction. Factories are industrial facilities, but are enclosed in buildings.

It would be impossible in a few pages to describe all of the possible kinds of industrial construction. However, the electrical power generation plant described in this chapter includes most of the major elements.

Environmental Concerns

Before any major industrial construction project, such as a power plant, can be started, the owners and their engineers prepare an **environmental impact statement**. This is a report of all important aspects of the environment where construction will take place

FIGURE 23–1 The Black River is important for recreation. *(Courtesy of James Besha Associates)*

and details concerning how construction will affect the environment. The environmental impact statement for a large civil or industrial project is similar to the one described and illustrated in Chapter 13, but contains more information. The review and approval process is often much more complex for a major project, because it has the potential to affect many more people. Preparing the environmental impact statement and negotiating with local, state, and federal environmental agencies is a major function of the engineering company in charge of the project.

What the Problems Were

The Black River is a fast-flowing, medium-size river in a wilderness area. The river is important to sport fishermen, white-water rafters, and sightseers, **Figure 23–1**. If the water from the river were completely diverted through a power plant, the river would be destroyed for all of these recreational purposes. The design of the power plant had to maintain a constant flow of water in the river.

In the early discussion with the state environmental agency, it was determined that this project might affect an **endangered species** (any animal or plant that may become extinct). The Indiana brown bat is an endangered species which lives in caves near this construction site. It was feared that blasting large amounts of rock for the project would disturb the sleeping bats. A method had to be found to keep track of the bats' sleep patterns and to blast the rock without disturbing them.

A Solution for the River. To protect the recreational value of the river, the power plant was designed in two parts, **Figure 23–2**. The water was diverted into the upper end of the canal. Five hundred cubic feet of water per second flows through the upper powerhouse and is returned to the river at point A. The remainder flows down the canal, nearly 1/3 mile long, to the lower powerhouse. Most of the power is generated at the lower powerhouse.

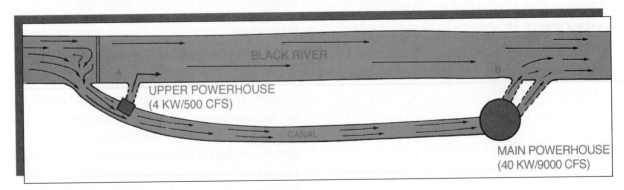

FIGURE 23–2 Overall plan of the power plant

The large flow, up to 9,000 cubic feet per second, is returned to the river at point B. At least 500 cubic feet (over 3,700 gallons) of water per second flows in the main river channel at all times.

A Solution for the Bats. Little information was available on the normal sleeping habits of Indiana brown bats. Therefore, the first step was to determine what the normal conditions for these animals are. Then, blasting was begun. Very small blasts were done at first, increasing the size of the dynamite charges slowly until the signs from the bats started to change. By decreasing the charge slightly from this level, the canal could be blasted and excavated without further endangering the bats.

The bats were monitored with infrared photography via video recorders and sensitive electronic instruments. Four television images were used. Three recorded movement among the bats, their pulse rate, and respiration, while the fourth recorded **seismic activity** (vibrations in the earth from blasting), **Figure 23–3**.

The Canal

It was a huge excavation project to create the canal that transports the water from the upper end to the main powerhouse, **Figure 23–4**. The canal is more than one-quarter mile long. It is more than 70 feet deep. About one-half million cubic yards of rock had to be excavated.

A team of surveyors staked out the boundaries of the canal. The **overburden**, or soil above the solid rock, was removed. Then groups of holes were drilled into the rock,

FIGURE 23–3 The bats were monitored with electronic instruments and video equipment. (A) video camera in cave; (B) video monitor *(Courtesy of James Besha Associates)*

FIGURE 23–4 One-half million cubic yards of stone had to be removed. *(Courtesy of James Behsa Associates)*

D. H. Patel

Occupation:

President, Atlantic Mechanical, Inc.

How long:

20 Years

Typical day on the job:

As president of his own contracting company, D.H. considers estimating and bidding new work a big part of his job. Estimating requires reading architectural plans, calculating the amount of material required to do the job, and calculating the amount of time it will take to complete the job. He is also involved in project management which means keeping in touch with the job superintendents on the job sites. Yet D.H.'s actual time on job sites is only a couple of hours per week. When business is booming, the office staff takes care of bookkeeping, banking, and running the office. When business is slow, D.H. does all of the jobs in the office. Bonding is a continuing concern. Contractors are required to be bonded, to guarantee their work. It is easy to have so much work going at one time that you exceed the limits of your bonding.

Education or training:

D.H. attended high school and started college in India, then completed his baccalaureate degree in mechanical engineering in the United States. Science and mathematics were the high school courses he considers most valuable to his career. In college, most of his courses were in mathematics and heat transfer.

Previous jobs in construction:

D.H. worked as a mechanical engineer for three contractors in Greenville, South Carolina; Traverse City, Michigan; and Galesburg, Illinois. This gave him a chance to see how mechanical systems are designed and installed in different parts of the country.

Future opportunities:

He would like to expand his company to do more and bigger projects.

Working conditions:

Most of his time is spent in the office.

Best aspects of the job:

Always a new challenge. Every job is different.

Disadvantages or drawbacks of the occupation:

The economy and your competition control your business. If competitors bid lower than they can afford to do the job, they often get the bid or at least force you to bid low. It takes 10 to 15 bids to get one contract.

FIGURE 23–5 Rock drills make holes into which explosive charges are placed. *(Courtesy of James Behsa Associates)*

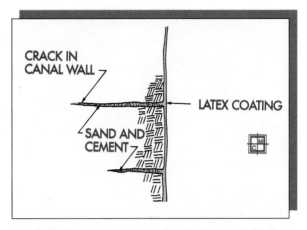

FIGURE 23–6 Cracks in the canal walls are sealed with sand and cement. A latex coating holds the sand and cement in place.

Figure 23–5. Carefully calculated charges of explosives were placed in the holes. When explosives are **detonated** (exploded) for rock excavation, the rock is crumbled in place without being thrown all over the construction site. Of course, in this case, the explosive charges had to be kept small enough so the bats were not disturbed.

With the rock broken into manageable pieces, it was loaded onto trucks and removed. One-half million cubic yards of rock would make a small mountain wherever it was dumped. A reasonable cost for removing and hauling away the rock was about twelve dollars per cubic yard. This cost included the price of buying a farm on which to dump the rock. However, the engineer on this project had good problem-solving skills and found a better solution. A nearby Army post was about to start a major construction project involving huge amounts of concrete. The power plant engineer made arrangements with the quarry that would supply the concrete aggregate to the Army project. The quarry hauled away the power plant stone for free. This saved the quarry the cost of blasting. The power plant was saved the cost of trucking. It finally cost about eight dollars per cubic yard to excavate the canal. This was a savings of four dollars per yard or two million dollars.

The walls of the canal are solid rock. There are deep cracks and crevices in the rock. These cracks would allow a large amount of the water to leak out of the canal into the surrounding earth. Lost water cannot be used to generate electricity, so the canal walls had to be sealed. To seal the walls, dry portland cement and fine sand were blown into the cracks under high pressure. The sand-and-cement mixture was forced at least two feet deep into the cracks. The entire surface was then sprayed with latex to hold the dry sand and cement in place, **Figure 23–6**. As the water seeps into the cracks, the sand and cement cures into hard concrete.

The Powerhouses

The two powerhouses are of different designs, but have much in common. Both are buildings with much reinforced concrete and some structural steel. Both exist only to house a turbine-powered electrical generator and the necessary controls. The main powerhouse is

FIGURE 23–7   Forms and reinforcing steel for the powerhouse *(Courtesy of James Besha Associates)*

described here because it is the more interesting design of the two powerhouses.

Like most large buildings, the main powerhouse is built on a foundation of reinforced concrete, **Figure 23–7**. The power plant project required that large amounts of concrete be placed in a variety of hard-to-reach places. To help place the concrete where it was needed, it was placed with a **concrete pump**, **Figure 23–8**.

The main powerhouse is a reinforced concrete shell with a metal geodesic dome roof,

FIGURE 23–8   A concrete pump delivers the concrete from the truck to the point where it is used. *(Courtesy George B. Gary and Sons)*

FIGURE 23–9   Geodesic-domed roof on the main powerhouse *(Courtesy of James Besha Associates)*

Figure 23–9. A **geodesic dome** is created by joining straight pieces into a network of triangles. This design was chosen for the powerhouse roof because it is a very efficient use of building materials and does not require support in the center.

There are five floor levels in the powerhouse. Each floor is about one-fourth the size of the powerhouse, **Figure 23–10**. These floors

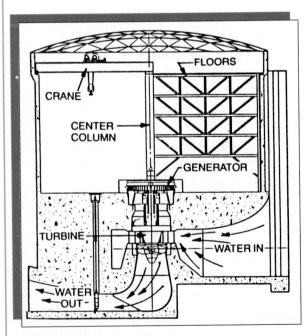

FIGURE 23–10   Cross section of the main powerhouse

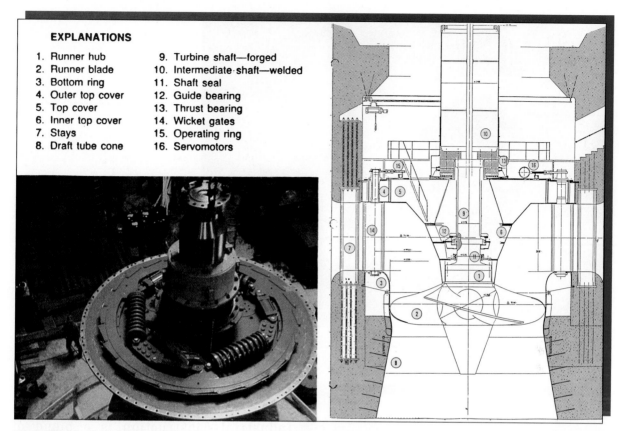

EXPLANATIONS

1. Runner hub
2. Runner blade
3. Bottom ring
4. Outer top cover
5. Top cover
6. Inner top cover
7. Stays
8. Draft tube cone
9. Turbine shaft—forged
10. Intermediate shaft—welded
11. Shaft seal
12. Guide bearing
13. Thrust bearing
14. Wicket gates
15. Operating ring
16. Servomotors

FIGURE 23–11 Two of these turbines drive the generators in the main powerhouse. *(Courtesy of James Besha Associates)*

will hold the computer that controls the generators, switches to start and stop the flow of electricity to the outside power company, and a small amount of storage space. The floors are made up of corrugated steel decking covered with concrete. The floors are supported in the center by a single column that also supports a crane.

The Equipment

The heart of the power plant is the turbine-powered generating equipment, **Figure 23–11**. Huge electric generators are turned by water passing over the turbine blades. The water then flows out of the powerhouse and back to the river through two large tunnels, **Figure 23–12**.

Both volume and pressure are needed for the water to turn the turbine. The volume of water is determined by how much water is flowing in the river. The pressure is determined by the depth of the water. This is called the **head pressure**. The turbines in the main powerhouse on the Black River have 66 feet of head. In other words, the turbines are 66 feet below the surface of the water, **Figure 23–13**.

The electrical generators themselves are so large that it would be impractical to build them in a factory and truck them to the site. Instead, they are built on site, **Figure 23–14**. It is fairly common in industrial construction to see large machinery assembled at the construction site. Usually, the workers who assemble such machinery are employees of the machinery manufacturer, not the construction companies or consulting engineer. This is a

FIGURE 23–12 Discharge tunnels from the main powerhouse *(Courtesy of James Besha Associates)*

major difference between industrial construction and other types of construction. Industrial construction is a blend of construction and manufacturing.

FIGURE 23–13 The surface of the water is 66 feet above the turbines. *(Courtesy of James Besha Associates)*

FIGURE 23–14 The generators are built on site by employees of the generator manufacturer. *(Courtesy of Niagara Mohawk Power Corporation)*

ACTIVITIES

Industrial Construction

Observe a nearby example of industrial construction or choose one described in a magazine. Write a short description of the major parts of the project. Use drawings as necessary. Then, list in one column all aspects of the project that might be found in building construction. In another column, list all aspects of the project that would only be found in industrial construction.

Careers

List all of the construction occupations you think would be needed to build the project described in the first activity.

Applying Construction Across the Curriculum

Social Studies, Science, and Communications

Interview someone who was involved in the design and construction of an industrial project. If there are no recent industrial projects in your area, you can obtain the same information by telephone interviews or writing letters to the owners or engineers on such projects. Your librarian can help you find an interesting project. What environmental concerns were there? What was the process for overcoming those concerns? What adjustments were made in the design or construction of the project as a result of the environmental concerns? Give a report to your class, explaining the above points.

REVIEW

The following is a list of operations performed in the construction of the Black River Power Plant. For each operation: (A) name the occupational level (professional, technical, skilled trade, or unskilled labor) that would perform the task and (B) write *building* if it is the type of job that might be done in building construction or write *industrial* if the type of job that would only be done in industrial construction.

1. Survey to determine boundaries of the project
2. Write an environmental impact statement
3. Drive sheet piles, forming a cofferdam to keep the river out of the site
4. Blast rock with explosives
5. Make drawings of the powerhouse roof with a CAD system
6. Place reinforcement steel for the powerhouse foundation
7. Clean concrete forms after the powerhouse walls are formed
8. Estimate the cost of excavating the canal
9. Connect the electrical cables from the generator to the switchgear
10. Assist the carpenters in building scaffolding

Section Six

Nonstructural Systems

The structural systems in a building create the basic shell. The nonstructural systems are the utilities and services that make the building functional.

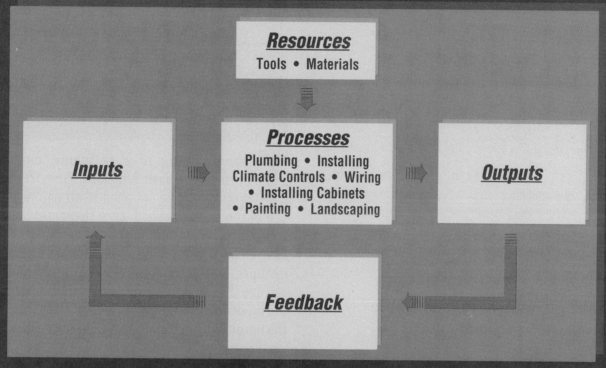

Resources
Tools • Materials

Inputs

Processes
Plumbing • Installing Climate Controls • Wiring • Installing Cabinets • Painting • Landscaping

Outputs

Feedback

Chapter 24
Plumbing

explains the function of supply plumbing and drainage, waste, and vent plumbing. This unit also surveys tools and materials and how they are used.

Chapter 25
Climate Control

discusses all aspects of climate control for a building. The major topics are thermal insulation, solar heating, forced-air and hydronic heating, and electrical air conditioning.

Chapter 26
Electrical Wiring

explains basic electricity_voltage, current, resistance, series circuits, and parallel circuits. The second half of

the chapter applies these principles to electrical systems in buildings.

Chapter 27
Interior Finishing

covers the most important interior finishing operations. Information is included on working with gypsum wallboard, installing flooring, and installing cabinets.

Chapter 28
Painting

discusses the different types of paints and finishing materials and how they are used.

Chapter 29
Landscaping

covers designing, installing, and maintaining the landscaping around a building.

CHAPTER 24

Plumbing

OBJECTIVES

After completing this chapter, you should be able to:

▼ identify the common tools and materials used in plumbing;

▼ explain the basic principles of plumbing design; and

▼ join plastic, copper, and cast-iron piping.

KEY TERMS

supply plumbing

sewage plumbing

DWV

flux

solder

tap

effluent

trap

vent

utility

rough plumbing

finish plumbing

Two Systems

The plumbing in a residence must perform two basic functions. First, fresh water must be supplied to all points of use in the house. Second, once the water has been delivered, it must be carried away after use. To accomplish this, a plumbing system includes two basic subsystems: supply and sewage. The **supply plumbing** includes all of the pipes and fittings to carry fresh water from the municipal supply or the well to the point of use. **Sewage plumbing** includes the pipes and fittings to carry the used water and waste to the septic system or municipal sewage system. Sewage plumbing is often called drainage, waste, and vent plumbing, or simply **DWV**.

Plumbing Materials

The materials most often used for plumbing are iron, copper, plastics, and cast iron.

Galvanized Iron

Galvanized iron piping was once the most widely used material for supply plumbing, but because it corrodes (rusts) more rapidly than

FIGURE 24–1   To join or separate screwed pipe fittings, one pipe wrench holds the pipe while the other turns the fitting. *(Courtesy of The Ridge Tool Co.)*

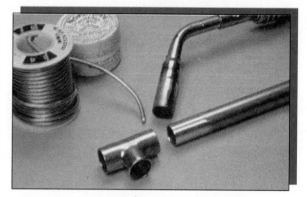

FIGURE 24–2   Equipment for soldering copper pipe *(Courtesy of The Ridge Tool Co.)*

other materials it is seldom used for supply piping in modern construction. Threaded iron pipe is, however, widely used for gas piping. Iron pipes and fittings are threaded, so they can be screwed together. Two pipe wrenches are used to join threaded pipe, **Figure 24–1**. The plumber holds the pipe from turning with one wrench and turns the fitting with the other one. Trying to join piping with only one wrench can do serious damage because there is danger of turning the entire assembly.

Copper

Copper is often used for plumbing because it resists corrosion. However, it is relatively expensive. Copper pipes and fittings can be threaded or unthreaded for soldered joints.

The tools needed for working with copper plumbing are a tubing cutter, steel wool or emery cloth, soldering **flux** (a very mild acid which chemically cleans the copper), solder, and a propane torch, **Figure 24–2**. **Solder** is a soft metal made by combining tin and another metal alloy. Most solder is an alloy of tin and lead, but even the small amount of lead that might be swallowed from drinking water that flowed through pipes soldered with lead tin and lead solder can cause brain damage. The

solder used to join supply piping must be lead-free. A plumber's torch is made up of a propane gas cylinder, a regulator to control the working pressure of the gas coming from the cylinder, a hose, and the torch. A small propane torch with a disposable cylinder can also be used.

To solder copper pipe and fittings:

1. Cut the pipe to the desired length.
2. Clean the end of the pipe and the inside of the fitting with steel wool or emery cloth.
3. Apply soldering flux to the parts to be joined, **Figure 24–3**.
4. Assemble the pipe and fitting.
5. Heat the pipe and fitting until the solder melts when touched to the joint, **Figure 24–4**.
6. Apply just enough solder to insure that the joint is completely soldered.
7. Clean any remaining flux from exposed surfaces with steel wool or emery cloth.

Plastic

Plastics for use in plumbing materials are lightweight, noncorrosive, and easily joined. However, most plastic plumbing materials cannot be used around heat above 200 degrees Fahrenheit (93°C). Plastics are not suitable for some applications where high strength is needed. They are common for general supply and DWV plumbing.

Clarence Luckey

Occupation:
Owner and President, Distinctive Plumbing

How long:
9 Years

Typical day on the job:
Clarence starts his day in the office at about 6:00 AM planning the day's work. He spends much of the day on the phone with suppliers, subcontractors, and bonding agents. Distinctive Plumbing specializes in infrastructure work—water mains, gas mains, fire mains, storm drains, etc.—so most of their work contracts are with government agencies. Government agencies usually require performance bonds, so a big part of Clarence's job as president is to ensure that the company has the necessary bonds. Distinctive is a union shop. They have few employees, but they get skilled labor from the union hall as they are needed for the work.

Education or training:
Clarence grew up in Watts with little family income and many personal and societal problems. He knew that he wanted more than he had as a child, so he attended a private trade school for 10 months where he learned to become a plumber.

Previous jobs in construction:
After trade school, Clarence did small repair jobs. He spent a great deal of time crawling in the mud under houses looking for leaks and repairing plumbing. He worked for himself and gradually built his business into what it is today. He has no formal education in running a business and says he made a lot of mistakes, but fortunately he was able to recover

from them. Today he is a better business manager because he learned from his mistakes.

Future opportunities:
Clarence looks forward to developing his company into an international contractor. Distinctive Plumbing is planning contracts for work in Mexico, Viet Nam, and Africa.

Working conditions:
He spends most of the day in an office now, but when Clarence was learning the plumbing trade he spent a lot of time in the mud, under houses and working in the rain or hot sun.

Best aspects of the job:
All of it! Clarence likes to do paper work in the office, but he also likes to get out in the field and actually help put a piece of plumbing together.

Disadvantages or drawbacks of the occupation:
There is no drawback. He is very happy with what he is doing and the progress of his company.

FIGURE 24-3 Applying soldering flux

FIGURE 24-5 Cast iron soil pipe is often joined with neoprene sleeves and stainless steel clamps.

FIGURE 24-4 Heat the joint until the solder melts and flows smoothly. *(Courtesy of The Ridge Tool Co.)*

Plastic pipe and fittings are joined with solvent cement. The cement is applied to the parts to be joined. The pieces are then quickly assembled and held in place for a few minutes while the cement cures. The solvent cement softens the surfaces. The softened surfaces fuse together and are held in position as the solvent evaporates.

Cast Iron

Cast iron is widely used for DWV plumbing because of its strength and resistance to corrosion. However, it is seldom used for supply plumbing. A common type of cast-iron pipe is called **hubless soil pipe**. Neoprene sleeves

and stainless steel clamps are used to join pipes and fittings, **Figure 24-5**.

Fittings

A wide assortment of fittings is available from manufacturers of plumbing supplies. These fittings are joined to the pipe to make turns at various angles, join additional pipes to the system for service, etc. Most fittings are made of the materials of which pipe is made. Plumbers must be familiar with all types of fittings so they can install their work according to the specifications of the engineer.

Couplings, **Figure 24-6**, are used to join two pipes in a straight line. Couplings are generally used only where a single length of pipe is not long enough.

Unions, **Figure 24-7**, allow piping to be disconnected easily. A union is made up of two parts, with one part being attached to each pipe. Then the two parts of the union are

FIGURE 24-6 Couplings are used to connect two pieces of pipe permanently.

FIGURE 24–7 A union allows the pipes to be disconnected.

screwed together. When it becomes necessary to disconnect the pipe, the two halves of the union are unscrewed.

Elbows, **Figure 24–8**, are used to make changes in direction of the piping. Elbows turn either 90 degrees or 45 degrees. Plumbers sometimes use two 45-degree elbows to produce a more gradual turn.

Tees and wyes, **Figure 24–9**, have three openings to allow a second line to join the first

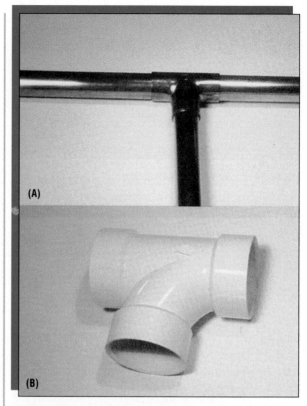

FIGURE 24–9 (A) Tee, and (B) Wye

from the side. Tees have a 90-degree side outlet. Wyes have a 45-degree side outlet.

Cleanouts, **Figure 24–10**, allow access to sewage plumbing for cleaning. A cleanout is made up of a threaded opening and a matching

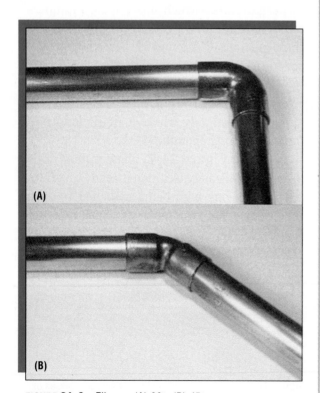

FIGURE 24–8 Elbows: (A) 90°; (B) 45°

FIGURE 24–10 A cleanout is a large plug that can be removed for access to the system.

Joe May

Occupation:
Counter Manager, Ocean City Supply (plumbing and heating supply store)

How long:
5 Years

Typical day on the job:
This business has regular hours, with the store opening at 7:30 AM and closing at 5:00 PM. During the first hour in the morning many of the plumbers from the Ocean City area come in to pick up the supplies they need for the day. It is a time when a lot of business gets done, but it is also an enjoyable time to visit with friends in the trade. Most of the day is spent answering questions about solving plumbing problems, filling orders, and checking inventory levels. Homeowners who want to do their own projects take a lot of time, but they do not make up as big a part of the business as contractors do.

Education or training:
Joe's father was a plumber and he got his start in the trade from him. He has no special training for this position, but he has learned a lot about new materials and equipment from sales people who represent manufacturers.

Previous jobs in construction:
Joe spent many years as a journeyman plumber.

Future opportunities:
After a full career as a plumber, Joe is now retired and does this as a second career.

Working conditions:
Most of the time is spent at the counter of the supply store.

Best aspects of the job:
Joe enjoys working with people. His occupation offers good future job security. There will always be a need for plumbers and plumbing supplies because people will always need water and will always need to get rid of their sewage.

Disadvantages or drawbacks of the occupation:
Ocean City is a resort area, so the business goes up and down with the tourist business. When tourism is slow, the construction business is slow. Working in a supply store does not pay as well as being a plumber or contractor.

plug. When cleaning is necessary, the plug is removed and a *snake* or *auger* is run through the line. Cleanouts are installed in each straight run of DWV.

Valves, **Figure 24–11**, are used to stop, start, or regulate the flow of water. The familiar faucets on a sink or lavatory are a type of valve.

Design of Supply Plumbing

In most communities water is distributed through a system of water mains under or near the street. When a new house is built, the municipal water department **taps** (makes an opening in) this main. The supply plumbing from the municipal tap to the house is installed by plumbers working for the plumbing contractor.

The main supply line entering the house must be larger in diameter than the individual *branches* running from the main to each point of use. There are two basic reasons for this. First, water develops friction as it flows through pipes, and the greater size reduces this friction in the long supply line. Second, when more than one fixture is used at a time, the main supply must provide adequate flow for both. Generally, the main supply for a house is 3/4-inch or 1-inch pipe.

At the point where the main supply enters the house, a water meter is installed. The water meter measures the amount of water used. The municipal water department relies on this meter to determine the proper water bill for that house.

The main water shut-off valve is located near the water meter. This is usually a *stop-and-waste valve*. A stop-and-waste valve has an opening in the side of the body that permits draining the system when the valve is closed, **Figure 24–12**.

From the water meter, the main supply continues to the water heater. Somewhere between the water heater and the meter, a tee

FIGURE 24–11 This valve is used to stop the flow of water into the water heater.

is installed to supply cold water to the house.

From the main supply, branches are run to each fixture or point of use. Generally, each branch has a shut-off valve so that the fixture can be repaired or serviced without shutting off the entire system.

FIGURE 24–12 A stop-and-waste valve has a small cap which can be unscrewed to drain the system.

Design of Drainage, Waste, and Vent Plumbing

The main purpose of DWV, as stated earlier, is to remove water after it has been used and to carry away solid waste. To do this, a branch line runs from each fixture to the main building sewer. The main building sewer carries the **effluent** (fouled water and solid waste) to the municipal sewer or septic system.

Traps

The sewer contains foul-smelling, germ-laden gases that must be prevented from entering the house. If a pipe were simply run to the sewer, then as the effluent emptied from the pipe, sewer gas would be free to enter the building. To prevent this from happening, a trap is installed at each fixture. Another trap is installed at the point where the sewer leaves the building. A **trap** in plumbing is a point in the system that naturally fills with water to prevent sewer gas from entering the building, **Figure 24–13**. Not all traps are easily seen.

Some fixtures, such as *water closets* (toilets), have built-in traps, **Figure 24–14**.

Vents

As the water rushes through a trap it is possible for a siphoning action to be started. To illustrate this siphoning action, a piece of

garden hose or tubing can be used to draw the water out of an open container, **Figure 24–15**. By sucking water through the hose and holding the discharge end at a point lower than the water level in the container, the water will continue to run through the hose without

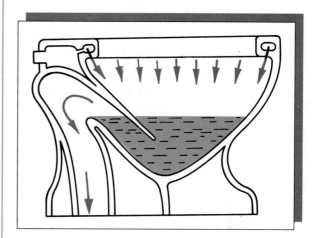

FIGURE 24–14   A water closet has a built-in trap.

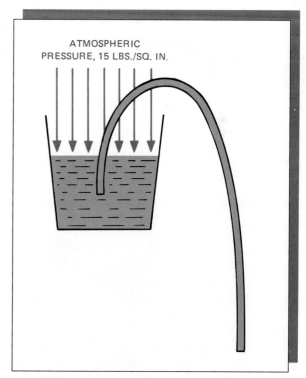

FIGURE 24–15   When a hose filled with water is suspended in a container of water with its outlet below its inlet, atmospheric pressure forces water from the container into the hose.

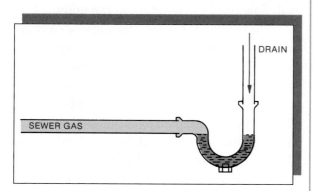

FIGURE 24–13   A trap prevents sewer gas from entering the building.

further sucking. This siphoning action can draw the water from the trap, leaving the sewer open to the inside of the building.

To prevent DWV traps from siphoning, a vent is installed near the outlet side of the trap. The **vent** is an opening that allows air pressure to enter the system and break the suction at the trap, **Figure 24–16**. Because the vent allows sewer gas to pass freely, it must be vented to the outside of the building. Usually all of the fixtures are vented into one main vertical pipe, called a *stack*. The vent stack extends up

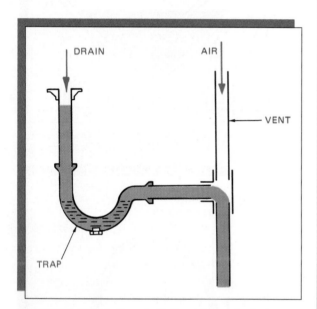

FIGURE 24–16 Venting a trap allows air to enter the system and prevents siphoning.

FIGURE 24–17 Plumbing vent. Notice the metal flashing to prevent leaks.

through the roof or through an exterior wall, **Figure 24–17**.

This is a simplified explanation of plumbing design. The design of a complete plumbing installation is highly technical. It must meet the plumbing codes of the community. Plumbing design in new construction is usually done by a plumbing engineer. The plumbers who install the system must know basic plumbing principles to understand the specifications and drawings developed by the engineer. Since plumbers often repair or modify existing systems, they must also understand plumbing codes and plumbing design.

Plumbing Utilities

The plumbing and electric wiring that join the building to the street-side sources are called **utilities**. Plumbing utilities are fresh water supply, gas supply, and sewer. The pipes for these utilities are often installed before the superstructure of the building is erected. They are discussed here because it is easiest to study all plumbing at one time.

Supply plumbing is fairly simple to install. Pipes are buried below the frost line. They are routed as directly as possible from the street main to the building. Because the water or gas is supplied under pressure, the pitch (levelness) is not important. Sewer pipes, on the other hand, must be pitched properly, so the sewage will flow from the building to the main.

The traditional method for aligning sections of pipe has been to use wire or string and a level. This method is slow and not always accurate. A laser can be used to do the job quickly and accurately, **Figure 24–18**. The laser is positioned at one end of the pipeline. If the pipeline is to be pitched (sloped up or down), the laser is tilted to that angle. The position of each length of pipe is adjusted so the laser beam strikes a target attached to the end of the pipe. The laser is placed inside large-

FIGURE 24–18   Lasers are used to align pipes. *(Courtesy of Spectra-Physics)*

FIGURE 24–19   Plumbing roughed in

diameter pipes. When the pipe is too small, the laser and target are positioned on top of the pipe.

Installation of Plumbing

When the basic structure of the house is completed, the plumbing installation begins.

This is done before any interior wall covering is applied. Plumbers install only the rough plumbing at this stage. **Rough plumbing** includes installation of main supply lines, main sewer lines, and all branch piping, **Figure 24–19**. Although they are considered part of the finished plumbing, bathtubs are generally installed at this time also. This is so the interior wall covering can be finished next to the tub.

When the interior of the house nears completion, the plumbers return to install the **finish plumbing**. This includes installation of all remaining fixtures, such as sinks, lavatories, and water closets.

ACTIVITIES

Joining Copper or Plastic Pipe

Equipment and Materials

Two 1/2-inch valves with hose thread
Two 1/2-inch 90-degree elbows
Two 1/2-inch tees
Two 1/2-inch caps
6 feet of 1/2-inch pipe
Solvent cement (for plastic)
Hacksaw (for plastic)
Propane torch and flint lighter (for copper)
Tubing cutter (for copper)
Solder and flux (for copper)
Steel wool
Tape measure or folding rule

Procedure

1. Cut all pipe the proper length to assemble as shown in **Figure 24–20**. Assemble pieces without solder or cement to check the fit.
2. If using plastic pipe, clean the surfaces to be cemented with steel wool. Apply cement to both parts and immediately join the parts with a twisting motion. Hold them in place for a full minute.
3. If using copper, clean the ends of the pipe with steel wool. Apply soldering flux, join the parts, and sweat solder the joint. Using the propane torch, heat the pipe and fitting just until the solder melts when touched to the joint. Apply just enough solder so that a line of solder appears all around the joint.

CAUTION: Work on a fireproof surface. Know the location of the fire extinguisher and how to use it. Allow the joint to cool completely before handling it.

4. When soldering valves, the stem should be removed to prevent damage to the seals, **Figure 24–21**.
5. When all fittings have been soldered, clean the excess flux from the pipe with steel wool.
6. Connect the soldered assembly to a faucet or hose bib with a hose. Turn on the water and open both valves. When water flows out of the assembly, close the outlet valve and check for leaks.

Applying Construction Across the Curriculum

Social Studies

Use the library to find out when major developments took place in plumbing (e.g., the first mechanical pump, construction of early aqueducts, pipes, metal pipes, fire sprinklers, etc.). Make a timeline of the development of modern plumbing on a poster.

Science

What are the advantages and disadvantages of the following plumbing materials?

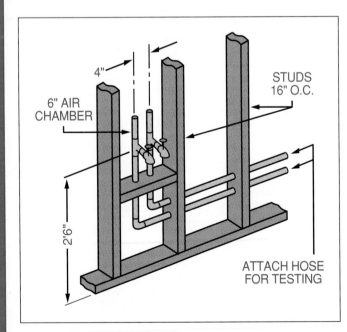

4"
6" AIR CHAMBER
STUDS 16" O.C.
2'6"
ATTACH HOSE FOR TESTING

FIGURE 24–20   Washing machine supply piping

FIGURE 24–21 Remove the valve stem before soldering.

lead solder
cast iron pipe
galvanized iron pipe
copper pipe
ABS plastic pipe
ceramic (terra cotta) drain tiles

Science

Using plastic pipe or tubing, demonstrate the function of a vent stack in the drain piping from a water closet. Give a science-based explanation of what the drain does and what would happen without it.

REVIEW

A. Multiple Choice.   Select the best answers for each of the following questions.

1. What does DWV mean?
    a. Drainage, waste, and vent
    b. Downward water vent
    c. Double waste vent
    d. A kind of double-weight material

2. Which of the following is part of the supply plumbing?
    a. Trap
    b. Vent
    c. Cleanout
    d. Air chamber

3. When do the plumbers normally begin installing piping in a house?
    a. As soon as the floor is framed
    b. As soon as the floor and walls are framed
    c. When the structure is enclosed, but before interior wall covering is applied
    d. When the house is nearly completed

4. What is a typical size pipe for a main house supply?
   a. 1/2 inch
   b. 3/4 inch
   c. 2 inches
   d. 4 inches
5. Which of the following materials is most apt to corrode?
   a. Copper
   b. Galvanized iron
   c. Cast iron
   d. Plastics
6. What is the purpose of soldering flux?
   a. To act as an adhesive in the joint
   b. To protect the joint from corrosion
   c. To help transfer the heat into the joint
   d. To chemically clean the joint
7. What is the purpose of a trap?
   a. To allow atmospheric pressure to enter the system
   b. To prevent sewer gas from entering the building
   c. To collect solid waste
   d. To create a siphoning action
8. What is the purpose of a vent?
   a. To allow atmospheric pressure to enter the system
   b. To prevent sewer gas from entering the building
   c. To prevent pressure buildup
   d. To act as an overflow in case of stoppage
9. Where is a stop and waste valve most apt to be located?
   a. At sinks
   b. At the inlet to the water closet
   c. In the waste plumbing
   d. At the lowest point in the supply plumbing
10. When do the plumbers install lavatories, sinks, and water closets?
    a. As soon as the floor is framed
    b. As soon as the floor and walls are framed
    c. When the structure is enclosed, but before interior wall covering is applied
    d. When the house is nearly completed

CHAPTER 25

Climate Control

OBJECTIVES

After completing this chapter, you should be able to:

▼ describe the operating principles of common heating and air-conditioning systems;

▼ explain the importance of thermal insulation; and

▼ calculate simple heating loads.

KEY TERMS

HVAC	solar collector
convection	Btu
radiation	U factor
evaporation	R value
gravity flow	design temperature difference
duct	greenhouse effect
combustion chamber	
thermostat	
evaporator	
compressor	
condenser	
expansion valve	
hydronic	
circulator	

Buildings of all types, especially buildings where people live, work, and play, use climate control systems. In almost all parts of the country some means of heating is needed in the winter. In most parts of the country cooling has become a necessity in the summer.

Heating, ventilating, and air conditioning (**HVAC**) is an important part of the construction industry. The services of mechanical engineers, electrical engineers, and heating and air-conditioning engineers are used to design the systems in most buildings. Electricians, plumbers, and HVAC technicians install the systems. All of these occupations must have some knowledge of the principles of climate control.

Heat Transfer

Controlling air temperature is a process of transferring heat into or out of a building space. This heat transfer can be done by convection, radiation, evaporation, or gravity flow.

▼ **Convection:** Heat flows from a warm surface to a cold surface. For example, heat flows from warm air to a cold wall.

359

- ▼ **Radiation:** The movement of heat by heat rays. This does not require air or other medium. The sun's heat travels through space by radiation.
- ▼ **Evaporation:** As moisture evaporates it uses heat, thereby cooling the surface from which it evaporated. This is how perspiration cools the body.
- ▼ **Gravity flow:** Cool air is more dense than warm air. Due to gravity flow, the air near a ceiling is warmer than the air at the floor.

Air Cycle

One of the most common systems for climate control circulates the air from the living spaces through or around heating or cooling devices. A fan forces the air into large sheet metal or plastic pipes called **ducts**. These ducts, which are installed by sheet metal technicians, connect to openings in the room. The air enters the room and either heats it or cools it as needed.

Air then flows from the room through another opening into the *return duct*. The return duct directs the air from the room over a heating or cooling device, depending on which is needed. If cool air is needed, the return air passes over the surface of a cooling coil. If warm air is needed, the return air either passes over the surface of a **combustion chamber** (the part of a furnace where fuel is burned) or a heating coil. Finally, the conditioned air is picked up again by the fan and the air cycle is repeated, **Figure 25–1**.

Furnace

If the air cycle just described is used for heating the air, the heat is generated in a furnace. Furnaces for residential heating produce heat by burning fuel oil or natural gas, or from electric heating coils. If the heat comes

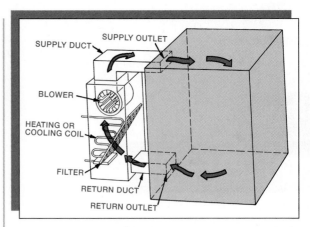

FIGURE 25–1 The air cycle

from burning fuel oil or natural gas, the *combustion* (burning) takes place inside a combustion chamber. The air to be heated does not enter the combustion chamber, but absorbs heat from the chamber's outer surface. The gases given off by the combustion are vented through a chimney. In an electric furnace the air to be heated is passed directly over the heating coils. This type of furnace does not require a chimney.

A **thermostat** senses the temperature at some point in the house. When the temperature drops below a preset level, the thermostat, which is an automatic switch, starts the furnace.

Refrigeration Cycle

If the air from the room is to be cooled, it is passed over a cooling coil. The most common type of residential cooling system is based on the following two principles:

- ▼ As liquid changes to vapor it absorbs large amounts of heat.
- ▼ The boiling point of a liquid can be changed by changing the pressure applied to the liquid. This is the same as saying that the temperature of a liquid can be raised by increasing its pressure, and lowered by reducing its pressure.

A Hole in the Sky

The earth's atmosphere is a relatively thin layer of air, which supports life and a layer of ionized air called ozone, which shields the earth from the sun's ultraviolet rays. In recent years scientists have discovered that the ozone layer is being destroyed near the North and South Poles of the Earth. It is likely that, if the ozone layer at the poles of the earth has been seriously depleted, there has been some thinning of the ozone layer over the rest of the earth. This destruction of the ozone layer can be traced in large part to our use of fluorocarbons. Fluorocarbon is a chemical composition that has been used as propellant in many aerosol sprays and as a refrigerant liquid in refrigeration and air conditioning systems.

The "hole in the ozone layer" over the North Pole has grown to several thousand miles wide. This is an area of thousands of square miles where the earth is subject to ultraviolet radiation and other forms of high energy radiation. Most scientists agree that this radiation is harmful to humans and other life forms on earth. There are many efforts to reduce the destruction of the ozone layer. Recently, the Federal government banned the use of dangerous fluorocarbon refrigerants. Most sprays that were once available only in aerosol cans are now available in pump-type sprays or with non-fluorocarbon propellants. ■

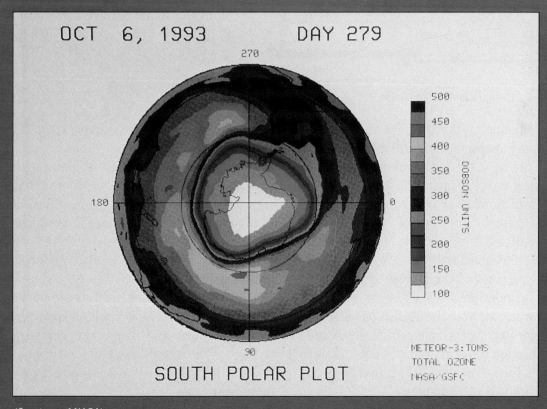

(Courtesy of NASA)

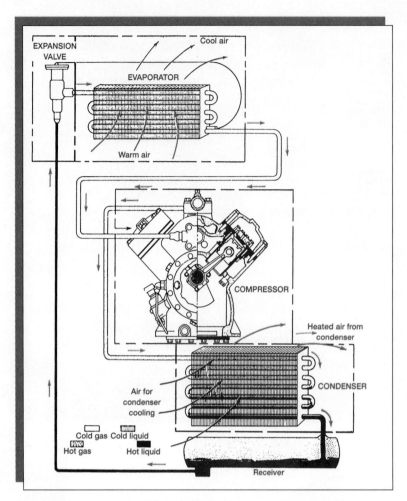

FIGURE 25-2 Parts of a refrigeration system

The principal parts of a refrigeration system are the **evaporator** (cooling coil), **compressor** (an air pump), the **condenser**, and the **expansion valve**, **Figure 25-2**.

Keep in mind that common refrigerants can boil (change to a vapor) at very low temperatures—some as low as minus 21 degrees Fahrenheit (−29°C). Also remember that a liquid boils at a higher temperature when it is under pressure.

The warm air from the ducts is passed over the evaporator. As the cold refrigerant liquid moves through the evaporator coil, it picks up heat from the warm air. As the liquid picks up heat, it changes to a vapor.

The heated refrigerant vapor is then drawn into the compressor where it is put under high pressure. This causes the temperature of the vapor to rise even more.

Next, the high-temperature, high-pressure vapor passes to the condenser where the heat is removed. In residential systems this is done by blowing air over the coils of the condenser. As the condenser removes heat, the vapor changes to a liquid. It is still under high pressure, however.

From the condenser, the refrigerant flows to the expansion valve. As the liquid refrigerant passes through the valve, the pressure is reduced. This lowers the temperature of the liquid still further, so that it is ready to pick up more heat.

The cold, low-pressure liquid then moves to the evaporator. The pressure in the

Linda S. Murphy

Occupation:

Energy Compliance Analyst

How long:

5 Years

Typical day on the job:

Linda works with owners, architects, and engineers to ensure that their plans are in compliance with California's Energy Code. To do this she must study the plans for a new construction and put information about the planned building into a computer system that compares the information with the Energy Code. The computer generates a report indicating any areas that do not meet code. The final product is a report to the city or county building department so that the owner can obtain a building permit.

Most of the day is spent doing area take-offs (calculating the area of walls, areas of windows and doors, areas of roofs etc.), doing data input on the computer, and writing reports.

Education or training:

Linda holds a state certificate for building codes. The California Building Codes Institute administers a voluntary examination and provides certification that tells clients that an analyst knows the codes and is up to date on construction knowledge. This certification is not required by California, but it is a valuable way to assure prospective clients that they are doing business with a qualified person.

Previous jobs in construction:

Linda worked five years for a mechanical and electrical engineering firm, first as a drafter and later, as a designer.

Future opportunities:

Linda works alone now, but she looks forward to expanding her customer base and adding employees as the business grows. She would like to start doing energy audits for small businesses, advising them concerning how they can reduce expenses by conserving electricity and fuel.

Working conditions:

Most of the work is in an office. Occasionally it is necessary to meet clients outside the office.

Best aspects of the job:

Linda especially enjoys working with clients and helping them solve their problems.

Disadvantages or drawbacks of the occupation:

Sometimes clients have to make changes in their building plans that they do not want to make. Although it is the state code, not her analysis that demands the changes, Linda is the one who has to tell them about the required changes.

Other:

Linda thinks people should realize that there is a lot more to the construction industry than just trades, laborers, architects, and engineers. A person who is interested in a career in construction should look for the niche that best fits his or her abilities and work preferences.

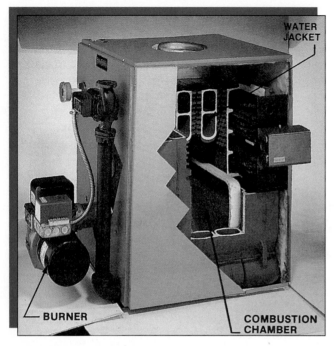

FIGURE 25–3 In a hot-water boiler, water is circulated around the combustion chamber. *(Courtesy of the Burnham Corporation, Hydronics Division)*

evaporator is low enough to allow the refrigerant to boil again and absorb more heat from the air passing over the coil of the evaporator.

Hot-Water Boiler System

Many buildings are heated by hot-water systems. A hot-water system is called a **hydronic** system. In a hot-water boiler system the water is heated in an oil- or gas-fired boiler. The water is then circulated through pipes to radiators or convectors in the rooms. The boiler is supplied with water from the freshwater supply for the house. The water is circulated around the combustion chamber where it absorbs heat, **Figure 25–3**.

In some systems, one pipe leaves the boiler and runs through the building and back to the boiler. In this type, called a *one-pipe system*, the heated water leaves the supply, is circulated through the outlet, and is returned to the same pipe, **Figure 25–4**. Another type, the *two-pipe*

system, uses two pipes running throughout the building. One pipe supplies heated water to all of the outlets. The other is a return pipe which carries the water back to the boiler for reheating, **Figure 25–5**.

Although several designs are used for hot-water heat outlets, most rely on convection for heat transfer, **Figure 25–6**. Cold air enters the bottom of the outlet. As it rises past the hot-water-filled pipes, it picks up heat. The heated air exits near the top of the outlet.

Hot-water systems use a pump, called a **circulator**, to force the water through the system. The water is kept at a temperature of 150 to 180 degrees Fahrenheit (66°C to 82°C) in the boiler. When heat is needed, the thermostat starts the circulator.

Electric Resistance Heat

There are a number of heating system designs that rely on electric heating elements located in each room. Some of these systems have electric heating elements imbedded in the floor or ceiling. In these systems the surface of the room is heated. The main method of heat transfer is radiation.

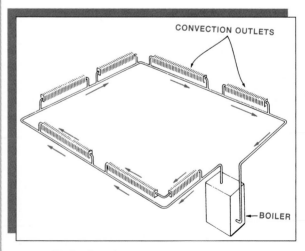

FIGURE 25–4 One-pipe system

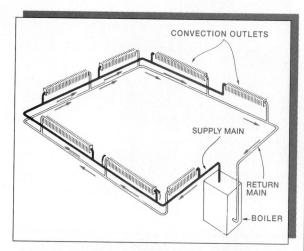

FIGURE 25–5 Two-pipe system

FIGURE 25–6 Convection draws cold air into the bottom of the heating unit and forces heated air out the top.

Another kind of electric heat has heating outlets similar to those described for hot-water heat. Heat transfer is by convection as the cool air passes the heating element.

Solar Heat

Greenhouse Effect

The simplest form of solar heating can be observed in greenhouses. Glass has a characteristic that is useful in solar heating. It allows the radiant heat of the sun to pass through easily, but does not allow reflected heat to pass. This is the principle that allows the sun's energy to warm a greenhouse. If the reflected heat passed through glass as easily as the incoming radiant heat, most of the heat that entered through the windows would be reflected back out. This use of solar heat is called the **greenhouse effect**.

Just as windows admit heat to a greenhouse, windows also admit heat to houses and other buildings. By planning the sizes and location of windows, this heat can be used to warm a building in winter and shade it in the summer. In the winter the sun is closer to the horizon for most of the day than it is in summer. By placing large window areas on the south side of the building, the winter sun shines through the windows, **Figure 25–7**. Because the sun is higher in the sky during the summer, it can be kept off the windows through most of the day by building a large overhang on the south side of the roof, **Figure 25–8**.

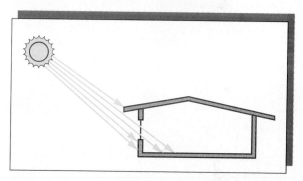

FIGURE 25–7 In the winter the sun is close to the horizon and shines in the windows

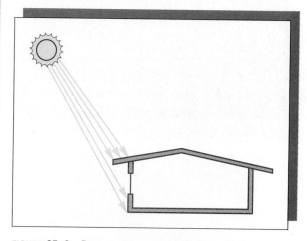

FIGURE 25–8 Because the sun is higher in the sky in summer, the windows can be shaded by the roof overhang.

Greenhouse Effect

Much of the sun's energy reaches the earth as shortwave and visible light. This shortwave energy easily passes through the earth's atmosphere. Some of the energy is absorbed by the earth's solid crust and some is reflected back into the atmosphere as long-wave radiation. The carbon dioxide and water vapor in the earth's atmosphere acts like glass in a greenhouse, absorbing some of the long-wave radiation and reflecting some of it back to the earth's crust again. Until this century, the earth maintained a natural balance of heat lost through radiation and heat gained by the greenhouse effect. The global temperature remained constant.

In the twentieth century we have burned huge amounts of fossil fuels: coal, gasoline, natural gas, etc. Also hundreds of square miles of tropical rain forests in South America have been cleared to create more agricultural land. The use of fossil fuels and the massive burning of the rain forests in tropical areas has increased the amount of carbon dioxide in the earth's atmosphere. The increase in carbon dioxide levels has resulted in increased greenhouse effect and global warming.

The destruction of the rain forests increases carbon dioxide levels in another way. Plants naturally absorb carbon dioxide and release oxygen—the opposite of what animals do. When a large amount of the earth's vegetation is destroyed, that absorption of carbon dioxide is lost.

Increasing the earth's temperature causes several undesirable effects. As the great bodies of water are warmed, the plant life in those waters grows much faster, sometimes at the expense of other aquatic life. Weather patterns are changed, sometimes causing a decline in available water resources in large areas. Some scientists predict that there will be severe droughts and flooding as a result of changing weather patterns. Can you think of other bad side effects of the greenhouse effect? ■

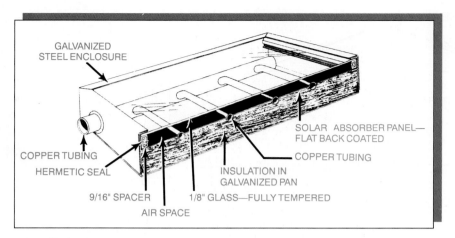

FIGURE 25-9 A commercially made solar collector

Solar Collectors

Solar energy can also be used as a source of heat for a hot-water system. All that must be done is to concentrate the sun's heat on pipes carrying water to the heating system. **Solar collectors** direct the sun's energy onto the heating system pipes.

A flat-plate solar collector is basically a box containing several pipes and having a light-transmitting top. The basic box can be of any material. Sheet metal is the most common, but wood is suitable. The pipes are connected to the heating system at each end by a manifold. The *manifold* is a large pipe with fittings to connect all of the smaller pipes. The top surface is made up of a layer of glass and a layer of plastic. The glass takes advantage of the greenhouse effect. The transparent plastic covers the glass to protect it. A layer of dull black or dark green sheet metal is placed between the glass and the pipes. This surface absorbs a large amount of radiant heat, **Figure 25-9**.

In operation, solar flat-plate collectors are placed where the sun's rays strike their surface in the winter, **Figure 25-10**. The sun's energy heats an antifreeze solution in the collector's pipes. This warmed liquid is pumped to a large tank near the regular boiler. The water entering the boiler is circulated through a coil in the tank. This preheats the water going into the

boiler, **Figure 25-11** Solar collectors do not usually provide 100 percent of the heat needed to heat a residence, but the use of solar collectors can reduce the fuel consumption of a boiler by as much as 60 percent.

FIGURE 25-10 (A) Photovoltaic flat-plate collectors absorb the sun's energy and convert it directly to electricity. (B) Some flat-plate collectors are used to warm liquids. *((B) Courtesy Advanced Energy Technologies, Clifton Park, NY)*

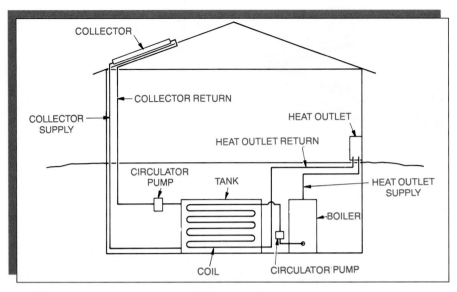

FIGURE 25–11 Solar-assisted hot-water heating system

Thermal Insulation

So far, the methods of supplying heated or cooled air to a building have been discussed. It is equally important to prevent heat from entering a cooled building in the summer and from leaving a heated building in the winter.

Architects and heating engineers measure heat by **Btu** (British thermal units). A Btu is the amount of heat needed to raise one pound of water one degree Fahrenheit. The rate of heat transfer is expressed at *Btuh* (British thermal units per hour).

All materials conduct heat. The rate at which a material conducts heat is its *K factor*. This is the number of Btuh that is conducted by one square foot of the material one inch thick with a difference of one degree Fahrenheit on each side. For example, consider a one inch thick piece of wood with one side being one degree colder than the other. If one Btu passes through one square foot of this wood every hour, its K factor is 1.

Since buildings are made up of an assortment of materials, the K factor is not a practical way to measure the amount of heat that is lost through a floor, ceiling, or wall. The combination of all of the K factors in a building section is the **U factor** for that section. **Figure 25–12** lists the approximate U factors for some common types of construction.

The purpose of thermal insulation is to resist the flow of heat. The resistance of a material to the flow of heat is its **R value**. The R value is the reciprocal of the U factor (R=1/U). In other words, by including thermal insulation with a high R value in a building section, the U factor, or ability to conduct heat, of that section is reduced.

The most common insulating materials for buildings are fiberglass, foamed polystyrene

Type of Building Section	U Value
Wood frame with plywood sheathing wood siding, and 1/2-inch drywall no insulation.	0.24
Wood frame with plywood sheathing wood siding, and 1/2-inch drywall 3-1/2-inch (R-11) insulation.	0.07
8-inch concrete block	0.25
Single-glazed window	1.1
Double-glazed window	0.6

FIGURE 25–12 U factors for some typical building sections

(Styrofoam®), and urethane foam. Fiberglass insulation is made in rolls and batts (pieces a few feet long), **Figure 25–13**. Foamed-polystyrene insulation is usually made in sheets from 1 to 4 feet wide, and 4 to 8 feet long. Urethane foam may be available in sheets or it may be sprayed onto the surface to be insulated, **Figure 25–14**.

The thermal insulation is installed in a building before the inside walls are covered. It is carefully fitted between the framing members, so that there are no openings through which heat can easily pass, **Figure 25–15**.

Estimating Heating Requirements

Design Temperature Difference

The difference between the inside and outside temperatures greatly affects the amount of heat that is lost through the shell of a building. The difference between the lowest probable outside temperature and the desired inside temperature (*design temperature*) is called the **design temperature difference**. The design temperature difference must be known in order to determine the heating needs of the building.

Determining Thermal Resistance of a Building

Properly installed insulation is important for comfortable, economical heating. Although it is impossible to stop the flow of heat through a building section completely, insulation can greatly reduce it. Insulating materials vary in

FIGURE 25–13 Fiberglass insulation is available in loose form for pouring as in (A), or in blankets as in (B).

FIGURE 25–14 Foamed-in-place polyurethane is an excellent insulation material. *(Courtesy of Foam Enterprises, Minneapolis, MN)*

thermal resistance of approximately 3.7 resistance units per inch of thickness. If all 3-1/3 inches of available stud space were insulated with fiberglass, 12.95 resistance units would be substituted for the .95 units provided by the air space (3.7 × 3-1/2"). This would increase the total R value to 16.59 (12.95 − .95 + 4.59).

Notice that the original resistance is less than one quarter of the insulated resistance. With good insulation, slight variations in the resistance of the structural and finish parts of the building have a minor effect on the overall resistance. In wall sections, the total resistance is the resistance of the insulation plus three units for the structural and finish parts. Similar reasoning can be applied to the floor and ceiling to arrive at values for these sections. Uninsulated floors have a resistance of approximately 2 units. Uninsulated ceilings have a resistance of approximately 1-1/2 units.

Windows and doors offer much less resistance to the flow of heat than do other building sections, so the values just mentioned do not apply to them. A single layer of window glass has an R value of approximately 0.88. However, trapped air offers good resistance to heat flow. Therefore, double glazing increases

FIGURE 25–15   Insulation installed between wall studs

their ability to restrict the flow of heat, depending on their type, density and other characteristics. For this reason, insulation is specified according to its R value, rather than thickness.

A building section is made up of the materials supporting the structure, the inside and outside surface coverings, and whatever material is used to minimize heat transfer. **Figure 25–16** shows a building section and the thermal resistance of each part. Notice that the total R value for the section is 4.59. Fiberglass insulation has a

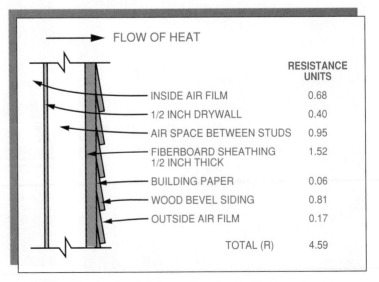

	RESISTANCE UNITS
FLOW OF HEAT	
INSIDE AIR FILM	0.68
1/2 INCH DRYWALL	0.40
AIR SPACE BETWEEN STUDS	0.95
FIBERBOARD SHEATHING 1/2 INCH THICK	1.52
BUILDING PAPER	0.06
WOOD BEVEL SIDING	0.81
OUTSIDE AIR FILM	0.17
TOTAL (R)	4.59

FIGURE 25–16   A typical building section showing the thermal resistance without insulation

Type of Building Section	R Value
Wood frame walls	3 plus the R value of the insulation used
Floors above unheated spaces	2 plus the R value of the insulation used
Ceilings	1-1/2 plus the R value of the insulation
Single-glazed windows	0.88
Double-glazed windows	1.67
Doors with glass	Use R value of the glass for the entire door
Doors without glass	1.67

FIGURE 25–17 R values for common building sections

the R value to approximately 1.67. The resistance values of several common building sections is shown in **Figure 25–17**.

The amount of heat lost through the various sections of the entire building can be found from the R value for each section and the exposed area for each section. By dividing the resistance (R) into the area, the heat transmission load is found in Btuh per degree Fahrenheit of design temperature difference. The *heat transmission load* is the amount of heat that is lost through the building materials.

The following four steps are used to find the transmission load:

1. Find the square-foot area of each outside section by multiplying its height by its length.
2. Divide the R value listed in **Figure 25–17** into the section area to find the heat transmission per degree difference for that section.
3. Add the transmission load per degree difference for all sections.
4. Multiply the total transmission load per degree difference by the number of degrees of design temperature difference. This is the total heat transmission load for the building.

ACTIVITIES

Solar Water Heater

Equipment and Materials

8 linear feet of 1" × 4" lumber
8 linear feet of 1" × 1" lumber
1' × 3' plywood or hardboard (any thickness)
1' × 3' glass
1' × 3' sheet metal
11 feet of 3/8-inch copper pipe
6 feet of 1/2-inch copper pipe
6 copper tees, 1/2" × 3/8"
Small water pump (the type that can be powered by an electric drill is suitable)
4 feet of 1/2-inch hose
3 hose clamps
Saw
Hammer
Supply of 4d common nails
Tape measure or folding rule
Drill and 3/4-inch bit
Thermometer
3 square feet of fiberglass insulation
Solder
Soldering flux
Torch

Procedure

1. Construct the solar collector as shown in **Figure 25–18**.
2. Position the collector where the glass surface gets direct sunlight. The collector should be tilted to face directly into the sun.
3. Connect the pump with a piece of the hose so that it pumps water into the collector.
4. Fill the pail with room-temperature water. Record the temperature.
5. Attach a short piece of hose to the pump inlet and another to the collector outlet. These hoses are to be kept in the pail of water.
6. Run the pump for 15 minutes and record the water temperature.
7. Try running the pump slower for 15 minutes and record the temperature. Experiment with different pump speeds to get the highest water temperature. If you cannot vary the speed of the pump, try installing a valve in the outlet hose so that the output flow can be restricted.

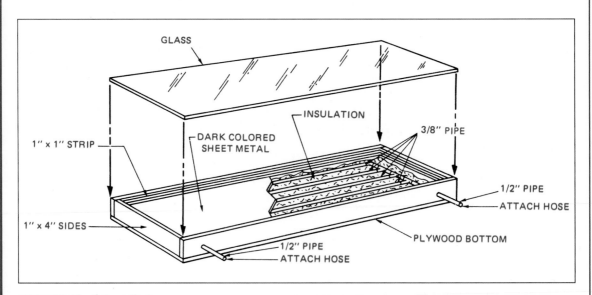

FIGURE 25–18   Solar collector

Estimating Transmission Loads

No special equipment or material needed.

Procedure

Find the total transmission load for a building with the following description:
24' × 40' frame construction with 8-foot ceilings
R-9 insulation in walls
R-21 insulation in floors and ceilings
2 doors, 3' × 7' each (no glass)
7 windows, 2' × 4' each (single glass)
1 window, 6' × 4' (single glass)
70°F design temperature/10°F outside temperature

1. Find the area of the floor in square feet.
2. Divide the R value shown in **Figure 25–17** into the area of the floor. This is the transmission load per degree for the floor.
3. The area of the ceiling is the same as the area of the floor, but its R value is different. Use the R value shown for ceilings to find the transmission through the ceiling.
4. Find the total area of all outside walls by multiplying the perimeter by the height of the ceilings.
5. Find the total area of all windows.
6. Find the total area of all doors.
7. Subtract the total area of the windows and doors from the total wall area.
8. Using **Figure 25–17**, find the transmission per degree Fahrenheit of the walls minus windows and doors.
9. Using **Figure 25–17**, find the transmission through the windows.
10. Using **Figure 25–17**, find the transmission through the doors.
11. Add the transmission of all individual building sections (steps 2, 3, 8, 9, and 10) to find the building transmission per degree Fahrenheit.
12. Multiply the transmission per degree Fahrenheit by the number of degrees of design temperature difference. This is the total amount of heat that will be lost through the surfaces of the building per hour.

Applying Construction Across the Curriculum

Science

Explain at least two specific examples of convection, radiation, evaporation, and gravity flow of heat which you have observed.

Science

Make a thermostatic switch, using a strip of copper and a similar strip of tin plate laminated together with solder. Test your thermostat with a flashlight battery, a bulb, and a heat gun or hair dryer for a heat source.

CAUTION: Do not use any electrical power source over 6 volts.

Science

Design an experiment to test the thermal resistance (the insulating value) of several small samples of available materials. Record the results of your tests and compare them with the relative values of your classmates or the values that you are able to look up in the library.

REVIEW

A. Questions. Give a brief answer for each question.

1. List three methods of heat transfer.
2. What is the name of the duct that carries cool air from a room to the furnace?
3. What device starts and stops the circulator in a hot-water system?
4. Which method of heat transfer does not require air or other medium?
5. What heat transfer method is used when heat is absorbed by boiling water?

B. Matching. Match the part in Column II with the correct function in Column I.

Column I	Column II
1. Cools warm air	a. Evaporator
2. Heated vapor is put under pressure	b. Expansion valve
3. Heat is removed from refrigerant	c. Condenser
4. Refrigerant picks up heat	d. Compressor
5. Pressure of refrigerant is reduced	

CHAPTER 26

Electrical Wiring

OBJECTIVES

After completing this chapter, you should be able to:

▼ describe the relationship between voltage, current, and resistance; and

▼ identify common electrical equipment.

KEY TERMS

current

ampere

volt

resistance

conductor

watt

circuit

open circuit

short circuit

series circuit

parallel circuit

overcurrent-protection
  device

service

ground

outlet

nonmetallic sheathed cable

armored cable

conduit

National Electrical Code

main disconnect

circuit breaker

ground fault circuit
  interrupter

wire nut

To understand the wiring system that provides electrical power throughout a building, it is necessary to know how electricity flows. Electricity is a form of energy that cannot be seen or touched. However, its principles can be demonstrated.

Current, Voltage, and Resistance

In order to do work, electric current must be made to flow through a device, such as a motor or lamp. **Current** is the movement of tiny particles called electrons. An electric generator provides a force which pushes the electrons through a wire. When more electrons flow, more work is done. The amount of current flowing through a device or wires is measured in **amperes**. Amperes are sometimes called *amps*.

If no force is applied by a generator, battery, or some other source, no current flows through the wires. The force that causes the current to flow is called **voltage**. The more voltage applied to a wire, the more current (amps) flows through it. Voltage and amperage (current flow) can be compared to a water system—the greater the water pressure, the more water flows through it, **Figure 26–1** and **26–2**.

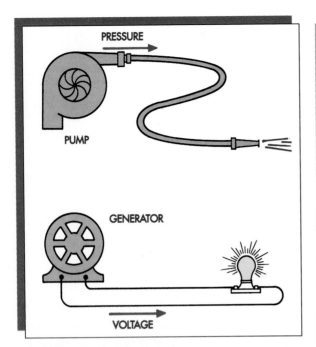

FIGURE 26-1 The force a water pump applies to a hose is called pressure. The force a generator applies to a lamp is called voltage.

There is one other force that affects the amount of current flowing through a system. If the current must do a great amount of work to flow through the system, less current will flow. If the current can flow without doing much work, more current will flow. This principle is easily seen in a water system. When the water must flow through a small nozzle, relatively little water flows. When the nozzle is opened, more water flows. The force that slows the flow of electric current is called **resistance**.

Some resistance is present in any electrical system. If only the wires (**conductors**) made up the system, there would be very little resistance. In this case a low voltage would cause high current flow. If the resistance is high, such as when the current flows through a heater, the same voltage is able to make less current flow.

The amount of work done by electric current depends on the amount of current flowing and the voltage causing it to flow. Electrical work is measured in units called **watts**. One watt is

the amount of work done by one ampere with a force of one volt. The utility company's electrical meter on the outside of a house measures the number of watts used and the number of hours for which they were used.

Circuits

A **circuit** is an arrangement of materials that allows current to flow. It includes an energy source, a device that makes use of the current, connecting wires, and a switch to stop and start the flow of current, **Figure 26-3**.

A complete circuit must provide a path for current to flow from the power source, through the device, and back to the power source. If current does not return to the source, the circuit will be unable to allow more electrons to pass and no current will flow. This is called an **open circuit**. When a switch is turned off, the circuit is opened.

If current is allowed to flow back to the source without overcoming the resistance of

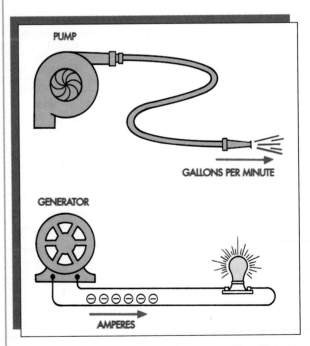

FIGURE 26-2 Water current can be measured in gallons per minute. Electrical current is measured in amperes.

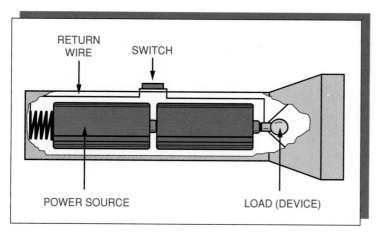

FIGURE 26-3 A flashlight is an example of a simple circuit.

the device, an excess of current will flow. This condition is known as a **short circuit**. A short circuit overloads the circuit wires and can either melt the conductors (wires) or start a fire in the insulation on the wires.

There are two basic kinds of circuits—series and parallel. In a **series circuit**, devices are wired in line with one another so that current must flow through each device to return to the source, **Figure 26-4**. In a **parallel circuit**, the current can take one of two or more paths to return to the source, **Figure 26-5**. Most wiring in buildings is parallel.

Every time a new device is added in a parallel circuit, the current has another path to return to the source. The effect of this is to lower the circuit resistance and allow more current to flow. A parallel circuit can be compared with a highway. Every time a new lane is added it is easier for electrons to get through, **Figure 26-6**. When too many devices are added the current becomes excessive and the wires overheat.

To prevent overloading a circuit, a fuse or circuit breaker is installed. These **overcurrent-protection devices** auto-

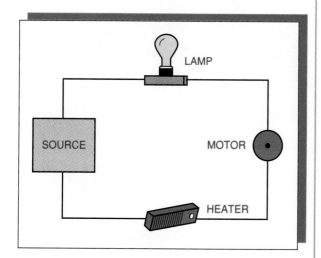

FIGURE 26-4 In a series circuit, the current must flow through all of the devices to return to the source.

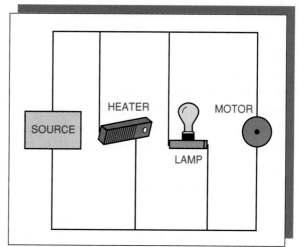

FIGURE 26-5 In a parallel circuit, the current can follow any one of two or more paths to return to the source.

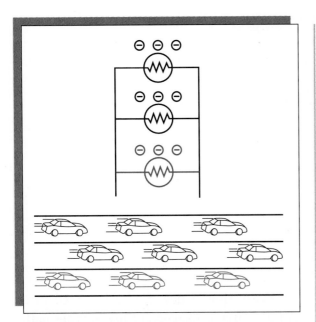

FIGURE 26–6   A parallel circuit is like a multilane highway. As more branches are added, it is easier for electrons to get through.

matically stop the current flow when a certain level is reached.

Power Distribution

Electrical power often must be transported great distances from the generating plant to the user. Due to the resistance of the miles of wire required for distribution, electricity is transported at as much as 600,000 volts. The voltage is reduced to 13,000 volts by transformers at substations. These substations supply distribution stations where the voltage is further stepped down to 2,200 volts. From the distribution station it is transported only a few miles to homes, businesses, and factories where it is stepped down to a more manageable voltage. Some industries use 480-volt electricity. Homes use 240 and 120 volts. From the last transformer, a cable, called a **service**, carries the current to the utility company's meter, **Figure 26–7**. This is where the utility company's responsibility ends and the customer's starts.

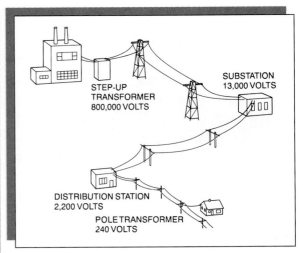

FIGURE 26–7   Electrical current is transported at a high voltage, then stepped down for use.

Grounding

Large power generating plants are connected to the ground by means of a metal rod driven into the ground. Each user of electrical energy is also **grounded**, **Figure 26–8**. In this manner, the earth provides a path for current to return to the source. Most electrical devices

FIGURE 26–8   Electrical system grounded through a ground rod

FIGURE 26–9 Electrical system grounded through a water pipe

have a means for connecting to this ground system. If a live conductor accidentally comes in contact with the frame of the device, the current is directed through the ground rather than through the user. Some devices are grounded by attaching a ground wire to a water pipe, **Figure 26–9**. In recent construction, plumbing may be done with plastic pipe. Plastic is a poor conductor of electricity, so plastic pipes should never be used for electrical grounding. Copper and steel water pipes conduct electricity quite well. Where they run through the earth, copper and steel pipes provide an excellent ground. The 3-prong plug on most small appliances provides grounding through a ground rod or water pipe where the electrical service enters the building.

Electrical Materials

The electrical system in a house is made up of the service and service panel, circuit wiring, and outlets. An **outlet** is a point in the system where equipment can be plugged in or permanently wired into the system, **Figure 26–10**. The most common outlets in a house are light fixtures, switches, and convenience outlets. Switches are commonly referred to as outlets because they are installed in electrical

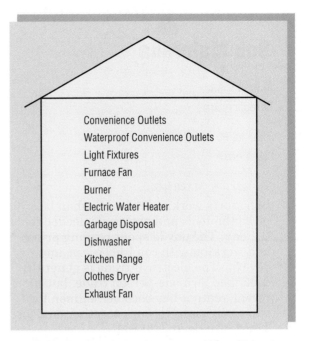

Convenience Outlets
Waterproof Convenience Outlets
Light Fixtures
Furnace Fan
Burner
Electric Water Heater
Garbage Disposal
Dishwasher
Kitchen Range
Clothes Dryer
Exhaust Fan

FIGURE 26–10 Electrical outlets commonly found in residences

boxes as are other devices used at outlets. However, since they do not draw current, electricians do not count them to determine the circuit load. *Convenience outlets* are the common outlets for plugging in various equipment, **Figure 26–11**

FIGURE 26–11 Convenience outlets are the common outlets for plugging in various appliances.

Bob Maiorana

Occupation:
Electrician

How long:
23 Years

Typical day on the job:
Bob starts work at 7:00 AM, when he picks the tools and materials needed for the day. The day is spent reading prints and working with conduit, wires, and devices. The electrical trade is generally described as a one-person trade, but in recent years it has become common for two electricians to work together on a job. The day ends at 3:30 PM.

Education or training:
Bob completed a five-year apprenticeship with the National Joint Apprenticeship and Training Committee. (This is the organization that manages apprentice training for the International Brotherhood of Electrical Workers.) The apprenticeship program consisted of a full day of work, followed by two hours in the classroom every night for five years.

Previous jobs in construction:
Bob worked as an electrician on a 60-story office building, on the renovation of an oil refinery, on an electrical generation power house, and on many smaller jobs.

Future opportunities:
Bob does not plan to change his job. He looks forward to retiring as an electrician.

Working conditions:
An electrician's work varies from indoor, air-conditioned sites to very hot or very cold outdoor construction sites. The work is often dirty and it is tiring to be on your feet all day.

Best aspects of the job:
There is always something new in the trade, making it an interesting job. You get a real feeling of accomplishment when you see the lights come on at the end of the job.

Disadvantages or drawbacks of the occupation:
Bob feels that fringe benefits are not as good today as they were 23 years ago.

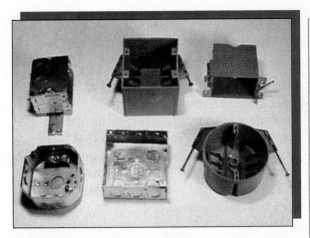

FIGURE 26–12 Electrical boxes are made of steel or plastic and come in a variety of shapes and sizes.

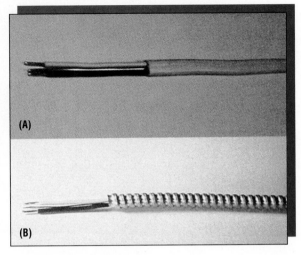

FIGURE 26–13 (A) Type NM nonmetallic sheathed cable; (B) BX armored cable

All connections in the wiring must be made inside a box. This includes connections with convenience outlets, switches, light fixtures, permanently connected appliances, and connections with other cables. Electrical boxes are made of steel or high-impact plastic. They protect the structure from fire in the event that an electrical spark occurs at the connection. Electrical boxes are made in many shapes and sizes for various uses, **Figure 26–12.**

Types of Wiring

There are three types of wiring commonly used in electrical systems, **Figure 26–13:**

▼ Nonmetallic sheathed cable
▼ Armored cable
▼ Conduit

Nonmetallic sheathed cable, commonly called *romex*, is made of copper or aluminum conductors covered with plastic insulation. It is lightweight, inexpensive, and easy to install.

Armored cable, commonly called *BX*, is made of separately insulated conductors encased in a spiral wound steel covering. It is flexible and easy to install. It provides more physical protection than romex, but cannot be used where moisture is present.

Conduit is metal or plastic tubing with conductors running inside of it. The conduit is installed first. Then the electricians pull the wires through it. Although conduit is more expensive and takes longer to install, it is commonly used in nonresidential construction.

Electrical cable is made in a range of sizes and with 2, 3, or 4 conductors. The size of the conductor is specified by American Wire Gauge (AWG). The higher the AWG number, the smaller the conductor. General purpose wiring in residences is generally 14 to 12 gauge. Cable is specified by gauge and number of conductors. For example, "14–3 w/ ground" indicates a cable with three 14-gauge insulated conductors and one uninsulated ground wire.

Designing an Electrical System

Architects and electricians consider many factors in designing the electrical system for a house. The system must provide electrical power for fixed appliances, such as electric heating systems, furnaces, and water heaters. It must also provide outlets at convenient locations for small appliances and lighting. It must be the proper size to be safe and to prevent overloading.

The National Fire Protection Association publishes the **National Electrical Code**

Smart House...

A conventional house has several separate electrical systems. The lighting and convenience outlets, telephone, and cable television often have cables running side by side throughout the house, but never connected. The dishwasher, clothes washer, VCR, and climate control systems probably all have microprocessors (computer chips) in their control units. None of these systems takes advantage of or in any way uses any of the components of the others and the controls for each are generally only located at the point of use.

There have been many advances in the area of home automation in recent years. Generally, they are the result of an innovative manufacturer or a hobbyist developing a means of controlling one of the electrical components in the home with a computer. The programmable control on a VCR is an example of automation. An exciting use of home automation has been connecting controls for home lighting, kitchen appliances, and even sound systems to a computer which can be accessed by telephone. With such a computer controlled system it is possible to use an office computer to send instructions to the home-automation computer to turn the connected systems on and off.

The Smart House system goes one step further. It uses a single wiring system, with only three cable types, to deliver all electrical energy required to control and operate the electrical devices throughout the house. The Smart House Limited Partnership is a partnership of the National Association of Home Builders Research Foundation and 55 manufacturers of electrical and control devices. Because Smart House L.P. includes manufacturers of all of the hardware and technology used in home electrical systems,

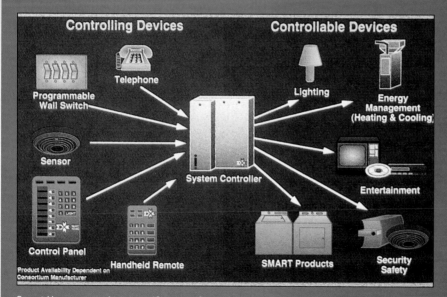

Smart House control system *(Courtesy Smart House, LP)*

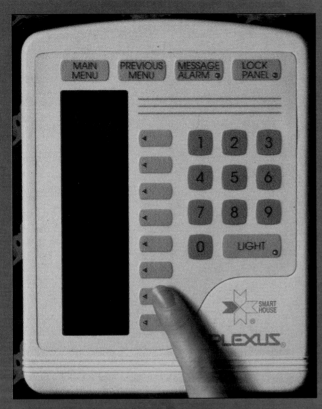

Smart House control panel *(Courtesy Smart House, LP)*

they have been able to agree on standards which allow coordination of the various components.

The Smart House service center is the combination service entrance and distribution point. Utility power, telephone, and TV enter at the service center and are distributed throughout the house. The service center includes the system controller (a computer) and circuit breakers. The service center acts like a switching station. It receives commands from control panels located throughout the house, directs the correct type and amount of electrical energy into the required cables, and signals the appropriate devices and appliances to carry out the desired action. When the user orders channel 12 on the television in the bedroom, the service center controller sends the necessary power to the television along with the cable TV signal for channel 12.

Because the Smart House controller only requires a simple numeric touch pad, it is possible to control the system from a touch tone phone. A Smart House owner can use a telephone from anywhere in the world to program lights to turn on and off at specified times, turn the heat up or down, check to see if the security system has been activated by an intruder, or start dinner cooking.

To install all of the wiring and outlets, making a house Smart-Redi adds about 10% to the cost of constructing a typical new home. Smart House appliances and other Full-Smart devices can be added later. ■

which specifies the design of safe electrical systems. Electrical engineers and electricians must know this code, which is accepted as the standard for all installations. Among the things the code covers are:

- ▼ Kinds and sizes of conductors
- ▼ Locations of outlets and devices
- ▼ Overcurrent protection (fuses and circuit breakers)
- ▼ Number of conductors allowed in a box
- ▼ Safe construction of devices
- ▼ Grounding
- ▼ Switches

The specifications for the structure indicate such things as the type and quality of the equipment to be used, the kind of wiring, and any other information that is not given on the drawings. However, electricians must know the *National Electrical Code* and any state or local codes that apply because specifications sometimes refer to these codes.

From the service entrance, large cables carry electricity to the service panel. The service or distribution panel contains a main disconnect and overcurrent-protection devices, **Figure 26–14.** Usually the overcurrent-protection devices are circuit breakers. The **main disconnect** is actually a large switch that allows all of the electrical power to the building to be disconnected. A **circuit breaker** is an automatic switch which opens the circuit if an excess of current tries to flow.

The service panel also splits the current up into several branch circuits. Each branch circuit has a circuit breaker or fuse. Branch circuits distribute the current to the various devices in the system. The designer of the electrical system must determine the expected load on each branch. The size of the conductors and the rating of the overcurrent protection depend on the expected load.

Another type of electrical protection device is the **ground fault circuit interrupter**

FIGURE 26–14 Installing branch circuit breakers in the service panel

(GFCI). A GFCI detects very small currents flowing from the live conductor (hot leg) to ground. When such an abnormal current is present, it indicates a dangerous path of current to ground—possibly a human body. The GFCI stops all current in the circuit instantly, usually before serious electrical shock occurs.

The set of working drawings for a building includes an electrical plan. The electrical plans for many single-family houses are included on the floor plans. On larger construction jobs,

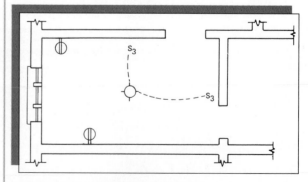

FIGURE 26–15 Electrical devices are shown on an electrical plan by the use of symbols.

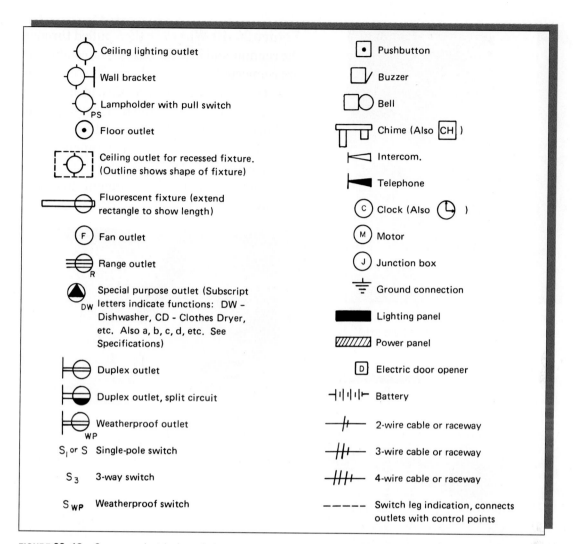

FIGURE 26–16 Common electrical symbols

there may be several pages of electrical plans. The plans include symbols to indicate where devices are to be located, **Figure 26–15**. Electricians must be able to read these plans. **Figure 26–16** shows some of the most common electrical symbols.

Roughing-In Wiring

After the building is framed and enclosed, and before the interior wall covering is applied, the wiring is roughed in. The electricians check the plans and specifications to determine the type and location of all materials.

The service panel is installed first. Then the locations of all devices are marked on the studs and joists. At each point where an outlet or switch is to be installed, a box is fastened to the building frame, **Figure 26–17**.

When all of the boxes are installed, the electricians drill holes through the framing members and install the cables, **Figure 26–18**. Where metal framing is used, the manufacturer provides holes through which wiring can be run. If romex or BX cable is used, it is stapled to the framing members. If rigid conduit is used, it is attached to the boxes with threaded connectors. Conduit is cut to length, then bent with a tubing bender to fit the installation,

FIGURE 26–17 Electrical boxes are fastened to the building frame wherever switches and outlets are to be located.

FIGURE 26–18 Electricians install electrical cable before the walls are enclosed.

Figure 26–19. Wires are then pulled through the conduit and into the boxes. The conductors are connected to one another with **wire nuts**, **Figure 26–20.** These are threaded plastic fittings that are screwed on to the bare ends of the conductors to make a connection.

Before the wall covering is applied, the wiring is inspected to make sure it meets the code requirements. Government agencies employ a large number of electrical inspectors.

FIGURE 26–19 Bending conduit with a tubing bender *(Courtesy of The Ridge Tool Co.)*

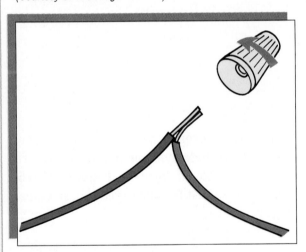

FIGURE 26–20 Two or more conductors can be joined by holding them together and twisting on a wire nut.

Wiring Branch Circuits

Construction electricians install electrical systems before the interior wall covering is applied. They consult working drawings and specifications, plan the layout, then install the electrical apparatus and wiring. When the system is completed, they test the circuits for proper connections and grounding. An electrical inspector also checks the electrician's work to see that it meets all regulations.

Equipment and Materials

2 electrical boxes, 2" × 3"
1 octagon box
Nonmetallic sheathed cable, 14–2 and 14–3
Duplex receptacle
2-way switch
Two 3-way switches
Light fixture

Procedure

First wire the circuits for a convenience outlet and a light fixture, **Figure 26–21**. Then wire the circuits for a light fixture controlled from two places, **Figure 26–22**.

CAUTION: Have the circuit checked by the instructor before connecting it to a power source.

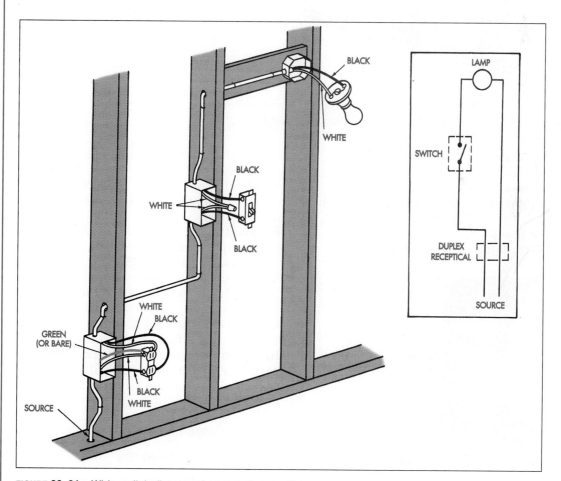

FIGURE 26–21 Wiring a light fixture and a convenience outlet

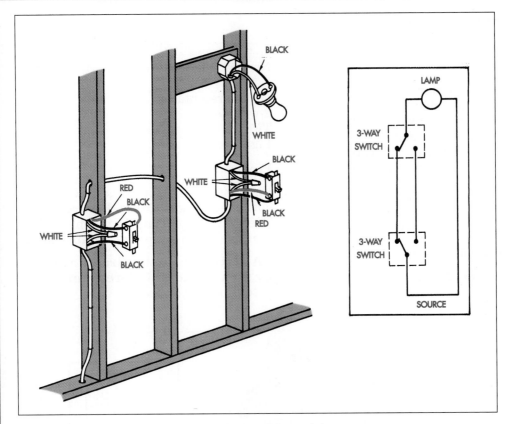

FIGURE 26–22 Wiring a light fixture controlled by two 3-way switches

1. Cut the cable to the proper length, allowing approximately 6 inches inside each box.
2. Strip 6 inches of sheathing from each end of the cable before inserting it in the boxes.
3. Strip approximately 5/8 inch of insulation from the end of each conductor.
4. Conductors that are to be connected to one another are connected with wire nuts.
5. To attach a conductor to a device, clamp it under the head of a terminal screw, **Figure 26–23**.
6. Be sure the wires are not kinked as the device is mounted in the box.

CAUTION: The black or hot conductor should be fastened under the brass-colored connector. The white or neutral conductor should be fastened under the silver-colored connector.

Applying Construction Across the Curriculum

Science and Mathematics

The relationship between current (I),

FIGURE 26–23 The conductor should be wrapped around the terminal screw in a clockwise direction to prevent it from slipping out.

voltage (E), and resistance (R) is stated in an equation called Ohm's Law. Ohm's Law can be used to calculate any one of the three values, if the other two are known:

$$I = E/4$$
$$E = I \times R$$
$$R = E/I$$

Use Ohm's Law to complete the following table:

Resistance (R)	Current (I)	Voltage (E)
200 Ohms	0.5 Amperes	—
—	1.6 Amperes	115 Volts
4,000 Ohms	—	13,000 Volts

Social Studies

Read article 90 of the National Electrical Code and write an explanation in your own language of what it means.

REVIEW

A. Questions.   Give a brief answer for each question.

1. How does adding devices in parallel affect the resistance of a circuit?
2. How does adding devices in parallel affect the current flowing in the circuit?
3. Are switches wired in parallel or series with the device they control?
4. How does adding devices in parallel affect the voltage on a circuit?
5. What publication indicates the requirements for safe wiring?
6. What is the purpose of an electrical box?
7. What is the proper name of the electrical cable commonly called romex?
8. What is the proper name of the rigid metal tubing sometimes used in electrical wiring?
9. At what stage during the construction of a residence do the electricians rough in electrical wiring?
10. While referring to **Figure 26–24**, answer the following:
    a. How many convenience outlets are there?
    b. How many 3-way switches are there?
    c. How many light fixtures are there?
    d. Where is the service panel located?

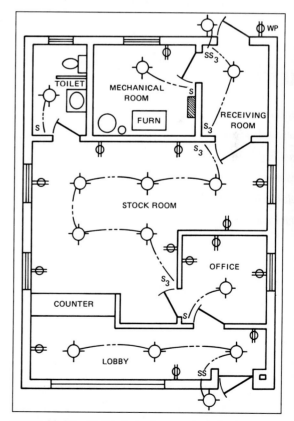

FIGURE 26–24   Electrical plan for an auto parts store

CHAPTER 27

Interior Finishing

OBJECTIVES

After completing this chapter, you should be able to:

▼ use common interior finishing materials; and
▼ consider the factors in finishing the interior of a residence.

KEY TERMS

gypsum wallboard

suspended ceiling

molding

coping

plastic laminate

backsplash

When all of the rough-in work for mechanical and electrical systems has been completed and inspected, the interior finish work is started. This includes installing the following:

▼ Ceilings
▼ Wall coverings
▼ Flooring or carpeting
▼ Molding and trim
▼ Cabinets and countertops
▼ Plumbing fixtures
▼ Electrical fixtures

Most of the interior finish is included in the subcontracts for the rough work. However, some trades are mainly concerned with finish work and others have specialists for this phase of construction. In general, workers who specialize in finish work must work to more precise dimensions and use more caution to protect surrounding work.

Ceilings

The ceilings are usually the first interior surface to be covered. The ceiling material is either attached to the ceiling joists or hung from the joists on steel wires. The most common type of ceiling is made of gypsum wallboard screwed

FIGURE 27–1 Screwing gypsum wallboard to the ceiling joists

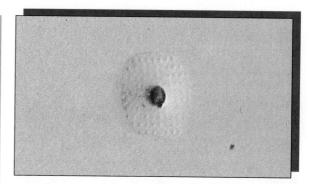

FIGURE 27–3 The dent caused by the hammer will be filled with joint compound.

or nailed to the ceiling joists, **Figure 27–1**.

Gypsum wallboard is made of a plaster core 3/8-, 1/2-, or 5/8-inch thick with a strong paper covering. Its name comes from the fact that gypsum rock is used to make the plaster. It is manufactured in sheets four feet wide and eight to sixteen feet long. The long edges are tapered to permit concealing the joints.

Gypsum wallboard is usually installed by drywall mechanics who specialize in this work. However, it can be installed by carpenters in smaller construction companies.

The wallboard is fastened to the joists with special nails or screws, **Figure 27–2**. Drywall nails have thin heads and ringed shanks. The thin head makes them easier to conceal. The ringed shank makes them less apt to pull loose. Drywall screws also have thin heads, making them easy to conceal. The heads of the nails or screws are driven slightly beyond the surface

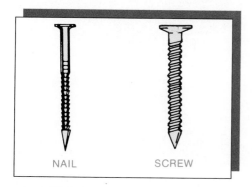

FIGURE 27–2 Drywall fasteners

of the wallboard. This leaves a shallow dent which is filled with joint compound, **Figure 27–3**.

The edges of the wallboard are concealed with paper tape and joint compound. A layer of joint compound is applied with a joint knife or trowel. Then the tape is pressed into it. This first coat is covered with a second coat of compound. When the second coat is dry, a finish coat is applied over a wider area, **Figures 27–4** through **27–7**. When the completed joints are dry, the surface is lightly sanded to remove any imperfections.

A **suspended ceiling** is made up of a metal framework hung on wires from the ceiling joists and fiber panels which fit into the framework, **Figure 27–8**. Drywall mechanics or carpenters attach the wires to the overhead framing, then hang the framework on these wires, checking to see that it is level.

Until recently the framework for suspended ceilings has been done with a spirit level and chalkline. Today, lasers are used to do the job more quickly and more accurately, **Figure 27–9**. The entire framework is installed first. Then the panels are set in place.

Walls

Gypsum wallboard is also a common wall covering material. It gives a sound surface for painting or papering. It also resists the spread

FIGURE 27–4 Applying bedding coat

FIGURE 27–5 Placing reinforcing tape

of fire. On walls, gypsum wallboard is usually applied horizontally so that only one horizontal joint results, **Figure 27–10**. When gypsum wallboard is used on the ceilings and walls, it is nailed to all surfaces first. Then the joints and nail heads are treated.

FIGURE 27–6 Applying a second coat

FIGURE 27–7 The third coat should be feathered out to very thin edges.

FIGURE 27–8 Suspended ceiling *(Courtesy of Celotex)*

FIGURE 27–9 The laser projects a horizontal reference line around the entire room. *(Courtesy of Spectra-Physics)*

FIGURE 27–10 Gypsum wallboard is applied horizontally so that only one joint results.

Another common wall covering is plywood or hardboard paneling. These materials are made up of sheets, usually 4 feet by 8 feet, of plywood or hardboard, with a decorative face. The decorative face may be hardwood veneer, wood grain printed on vinyl, or any attractive pattern on plastics. Wood-grained paneling creates a warm, natural atmosphere in dens, family rooms, living rooms, offices, and kitchens. Plastic-faced paneling provides an easy-to-clean, water-resistant surface in bathrooms and laundry rooms.

Wall paneling can be nailed to the wall framing, glued to the framing, or cemented to gypsum wallboard. Special colored nails are available for nailing wood-grained paneling and its trim. Water-resistant, plastic-faced paneling is usually cemented in place with special adhesives, **Figure 27–11**

Wall paneling is cut and installed by finish carpenters. They carefully measure the location of windows, doors, and electrical outlets. Then the carpenters cut the pieces with common woodworking tools.

Floors

Hardwood flooring is laid by finish carpenters. Some carpenters specialize in laying flooring.

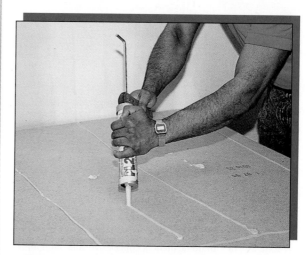

FIGURE 27–11 Plastic-faced panels are cemented in place using a special adhesive.

FIGURE 27–12 Installing hardwood strip flooring
(Photo by John Ewald, Ewald Flooring)

One type of hardwood flooring is made of narrow strips of oak or maple. These strips have tongue-and-groove joints, **Figure 27–12**. As each piece is installed over the subfloor, it is driven up tight against the preceding one. It is nailed through the edge at the base of the tongue. The completed floor is sanded with a floor sander to prepare it for varnishing.

When the thickness of the floor will not be built up by applying wood flooring, underlayment is applied over the subfloor, **Figure 27–13**. Floor underlayment is either plywood or particleboard. It strengthens the floor as it builds areas up to the thickness of the hardwood floors.

Carpenting is also a common floor covering. Carpeting and the pad used under it are purchased by the square yard. Floor covering installers stretch the carpet to the edge of the room. The carpet is held in place there by carpet grippers, **Figure 27–14**.

Seamless vinyl and vinyl tiles are often used where floors may get wet. Seamless vinyl (sometimes incorrectly called linoleum) is measured and cut much like carpeting. It is then cemented to underlayment with a special adhesive. Vinyl tiles are cemented individually.

Ceramic tile is also popular for use on bathroom floors. Ceramic tiles are manufactured in a variety of sizes and colors. They are first cemented to the floor. Then the joints between the tiles are filled with a mortarlike material called *grout*. Ceramic tile is usually installed by a tile setter.

Molding

Wood is machined into a variety of shapes, called **molding**, for use as trim. Molding is used to create special effects on paneling, to cover joints between building parts, and to

FIGURE 27–13 Underlayment is applied over the subfloor to provide a rigid, smooth surface for vinyl, carpeting, or tile. *(Courtesy of the American Plywood Association)*

FIGURE 27–14 Installing carpeting *(Courtesy of Larry Jeffus)*

Elizabeth K. Favreau

Occupation:

Interior Designer

How long:

3 years

Typical day on the job:

Many people confuse interior decorators and interior designers. An interior decorator is primarily concerned with finishes, such as wallpaper and furniture, and usually spends most of the time in homes. Interior designers do mostly commercial work and plan finishes plus doing architectural space planning. They look at traffic patterns in an office and plan where walls should be placed. The designer surveys the requirements of the customer either by interviews or analyzing the uses of the space. The designer then decides, with the client, who should have windows in their offices, who will work in a temporary cubicle, how many conference rooms will be needed, whether or not libraries will be needed, etc. The designer makes drawings like architectural floor plans to describe the general layout. When the layout is finished, the designer makes partition drawings, reflected ceiling plans (to show where light fixtures will be placed), furniture plans, electrical plans, etc.

Education or training:

Liz attended the High School of Art and Design in New York City. She took courses in drafting, architectural history, and physics in high school. She also attended Rhode Island School of Design and completed the first 2 years of the 5-year program. She received a bachelor's degree in Professional Studies, specializing in Architecture, from the State University of New York at Buffalo. Drafting ability is essential in this profession.

Previous jobs in construction:

Liz has worked as an assistant interior designer, where she learned library and pricing, and gained experience in drafting.

Future opportunities:

Liz hopes to advance to designing larger spaces and entire offices. Eventually, she wants to become a principal interior designer in a major design firm.

Working conditions:

She mostly does office work, either in her office or in her clients' offices.

Best aspects of the job:

The end result is a product everyone can see and appreciate. There are always new challenges.

Disadvantages or drawbacks of the occupation:

There is less work available in the winter, because there is less new construction in the winter.

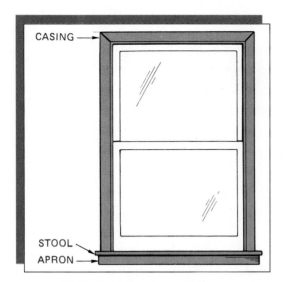

FIGURE 27–15 Trim around a window

protect areas. When the interior walls are covered with wood-grained paneling, special colored molding is used to match the paneling. Molding is installed by finish carpenters.

Window and door frames are trimmed with molding called *casing*, **Figure 27–15**. The casing is mitered (cut at a 45-degree angle) at the corners. It is nailed through the wall covering into the wall framing. Often, windows also have a stool and apron at the bottom.

To protect the walls from floor cleaning equipment and furniture legs, base molding is installed. Base molding is similar in shape to casing, but wider.

Corners of molding are mitered or coped to

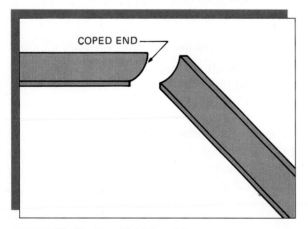

FIGURE 27–16 Coped joint in molding

give a neat appearance. **Coping** means to cut the end of one piece so that it can be butted against the face of the other, **Figure 27–16**. A miter box is used to cut accurate miters. The molding is fastened in place with finishing nails. The finishing nails are then set and covered with wood putty.

Cabinets

Cabinetmaking is a specialized field. Some custom cabinetmaking is done by independent cabinetmakers. Most cabinets are mass-produced by large companies. Engineers and designers develop various styles and include special features, such as revolving shelves, special drawers, and attractive door designs. Drafters make working drawings. These drawings are used by the production department to produce hundreds of cabinets. The completed cabinets are delivered to the construction site ready for installation by the finish carpenters.

The specifications for a residence indicate the brand and style of the cabinets, **Figure 27–17**. The working drawings indicate where the cabinets are to be installed, **Figure 27–18**. Cabinets can be included in several rooms, but

---

DIVISION 10

C. Cabinets

1. Kitchen cabinets shall be Kingswood Oakmont, as manufactured by the B. J. Sutherland Company of Louisville, Kentucky, or equal. Sizes and styles are to be as shown on the special detail drawings.

2. Bathroom vanity shall be RV-48 Moonlight, as manufactured by the B. J. Sutherland Company of Louisville, Kentucky, or equal.

3. Kitchen cabinet and vanity countertops shall be 1/16" laminated plastic bonded to 3/4" plywood. Countertops shall be of one-piece molded construction, with 4-inch backsplash and no back seams. The color and pattern are to be selected by the owner.

FIGURE 27–17 Cabinet specifications

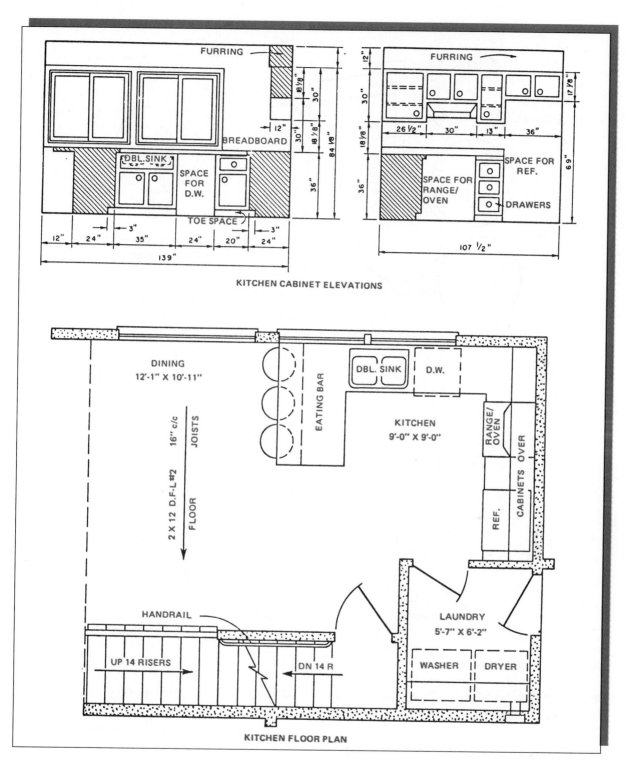

FIGURE 27–18 Kitchen floor plan and cabinet elevations *Courtesy of Home Planners, Inc.)*

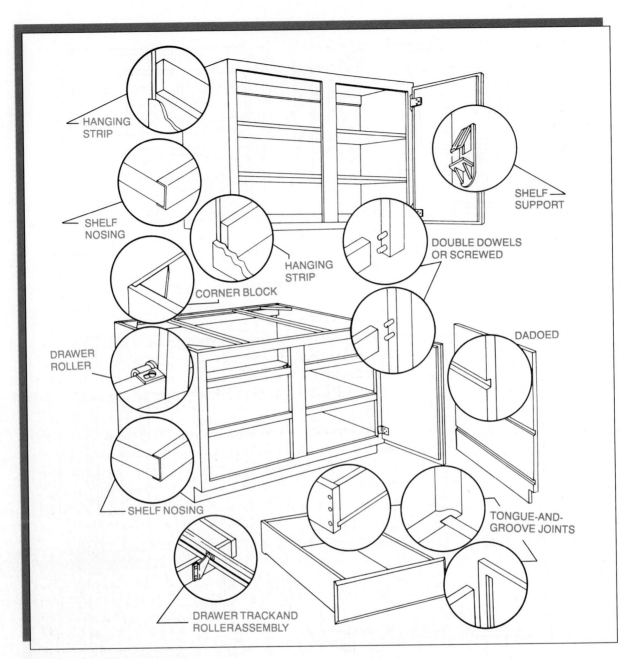

FIGURE 27–19 Cabinet construction

most are used in kitchens. The kitchen cabinet layout is carefully planned for convenience. The cabinets and appliances used for food preparation are close to one another and allow for easy serving. Usually the kitchen includes base cabinets with a countertop and wall cabinets above.

Good quality cabinets have strong, glued joints, **Figure 27–19**. Although most cabinets

are factory made, good quality cabinets can be constructed using carpentry tools.

The cabinets are carefully uncrated, set in place, and shimmed with thin pieces of wood to level them. The backs of the cabinets are often made of thin hardboard with one solid wood crosspiece included to screw the cabinet to the wall framing, **Figure 27–20**. The base cabinets rest on the floor and the wall cabinets are hung

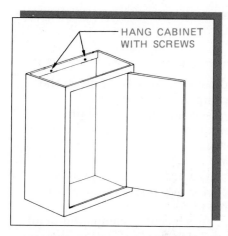

FIGURE 27–20 A strip of solid wood is built into the back of the cabinet for hanging it on the wall.

14 to 24 inches above the base cabinets.

Countertops are made of particleboard or plywood covered with **plastic laminate**. Countertops for most kitchen base cabinets are 25 inches wide with a 3- or 4-inch **backsplash** to protect the wall. Preformed countertops are available with the plastic laminate molded over the edge and the backsplash, **Figure 27–21**. This type of countertop is simply cut to length and attached to the cabinets with screws.

Countertops can also be made by the carpenter. The plywood or particleboard is fastened in place. Then the plastic laminate is cut slightly larger than the countertop. Contact cement is applied to the plywood or particleboard and the back of the plastic laminate. When the contact cement is dry to the touch, the plastic laminate is pressed into place. When properly used, contact cement bonds immediately when the two surfaces touch. After the edge has been covered in the same manner, the laminate is trimmed with a special bit in an electric router, **Figure 27–22**. Carbide-tipped bits are used for trimming plastic laminates. Either a bit with a ball-bearing pilot is used, or a special guide is attached to the router base. All adjustments must be carefully made to regulate the amount that is trimmed.

FIGURE 27–21 Preformed plastic countertop. *(Courtesy of Wood-Mode, Inc.)*

FIGURE 27–22 Laminate can be trimmed with an electric router or laminate trimmer. *(Courtesy of Porter-Cable Corporation)*

The final important step in interior finishing is cleaning. After the finish work has been completed by each trade, all debris must be removed. This step is important enough that it is sometimes included in the specifications.

ACTIVITIES

Gypsum Wallboard Application

Installing drywall panels and preparing them for painting may seem easy. However, small defects on the finished surface will greatly affect the final appearance of the wall or ceiling. Drywall mechanics and finishers must work carefully to insure a smooth surface.

Equipment and Materials

4-foot section of wall frame with one opening
3/8-inch or 1/2-inch gypsum wallboard to cover wall frame
Supply of ring-shank drywall nails
1/4-gallon wallboard joint compound
100-grit abrasive paper
Perforated tape
Steel square
Utility knife
Tape measure or folding rule
Hammer
Joint knife

Procedure

1. Measure the area to be covered and the exact location of any openings.
2. With a pencil, lay out any necessary cuts on the best face of the wallboard.
3. Straight cuts are made in wallboard by first cutting the paper on the face, then folding it to break the gypsum core. The paper is then cut on the back, **Figure 27–23**.
4. Nail the wallboard to each stud, using drywall nails spaced approximately 10 to 12 inches apart. Nails should be driven just far enough to create a slight dent without tearing the paper on the face of the sheet.
5. Use a joint knife to apply a thin coat of joint compound to fill each nail dimple. Apply a thin (approximately 1/8 inch thick by 4 to 5 inches wide) coat of joint compound to all joints.
6. Embed perforated paper tape in the fresh compound at the joints.
7. Apply a second thin coat of joint compound over the perforated tape. Feather the edges out to approximately 8 inches wide.
8. After the compound is completely dry, remove any large bumps with abrasive paper.
9. Apply a finish coat of compound to all joints and nail heads. Feather the edges of the joints 12 to 15 inches wide.
10. After all of the compound is dry, smooth all surfaces with abrasive paper.

Building Countertops

Equipment and Materials

Plywood or particleboard, 2 feet square
1" × 1" × 24" piece of lumber
Three 1-1/4 × 8 flathead steel screws
Electric or hand drill and selection of bits
Screwdriver
Contact cement
Plastic laminate to cover surface
Plastic laminate edge banding
Electric router and laminate trimmer bit
Waxed paper

Procedure

1. To create an edge with the appearance of thicker material, attach a piece of solid wood to the underside of the surface at the edge, **Figure 27–24**.
2. Brush a uniform coat of contact cement on the back of the laminate and on the surface to be covered.

CAUTION: Some contact cement is highly flammable. Do not use contact cement near an open flame. Use adequate ventilation.

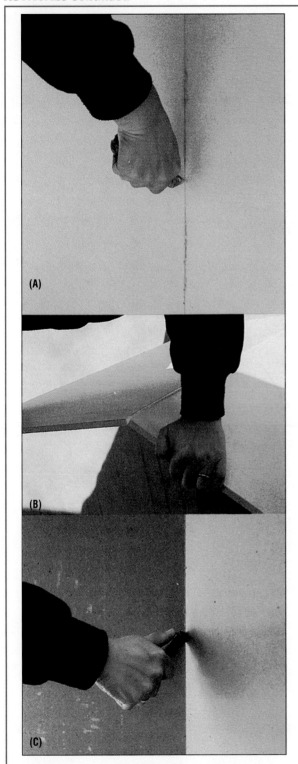

FIGURE 27–23 (A) The paper on the face of the wall-board is cut with a knife. (B) The gypsum core is broken where the paper was cut. (C) The paper on the back is cut.

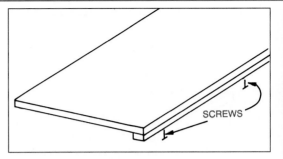

FIGURE 27–24 Attach a piece of solid wood to build up the edge of the counter.

3. When the contact cement is no longer sticky to the touch, cover the surface with two pieces of waxed paper. Overlap the pieces of paper near the center. The waxed paper prevents the laminate from sticking until it is in position.

4. Position the laminate on top of the waxed paper. Allow a slight overhang at the edges. Raise one side of the laminate enough to remove one piece of waxed paper. Raise the other side and remove the remaining piece of waxed paper, **Figure 27–25**.

5. Apply pressure all over the surface to insure good contact at all points. This can be done with a soft-rubber mallet or by rubbing the surface with the corner of a piece of soft pine. Be careful not to break the overhanging edges.

6. Cement the edge banding to the counter edge in the same way. It is not necessary to use waxed paper with the edge banding. Be sure the top edge of the edge banding is against the underside of the top laminate before allowing the cemented surface to touch.

7. Insert the laminate trimmer bit in the router. Adjust the depth of cut as shown in **Figure 27–26**. It is best, at first, to adjust the depth of cut slightly high. Then make a trial cut and readjust the depth.

8. Trim the overhang from the top laminate with the router base on the countertop and the trimmer-bit pilot against the edge banding.

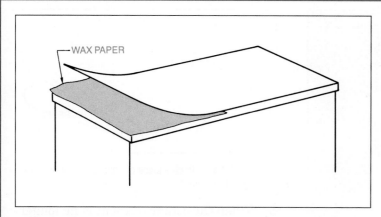

FIGURE 27–25 Raise the end of the laminate and pull out the second piece of waxed paper.

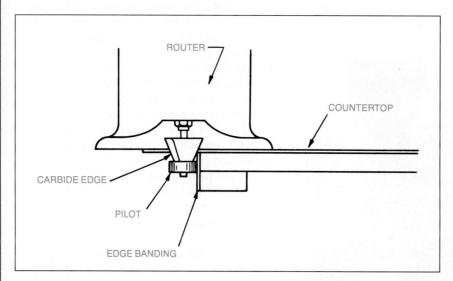

FIGURE 27–26 Adjust the router depth of cut so that the countertop will be flush with the edge banding.

Applying Construction Across the Curriculum

Communications

Choose one room of a home that you would like to design. Make a list of the major items or materials necessary to finish that room, then write to at least two manufacturers or distributors of each of those items requesting literature about the items or materials. When the literature arrives, select one of the items about which you received good information. Write a detailed comparison of the material from each of the two sources, explaining which you would prefer to use and why.

REVIEW

A. Identification. Identify the trade or profession that performs the interior finishing jobs listed.

1. Installs window casing
2. Installs carpeting
3. Installs ceramic tile floors
4. Applies gypsum wallboard
5. Installs plumbing fixtures
6. Installs suspended ceilings
7. Installs kitchen cabinets
8. Plans the kitchen layout
9. Installs base molding
10. Installs lighting fixtures

B. Identification. List the items used for interior finish in **Figure 27–27**.

FIGURE 27–27 *(Courtesy of Marvin Windows, Warroad, MN 56763)*

CHAPTER 28

Painting

OBJECTIVES

After completing this chapter, you should be able to:

▼ explain the purposes of paint and clear wood finishes;

▼ describe the differences between the various kinds of paint and finishing materials;

▼ prepare surfaces for paint or clear finish; and

▼ apply paint and varnish.

KEY TERMS

pigment

vehicle

thinner

acrylic latex

alkyd resin

flat paint

primer

penetrating finish

stain

shellac

lacquer

varnish

tack rag

Surfaces are painted for several reasons. Decoration is the most obvious reason. Paint also gives protection from the sun, wind, and rain, **Figure 28–1**. Steel and iron rust unless they are protected from moisture by paint or some other coating. Wood will warp, crack, and decay if allowed to absorb too much moisture. Painted surfaces are also easier to clean than unpainted surfaces.

There are several kinds of paint for interior and exterior uses, **Figure 28–2**. Painters must know the differences between them and how to use each. Painters must also know how to use wood stains and several kinds of clear coating materials.

Paint

Paint is made of solid pigments ground into a fine powder, a vehicle or liquid to hold the pigment, driers, and thinners. The **pigment** is the coloring material in paint. White lead, zinc oxide, and titanium dioxide are common pigments. The **vehicle** is the liquid in which the other ingredients are mixed. The kind of vehicle used depends on the kind of paint.

FIGURE 28–1 Paint provides protection and improves appearance.

Driers are substances added to paint to speed its drying. Without driers, paint would dry very slowly, if at all. **Thinners** are chemical solvents added to make the paint more liquid and easier to apply. The thinner evaporates from the paint after it is applied to a surface.

CAUTION: Many paint, varnish, and lacquer fumes are harmful to breathe. Always read the container label before opening a container of any finishing, painting, or solvent product. If you want more information, read the Material Safety Data Sheet for that product.

FIGURE 28–2 Painters must be familiar with a wide range of products.

Water-Base Paint

Water-base paint has a water-soluble, synthetic-resin vehicle. The most common synthetic resin in this type of paint is **acrylic latex**, often called *acrylic latex paint*, **Figure 28–3**. The thinner for water-base paint is water. This paint dries quickly (usually within 30 minutes) and covers well.

Alkyd-Resin Paint

Alkyd-resin paints have a vehicle of soya and alkyd resin (a type of plastic), **Figure**

COMPOSITION OF PAINT BY WEIGHT
Pigment 35%
Black Iron Oxide 57.2%
Barium Sulfate 36.5%
Silica and Silicates 6.3%
Vehicle 65%
Nonvolatile
Plasticized Acrylic Resin 34.8%
Volatile
Water 65.2%

FIGURE 28–3 Ingredients of typical water-soluble paint

COMPOSITION OF PAINT BY WEIGHT

Pigment 34%

Titanium Dioxide	11.4%
Calcium Carbonate	85.0%
Silicates	3.6%

Vehicle 66%

Nonvolatile
Soya Alkyd Resin	42.2%

Volatile
Mineral Spirits	57.4%
Driers	0.4%

FIGURE 28–4   Ingredients of typical alkyd-resin paint

28–4. Alkyd resin is made by combining alcohol and acid. Alkyd-resin paints produce an extremely hard surface which is very water resistant. Although it is not a true enamel, alkyd-resin paint is sometimes called enamel because its hard surface resembles that of enamel. These paints are thinned with mineral spirits or turpentine.

Interior Paints

Paint that is used on interior surfaces must be able to be easily cleaned. It must produce a smooth, uniform surface. It also must have good covering ability. Oil-base and alkyd-resin paints for interior use are available in gloss, semi-gloss, and flat paint. Water-base paint is available in semi-gloss and flat. Most interior walls and ceilings are painted with flat paint. A **flat paint** is one that has no gloss when dry. Kitchen and bathroom walls are sometimes painted with semi-gloss paint. Woodwork is usually painted with gloss or semi-gloss paint.

Exterior Paint

Paint for the outside of a building must have different properties than that intended for indoor use. Exterior paint gives greater protection from sun, snow, sleet, and rain. Some white exterior house paint also has a self-cleaning property. This paint is chalky so that the surface cleans itself when it rains. Colored paints must be nonchalking to resist fading. Paint for exterior trim is usually glossy and nonchalking.

Painting

To get good results from any paint job, the surface to be painted must be properly prepared. The surface should be clean, dry, and free of old loose paint, **Figure 28–5**. Wood surfaces should be sanded where necessary to smooth rough spots. Sanding is especially important on interior woodwork. Any defects, such as split boards, rusty metal, and loose fasteners, must be repaired before painting begins. Nail heads should be set and puttied. When painting the exterior of a building, painters protect nearby shrubs from dropped paint. When painting the interior, painters use drop cloths to protect plumbing fixtures, finished floors, and other unpainted work, **Figure 28–6**.

Most paint manufacturers recommend applying one coat of primer before the regular paint. **Primer** is a special paint that sticks to

FIGURE 28–5   To get good results from any paint job, the surface should be clean, dry, and free of old, loose paint.

FIGURE 28–6 A drop cloth protects fixtures and finished surfaces from paint spills.

the surface better than regular paint. The primer should be the one recommended by the paint manufacturer.

In general, painting should proceed from top to bottom and from large areas to trim. In painting the exterior, painters complete the siding from the roof to the foundation. Then they paint the windows, doors, and other trim. If the structure is made of masonry or masonry veneer, the painters are particularly careful not to drop paint on the masonry work. When painting the interior, the ceilings are painted first. Then the walls, windows, doors, and trim are painted.

Even though the paint may have been mixed when it was purchased, it should be stirred before it is used, **Figure 28–7**. It can be applied with a brush, roller, or sprayer.

To brush paint on, use a good quality brush as wide as is convenient on the surface being painted. Dip the brush about one-third the length of the bristles into the paint. Remove the excess paint by tapping the bristles against the inside of the can, **Figure 28–8**. Flow the paint on the surface with long, full strokes.

To use a paint roller, pour a little paint into a roller pan. Work the roller back and forth in the pan until the roller cover is evenly saturated with paint. Roll the paint in several directions on the surface, then finish by rolling in one direction, **Figure 28–9**.

Regardless of whether painting with a brush or roller, paint an area 2 or 3 feet wide across the ceiling or down the wall. Before the paint in that area dries, paint another area, overlapping the first slightly. If paint is allowed to overlap an area that is already dry, a line may show.

FIGURE 28–7 Always stir finishing material before using it.

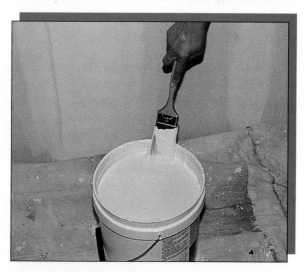

FIGURE 28–8 Tap the bristles of the brush on the inside of the can to remove excess paint.

FIGURE 28–9 Large surfaces can be painted with a roller.

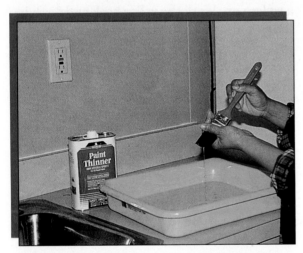

FIGURE 28–10 Clean painting equipment with the proper thinner for the paint used.

FIGURE 28–11 Clean all traces of thinner out of the brush with soap and water.

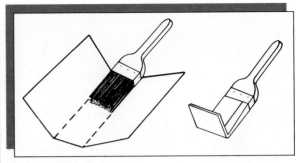

FIGURE 28–12 When the brush is thoroughly cleaned, it should be wrapped in paper and stored.

When the job is finished, or at the end of the day, the painters clean their equipment. Brushes and paint rollers are first cleaned with the thinner for the paint being used, **Figure 28–10**. The thinner and any remaining traces of paint are washed out with soap and water, **Figure 28–11**. Rollers are hung to dry. Brushes are wrapped in paper to protect the bristles, **Figure 28–12**. Store paint brushes flat so that the bristles will not be bent.

CAUTION: It is dangerous to use gasoline or kerosene to clean painting equipment. These fuels present a fire hazard and may cause severe skin irritation.

Stains and Penetrating Finishes

Some surfaces are not painted. Instead, the grain figure of the wood is used for a decorative effect. These surfaces still need protection from the weather. A **penetrating finish** is one which is absorbed into the surface of the wood. The wood is protected while retaining the appearance of natural wood.

Often it is desirable to darken or change the color of wood. *Wood stain* is made from natural colors mixed with a penetrating vehicle and drier. Exterior stains provide a penetrating finish and stain in one operation. *Exterior stain* is applied in the same way as exterior paint.

Stains for interior uses are similar to exterior stains, but they generally do not have as much body and do not protect the wood. On interior trim and cabinetwork, stain is used to change the color of the wood before a clear finish is applied. There are several kinds of stain for interior use. One of the most common is pigmented oil stain. Pigmented oil stain is applied to the surface, then any stain which is not absorbed by the wood is wiped up with a rag. Stained surfaces should be allowed to dry before a clear finish is applied. Follow the manufacturer's instructions for drying times.

Clear Finishes

There are many clear finishing materials available. Each has different properties and is used in a different way. Only a few of the most common types of clear coatings for wood are discussed here. In selecting and using any paint or finishing material, it is always wise to ask for the assistance of the paint dealer and follow the manufacturer's instructions.

Shellac

Shellac is one of the oldest finishing materials in use. The basic ingredient in shellac is the secretion of an insect which is found in India. This solid material is dissolved in denatured alcohol to make shellac finishing material.

Shellac produces a very fine finish on wood. However, it does not withstand heat, direct sunlight, or water spills well. An important use of shellac is for sealing knots in pine and other resinous woods before painting, **Figure 28–13**. The shellac prevents the resin in the knot from discoloring the paint.

Lacquer

True **lacquer** has a nitrocellulose base in some kind of fast-drying vehicle. It is common to refer to any finishing material that dries very quickly through evaporation as lacquer. Many of the modern coatings in this category produce very tough, water-resistant, alcohol-resistant surfaces. They are excellent for spray application because of their very short drying time, **Figure 28–14**. Brushing lacquers are specially formulated to dry more slowly.

FIGURE 28–14 Spraying the finish on cabinet parts on a production line *(Courtesy of Binks Manufacturing Corp.)*

FIGURE 28–13 Shellacking knots

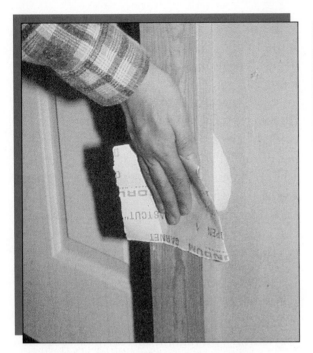

FIGURE 28–15 Sand woodwork before varnishing.

Varnish

Varnish is made of natural or synthetic resins in an oil base. Varnish also contains driers and thinners, much like paint. There are several kinds of varnish, depending on the

FIGURE 28–16 Clean the surface with a tack rag.

resin used, the kind of oil, and other ingredients. Varnish produces an extremely tough, clear finish.

A disadvantage of varnish is its slow drying time and tackiness (sticky quality). Varnish is difficult to apply without getting dust bubbles in its surface. The tough, durable surface produced by varnish makes it a popular finishing material.

Applying Clear Finishes

As there are a wide range of products available, no one set of instructions applies to all clear finishing materials. All manufacturers print instructions on their labels. However, the general procedure is the same for many of these products.

As with painting, the surface should be dry, clean, and smooth. The finish can only be as smooth as the surface on which it is applied. Surfaces should be thoroughly sanded with 150- to 180-grit abrasive paper, **Figure 28–15**. Any scratches will show more clearly after the finish is applied. The surface should be dusted with a tack rag. A **tack rag** can be made by working a small amount of mineral spirits and a few drops of varnish into a clean, dust-free rag. This will remove all traces of dust, **Figure 28–16**.

Before applying a clear finish, clean up all dust and, if possible, mop the floor. Do not shake or stir varnish as this will introduce bubbles. Dip the bristles of a good-quality brush one-third their length into the finishing material. Remove the excess by tapping the bristles against the inside of the can. Flow the finish on with long strokes in the direction of the grain, **Figure 28–17**. Before the finish begins to dry, brush across the grain to spread the finishing material, **Figure 28–18**. Finally, brush in the direction of the grain using only the tips of the bristles to remove brush marks,

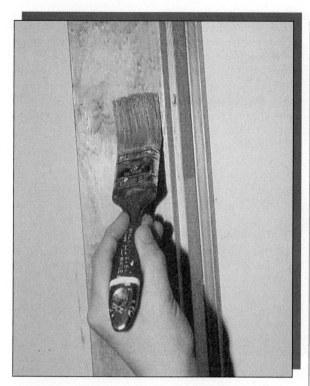

FIGURE 28–17 Flow the finish on with long strokes in the direction of the grain.

FIGURE 28–18 Brush across the grain to spread the finishing material.

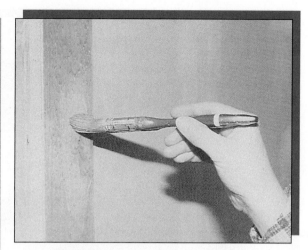

FIGURE 28–19 Brush in the direction of the grain using only the tips of the bristles to remove brush marks.

Figure 28–19. Allow the finish to dry thoroughly, then lightly sand with the fine abrasive to remove any imperfections. Apply a second and third coat following the same procedure. Generally, at least three coats are recommended for brushed-on finishes.

Equipment should be cleaned using the recommended thinner and soap and water. The final coat may be rubbed with fine steel wool, pumice and water, or rubbing compound when completely dry.

Wood Preservatives

Although they are not coatings like paint and clear finishes, wood preservatives are another important wood treatment. Wood that is exposed to water and certain insects and fungi decays very quickly unless it is treated. Wood preservatives prevent the decay which would otherwise ruin the wood.

Preservatives can be applied by dipping or brushing, but only a thin surface layer is treated in this way. A more effective treatment results from pressure treating the wood. The wood is loaded into large chambers where the

FIGURE 28–20 Lumber being loaded into a pressure-treating chamber. *(Courtesy of Wolmanized Pressure-Treated Lumber)*

preservative is forced into the cells of the wood under pressure, **Figure 28–20**. Utility poles, bridge timbers, and piers are examples of the uses of pressure-treated timbers.

ACTIVITIES

Painting

Select a small project and apply either one coat of primer and two coats of paint or three coats of clear finish. Be prepared to explain why you chose the particular coating.

Identification

Study the labels on three different types of paint and two types of clear finishing materials. Give the following information about each:

1. Brand name
2. Type of paint or clear finish
3. Percent of pigment
4. How much new wood will one gallon cover?
5. Is primer or sealer recommended? If so, what kind?
6. What thinner should be used?
7. How many coats are recommended?
8. Drying time

Applying Construction Across the Curriculum

Science

List the ingredients from the label of a can of paint, stain, or other finishing material. Look up each of the ingredients in an encyclopedia or science book. If there are ingredients that you cannot find, your science teacher may be able to help you. For each of those ingredients, write a short description of what it is, where it comes from, and what it contributes to the paint, stain, or finishing material. How would the product be affected if that ingredient were left out?

Communications

Write to the manufacturer of a particular brand of paint, stain, or other finishing material and request a Material Safety Data Sheet for that product. Explain what hazards might be associated with the use of that product and what safety precautions should be taken.

REVIEW

A. Multiple Choice. Select the best answer for each of the following questions.

1. Why are surfaces painted?
    a. To prevent decay
    b. For protection from the weather
    c. To prevent rusting
    d. All of these
2. Which of the following is not used in paint?
    a. Oil
    b. Nitrocellulose
    c. Titanium dioxide
    d. Mineral spirits

3. What is an advantage of alkyd paints?
   a. They can be thinned and cleaned up with water.
   b. They produce a very hard surface.
   c. They can be applied by brushing or spraying.
   d. They have an oil vehicle.

4. What should be used to thin water-base paint?
   a. Gasoline
   b. Mineral spirits
   c. Turpentine
   d. None of these

5. What is the purpose of pigment in paint?
   a. It makes the paint more durable.
   b. It makes the paint easier to apply.
   c. It gives the paint color.
   d. None of these

6. Which of the following surfaces would probably be painted with semi-gloss paint?
   a. Living room ceiling
   b. Living room walls
   c. Bedroom walls
   d. Kitchen walls

7. What is the purpose of pigmented oil stain?
   a. To protect the wood from water
   b. To color the wood
   c. To prevent scratches
   d. All of these

8. Which of the following is an ingredient in shellac?
   a. Nitrocellulose
   b. Tun oil
   c. Denatured alcohol
   d. Acrylic resin

9. What is the greatest disadvantage of varnish?
   a. It is difficult to apply without picking up dust.
   b. It is not water resistant.
   c. It is not durable.
   d. It deteriorates with age.

10. What is the most effective way to apply wood preservative?
    a. Pressure treating
    b. Spraying
    c. Brushing
    d. Dipping

CHAPTER 29

Landscaping

OBJECTIVES

After completing this chapter, you should be able to:

▼ list the important considerations in landscape design; and

▼ outline the procedure for landscaping a typical building site.

KEY TERMS

landscape plan

grading

surface water

topsoil

sod

mulch

Designing the Landscape

One of the final steps in completing any construction project is landscaping. This may be the simple grading and planting of a roadside, **Figure 29–1**. It may involve elaborate gardens and constructed features, **Figure 29–2**. Landscaping improves appearance, holds the soil in place, and provides access to structures.

The landscape design begins early in the planning of the project. On small residential jobs the architect for the building designs the landscaping. On larger projects the planners rely on a landscape architect for this part of the design. Four or five years of college preparation are needed to become a landscape architect. The landscape architect considers the needs of the people served by the structure, the environmental surroundings, and advice from other professionals and experts in designing the landscape.

All of the features of the design are included on a working drawing called a **landscape plan**, **Figure 29–3**. This plan shows the location of buildings; the locations and designs of driveways, parking lots, patios, and walks; and the kind and location of all vegetation.

FIGURE 29–1 Landscaping is an important part of highway construction.

FIGURE 29–2 Landscapes may include elaborate gardens. *(Courtesy of National Landscape Association)*

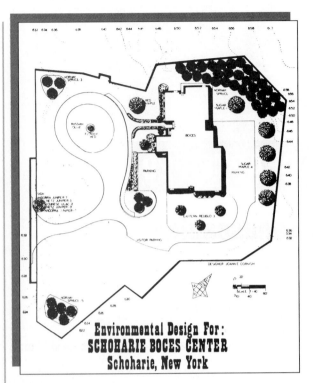

Environmental Design For:
SCHOHARIE BOCES CENTER
Schoharie, New York

FIGURE 29–3 Completed landscape plan

When the building is completed, the landscaping begins. The landscape design for a small project may be done by the general contractor or a landscaper hired by the general contractor. On larger jobs the landscaping is usually done by a separate subcontractor who is selected by competitive bidding in the same manner as other contractors.

Constructed Features

The first step in landscaping the site is to complete constructed features. These include such things as driveways, parking lots, patios, and fountains. Most of these features are constructed of materials that might also be used in a building. For example, driveways and walks can be concrete and patios can be stone, wood, or brick. The details of such features are shown on drawings like those used for the main structure. Where a great amount of construction is included in the landscape, the landscape contractor works with contractors from the appropriate trade to complete the work. **Figures 29–4** through **29–6** show examples of constructed landscape features.

Grading

The contour of an area is basic to the overall landscape. **Grading** is the shaping of the contour of the site. Grading affects the appearance and controls the runoff of water. Rain and melting snow run downhill. Eventually this **surface water** finds its way

Jerome Edmunds

Occupation:

Landscape Manager, Landscape Division, Hewitts Garden Centers

How long:

10 years as landscape helper, landscaper and landscape manager

Typical day on the job:

One-third of Jerome's day is spent sketching and drafting landscape plans and itemizing plant lists. The rest of the day, Jerome's time is spent meeting with customers, discussing their needs, planning their landscapes, and building or maintaining landscapes. This job requires a lot of time devoted to setting up and organizing materials, checking plants, and organizing workers. Most landscape work today is exterior, but interior plantscapes are growing. Jerome supervises 3 or 4 jobs at once.

Education or training:

Jerome had an interest in art and design in high school, but had no formal preparation for the job until college. He received a landscape degree from Cobleskill College.

Previous jobs in construction:

Experience is extremely important in this business. Jerome started working as a landscape helper while in high school and has stayed involved in the landscape business since. After college, he started working as a landscaper for Hewitts, a full service nursery.

Future opportunities:

Jerome hopes to expand the Hewitts business into franchises throughout the United States.

Working conditions:

The working conditions are good, with lots of time outdoors in good weather.

Best aspects of the job:

You can see the results of your work as the landscape is completed and becomes established. Jerome feels good about making a positive contribution to the environment. The environmental aspects of the career are very important.

Disadvantages or drawbacks of the occupation:

Long hours, sometimes as much as 90 hours per week.

to the sea. On a properly graded site, water runoff is gradual until it reaches storm drains or streams. If low spots remain on the site, water will collect in these spots until it seeps through the soil. If the land is contoured too steeply, or if there are gullies, the runoff is fast. This can result in soil erosion.

Earth-moving equipment is used where large amounts of earth must be rearranged, **Figure 29–7**. This operation is done when the initial site work is completed. Most of the

FIGURE 29–6 Patios are constructed landscape features. *(Courtesy of the National Landscape Association)*

earth's surface has a thin layer of **topsoil**. Where the topsoil must be scraped away, it is piled up so that it can be spread over the contoured site later.

Rough grading does not prepare the site for planting. It simply contours the site. *Finished grading* often involves a great amount of hand work. Large stones and debris can be removed from the topsoil by rakes mounted on tractors. However, to prepare the soil for a fine lawn, landscapers hand rake the topsoil, **Figure 29–8**. The raked surface must be smooth and completely free of unwanted stones.

FIGURE 29–4 This landscape design includes many constructed features. *(Courtesy of the Indiana Limestone Institution of America, Inc.)*

FIGURE 29–5 This walkway is an important part of the landscape. *(Courtesy of the Professional Grounds Management Society)*

FIGURE 29–7 Rough grading is done with machines. *(Courtesy of Deere and Co.)*

FIGURE 29–8   The topsoil is raked by hand.

Plantings

Trees

Trees are a valuable part of a landscape design. They provide shade, break the force of harsh wind, and create a natural, attractive appearance. Their roots are also valuable in controlling soil erosion.

Landscape designers try to include existing trees in their plans. Many of the trees used are grown in nurseries and transplanted on the site, **Figure 29–9**. The landscape architect specifies the kind of trees to be planted. Landscape gardeners must know how to transplant all of the trees commonly grown in their area.

Shrubs

Shrubs are smaller than trees and usually have several woody stems. They are attractive when used as part of a total design. They are also

FIGURE 29–9   A large tree is being transplanted with a tree space. *(Courtesy of Vermeer Mfg.)*

valuable for controlling traffic. A row of shrubs (hedge) planted near a walkway keeps people from straying off the walk, **Figure 29–10**.

Shrubs, like trees, are grown in nurseries. The landscape contractor buys shrubs from the nursery. Shrubs and trees are sold in pots or with their roots balled and wrapped in burlap, **Figure 29–11**.

Grass

Large areas of landscape are usually covered with grass. This provides an attractive, easily maintained ground cover and prevents

FIGURE 29–10   Plantings are effective for stopping people from cutting across the corners of lawns.

FIGURE 29-11   The roots of this tree are balled in burlap.

FIGURE 29-12   Fertilizer is spread with a lawn spreader.

FIGURE 29-13   Placing sod

soil erosion. The most familiar use of grass is around homes. Grass is also valuable for roadsides, parks, and lawns around commercial buildings. The landscape designer specifies the type of grass that is suited to the particular site.

Before grass is planted, the topsoil is prepared. If the soil is compacted, it is cultivated to break it up. The soil is analyzed to determine what fertilizer is needed. Fertilizer is spread with a lawn spreader, **Figure 29-12**.

There are three methods of planting grass: sodding, plugging, and seeding. **Sod** is a blanket of existing grass that is grown on a sod farm. A sod cutter removes the growing grass along with a thin layer of soil. The rolled up strips of sod are transported to the site where they are carefully placed on the prepared topsoil, **Figure 29-13**. In some parts of the country, *plugging* is a common method for planting grass. With this method, small plugs of grass are inserted into holes in the topsoil. Most grass, however, is planted by *seeding*. Grass seed is spread evenly over the topsoil, then the seeds are forced into the soil with a roller.

Freshly planted grass must be treated with care until it is well established. Lawns near walks and buildings are roped off to stop people from walking on them. A **mulch** of clean straw or hay is spread over the surface. Mulching protects the new grass from direct sunlight and helps the soil holds its moisture. Later, the mulch decays and provides organic fertilizer. On steep slopes netting is used to hold the soil and grass seed in place. **Figure 29-14** shows what happens when netting or mulch is not used.

FIGURE 29–14 Erosion netting or mulch might have prevented this erosion.

Landscape Maintenance

Landscaping involves the planting and growing of many living plants. Like any living thing, the landscape requires care. Landscape gardeners mow lawns, prune trees and shrubs, and care for gardens. From time to time, these plants need fertilizing and other special care. Landscape gardeners analyze the needs of this expensive part of the owner's investment and provide the necessary care.

ACTIVITIES

Designing Landscape

Landscape architects arrange trees and shrubs with such constructed features as walkways, patios, terraces, and fences. They often supervise the necessary grading, construction, and planting. In order to do this, landscape architects must study construction techniques as well as horticulture and art.

Equipment and Materials

Architect's scale
Pencils and paper
Straightedge
Compass

Procedure

Using the symbols shown in **Figure 29–15**, draw a landscape design for your home. You may draw a plan of the existing landscape, **Figure 29–16**, or completely redesign it. Your design should include the following:

1. General shape and location of buildings
2. Approximate size and shape of the area to be landscaped
3. Constructed features
4. Trees
5. Shrubs
6. Lawn
7. At least one ornamental garden
8. It is not necessary to include contour lines on this drawing. Indicate the direction of water runoff by arrows labeled "runoff."

Landscaping Procedures

Study the landscape around your school or another commercial structure. List all of the features of the landscape. Make an outline of the tasks involved in producing that landscape.

Applying Construction Across the Curriculum

Science

Visit a landscaped site with several species of plants. Take careful notes and

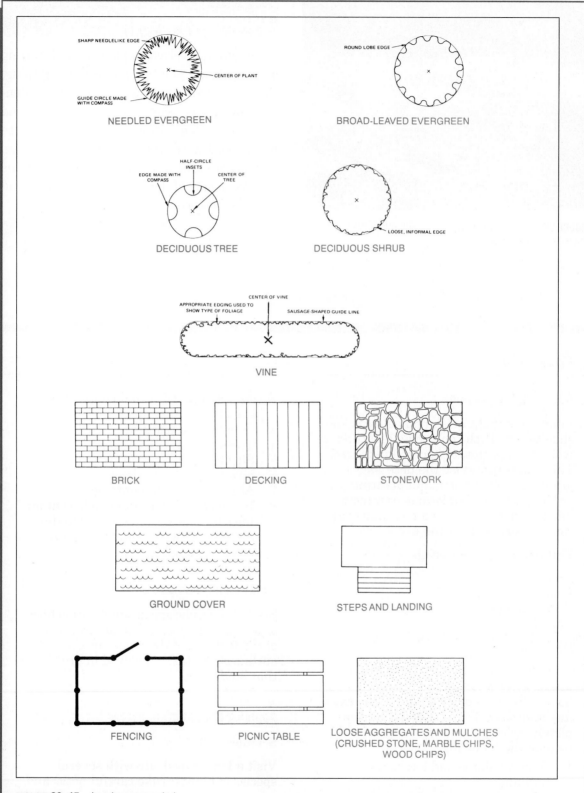

FIGURE 29–15 Landscape symbols

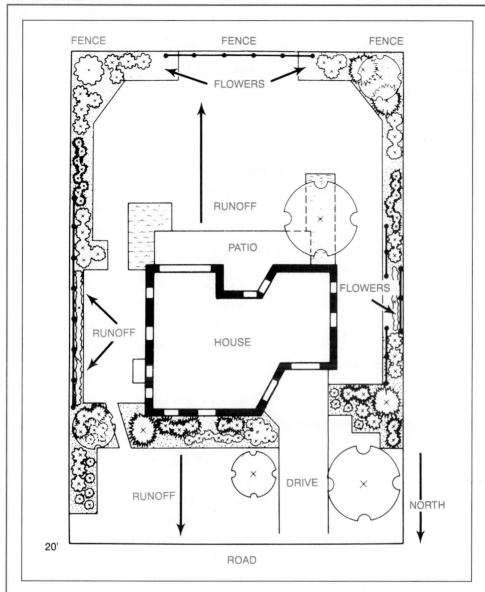

FIGURE 29–16 Sample landscape plan

draw sketches of the overall shape of the plants and details like leaves and flowers. Include grass, woody shrubs, trees, and flowers. Use reference books to identify at least 5 different plants you described. List the following information about each of the identified plants: botanical name of the species, identifying characteristics, geographic area in which it commonly grows, preferred growing conditions (e.g., soil type, water requirements, light requirements, climate, etc.).

REVIEW

A. Multiple Choice.   Select the best answer for each of the following questions.

1. When is the landscape designed?
   a. Before the main structure is designed
   b. Before construction begins
   c. When construction is nearly finished
   d. As soon as construction of the main structure is completed

2. What professional usually designs the landscape for a major project?
   a. Landscaper
   b. Civil engineer
   c. Landscape architect
   d. Landscape contractor

3. Which of the following steps is normally done first?
   a. Finished grading
   b. Construct patio
   c. Transplant trees
   d. Seeding

4. Who would normally construct a concrete driveway?
   a. General contractor
   b. Road builder
   c. Landscaper
   d. Cement mason

5. What can result from improper grading?
   a. Water collects in puddles
   b. Soil erosion
   c. Unsightly appearance
   d. All of these

6. Which operation is usually done with machines?
   a. Rough grading
   b. Seeding
   c. Planting shrubs
   d. Finished grading

7. Which of the following is not a method of planting grass?
   a. Seeding
   b. Sodding
   c. Spraying
   d. Plugging

8. What is the minimum education required to become a landscape architect?
   a. High school
   b. 2-year college
   c. 4-year college
   d. No minimum requirement

Section Seven

Advanced Construction Systems

This section includes two relatively recent developments in construction: manufactured construction and construction in space. Manufactured construction bridges the gap between construction and manufacturing. Buildings that are largely assembled in a factory do not fit the most common definitions of construction, but they are often included in the study of construction. Construction in space includes the materials, techniques, and special concerns for building large space structures. These structures clearly fit the definition of construction, but they are very different from structures built on earth.

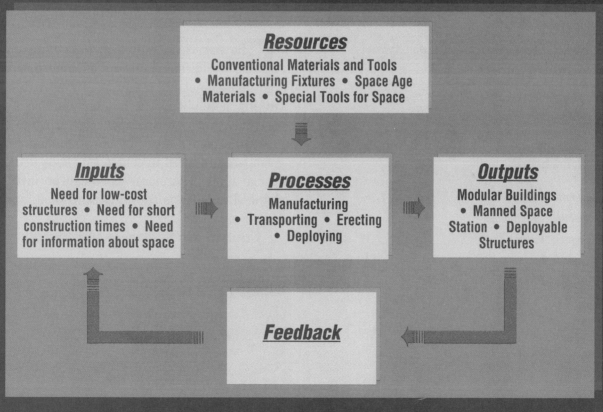

Resources
Conventional Materials and Tools
• Manufacturing Fixtures • Space Age
Materials • Special Tools for Space

Inputs
Need for low-cost structures • Need for short construction times • Need for information about space

Processes
Manufacturing
• Transporting • Erecting
• Deploying

Outputs
Modular Buildings
• Manned Space Station • Deployable Structures

Feedback

CHAPTER 30
Manufactured Construction
describes the manufacturing systems used to produce buildings in a factory and the delivery and installation of these units.

CHAPTER 31
Construction in Space
discusses the developments that led up to our plans for a manned space station, the special requirements for construction in space, and the materials used in space construction.

Manufactured Construction

OBJECTIVES

After completing this chapter, you should be able to:

▼ describe various manufacturing systems used in construction; and

▼ compare manufactured construction with conventional construction.

KEY TERMS

manufacturing

panelized construction

sectional house

modular construction

core unit

jig

plumbing tree

Development of Manufactured Construction

The American manufacturing industry is world famous for its efficient use of personnel and materials. In **manufacturing**, assemblies are put together on an assembly line by workers who specialize in a particular part of the total job. Manufactured goods are produced in controlled shops using special tools. Therefore, parts can be interchanged, quality is carefully controlled, and time is saved.

Until the middle of the twentieth century,

FIGURE 30–1 With trusses this large, a roof can be framed in a few hours. *(Courtesy of Weyerhaeuser Co.)*

nearly all construction was done using the same methods that had been used for centuries. Each piece of a structure was delivered to the construction site and installed separately. In the 1950s, lumber dealers devised a system to save valuable time. Roof trusses were assembled quickly on special fixtures. A construction crew that would normally have spent a whole day framing a roof could set roof trusses in two or three hours, **Figure 30–1**. Soon after the introduction of prefabricated trusses, similar methods were used to manufacture prehung doors, prefabricated wall panels, and other items.

Panelized Housing

Major home builders were quick to recognize the advantages of the use of manufactured building components. By the 1960s, houses could be enclosed in a few days with **panelized construction**. Using this method, the excavation and foundation are completed first. Then the panelized components are delivered to the site. These parts include factory-assembled floor, wall, and roof panels with exterior sheathing in place, **Figure 30–2**. Panelized construction results in an enclosed shell ready for installation of utilities and finishing.

Sectional and Modular Housing

Panelized construction reduces the time it takes to erect the basic shell of a house. However, the shell only accounts for about one-third of the cost of a house. To take further advantage of the benefits of manufactured housing requires more complete finishing and installation of utilities.

Two systems are used to manufacture completed building units. **Sectional houses** are completely built on assembly lines, **Figure 30–3**. They are made with the same materials

FIGURE 30–2 This wall was assembled (panelized) in the factory, then shipped to the site. *(Courtesy of Northern Homes)*

and types of construction as site-constructed houses. A completed house is made up of two or more sections which are placed together on the foundation. These sections need only a small amount of exterior finishing where they are joined and hook up to water, sewage, and electrical service. A sectional home can be ready for occupation the day after it is delivered to the site, **Figure 30–4**.

Modular construction allows more flexibility than sectional construction. A *module*

FIGURE 30–3 Sectional house on assembly line *(Courtesy of Cardinal Industries, Inc.)*

The House That Sven Built

A Swedish home factory looks nothing like a traditional building site. At the Anebyhus plant in Aneby, construction begins with logs from nearby forests. At the sawmill, sensors scan each log, measuring its length and diameter; using this information, a computer calculates the optimum way to cut in order to yield the greatest supply of usable lumber. Just before the cut begins, a laser projects a line of light indicating the saw's path, and an operator adjusts the cutting blades accordingly. Machines sort the cut lumber and deliver it to the kilns, where it is dried to a moisture content of 10–12%—resulting in less warping and splitting than the 15–19% water levels common in U.S. lumber.

In the factory, boards are fed automatically into a jib that holds them in place for the assembly of wall panels. Automatic pneumatic screwdrivers descend and quickly secure the frame. Mounted on an automated tilt table, the jig moves down the assembly line, where workers add doors and windows that are sealed with gaskets for an airtight fit. Other workers slip precut pieces of insulation into the walls. The conveyor belt moves the jig to the next station, where two workers attach a plastic vapor barrier. An inner layer of framing adds space for additional insulation and makes it possible to install wiring without piercing the vapor barrier. Finally, robot screwdrivers attach the exterior siding and interior gypsum board. In other parts of the factory, similar assembly lines produce floors and roofs.

Next comes the most distinctive characteristic of Swedish manufactured housing. Skilled woodworkers use precision tools to complete the trim and finish. The automated jigs and the comfortable indoor conditions make it easier to do careful work. The resulting quality shows: doors close with the reassuring whoosh of an airtight seal, and windows have essentially no leakage, even in winds up to 50 mph. Dormers and spiral wooden stairways give the feel of a custom-built house.

Wall, roof, and floor panels are tightly wrapped in plastic for delivery to the building site. A crane lifts the panels onto a concrete foundation topped with a rubber gasket, which forms an airtight seal. A three-person crew along with a crane operator assembles the shell of the house in one day. Some Swedish companies produce smaller wall panels that enable the do-it-yourself builder to erect a house in about three days, without a crane. Kitchens and bathrooms are supplied as complete room modules for quick installation. ■

(Reprinted with permission, High Technology magazine, November, 1986. Copyright 1986 by High Technology Publishing Corporation, 38 Commercial Wharf, Boston, MA 02110.)

FIGURE 30–4 Completed sectional house *(Courtesy of Cardinal Industries, Inc.)*

FIGURE 30–5 A completely finished motel unit being set in place. *(Courtesy of Cardinal Industries, Inc.)*

is a boxlike unit that includes several rooms. Modules are combined to make homes, motels, and office buildings, **Figure 30–5**.

A special type of module, called a **core unit**, contains most of the utilities for the finished building. By making three connections (water, sewage, and electrical), the core unit provides complete utilities for the structure. This greatly reduces the time needed for many of the subcontractors to complete the house on the site.

Manufacturing Core Units

This discussion of how core units are manufactured by a typical company shows how much manufactured construction is done. The main parts of the core are constructed at the same time in different parts of the factory. Materials are cut to size for the various assemblies. They are then stockpiled near where they will be used, **Figure 30–6**.

The floor is framed using conventional design with joists and headers. To save time and ensure greater accuracy, jigs are used to position the members. A **jig** is any simple device to position parts for assembly. Jigs for construction can be made up of blocks of wood nailed to a work surface. Carpenters nail the members with pneumatic (air-powered) nailing machines, **Figure 30–7**. The sections are mounted on wheels so they can be moved along the assembly floor. The subfloor and finished flooring are installed before the unit goes to the final assembly area.

Walls are framed in a similar manner. Most walls have 2" × 4" × or 2" × 6" studs spaced 16 inches or 24 inches on centers. Bathroom walls are framed with 2" × 6" studs to provide

FIGURE 30–6 Precut materials are stored near where they will be used. *(Courtesy of Cardinal Industries, Inc.)*

FIGURE 30–7 Carpenters use pneumatic nailers to assemble the units. *(Courtesy of Cardinal Industries, Inc.)*

FIGURE 30–8 Walls are assembled on a raised surface. *(Courtesy of Cardinal Industries, Inc.)*

FIGURE 30–9 The partially completed walls are stored on carts, then lifted into place with a crane. *(Courtesy of Cardinal Industries, Inc.)*

room for plumbing. The walls are assembled on a raised work surface so that the carpenters can work in a comfortable upright position, **Figure 30–8**. Electric hoists are used to turn framed components over. One surface is covered with wallboard or prefinished paneling. The partially completed walls are stored on rolling carts, **Figure 30–9**.

The ceiling is built in the same way as the floor and walls. The ceiling, which is a single panel, can be painted prior to installing on the core unit.

The plumbing systems are installed in much the same manner as they would be installed in a site-built house. When the structural parts of the building or module are in place, plumbers install pipes and fittings one piece at a time, **Figure 30–10**. In the early days of manufactured housing, plumbers used to prefabricate assemblies of pipes and fittings to be installed later, but most factory built homes today are designed to the customer's specifications, so it is no longer practical to presassemble plumbing systems. However, there still is an advantage in plumbing the manufactured unit in the controlled conditions of a factory.

In the final assembly area, carpenters nail the walls to the floor assembly and to one another, **Figure 30–11** Next the ceiling is nailed on. Plumbers install the plumbing tree and fixtures. Electricians install the wiring and electrical fixtures, **Figure 30–12**. Insulation is placed between the framing members of exterior surfaces. Finally, the interior is painted and trimmed according to specifications and cabinets are installed. Each core unit is hooked up and tested before it leaves the factory, **Figure 30–13**.

FIGURE 30–10 Plumbers install pipes and fittings, as in site-built construction. *(Courtesy of Copper Development Association, Inc.)*

FIGURE 30–11 Final assembly of a unit *(Courtesy of Cardinal Industries, Inc.)*

FIGURE 30–12 Assembling core units, complete with plumbing fixtures and cabinets for the kitchen and bathroom. *(Courtesy of Cardinal Industries, Inc.)*

FIGURE 30–13 Final inspection is done just before the unit leaves the factory. *(Courtesy of Cardinal Industries, Inc.)*

Transportation and Placement

The greatest restriction on manufactured construction is the size that can be transported, **Figure 30–14**. In most states, objects over 12 feet wide are not allowed on the highways. This means that sections and modules cannot be greater than this size.

FIGURE 30–14 Transporting the manufactured house to its site

When the manufactured building components arrive at the site, they are lifted from the truck to the foundation with a crane, **Figure 30–15**. The foundation has anchor bolts that fit into predrilled holes in the house frame. A sectional house can easily be installed on a foundation by a crew made up of:

▼ An operating engineer
▼ A rigger who also functions as a signaler
▼ An electrician to make the electrical hookup
▼ A plumber to make water and sewage connections

These persons may be regular employees of a manufactured housing contractor. Their skills and training are similar to those required in traditional construction.

FIGURE 30–15 Lifting a manufactured house to its foundation *(Courtesy of Cardinal Industries, Inc.)*

ACTIVITIES

Model Sectional House

Draw a floor plan for a three-bedroom house to be constructed in two or three sections. On a scale of 3/8" = 1'-0", construct a simple cardboard model of the floor and walls of that house. Do not fasten the section together. Remember that sections over 12 feet wide cannot be transported.

Working Conditions in Manufactured Housing

Make a list of all of the trades involved in the construction of a typical residence. For each of the trades listed, compare the working conditions for site-constructed and manufactured housing. List everything that helps describe the nature of the work performed by each of the trades on each type of construction. It may be easier to make the comparisons under two columns: one for site-constructed and one for manufactured construction.

Applying Construction Across the Curriculum

Communications

Visit a manufacturer of houses or housing components, write to a manufacturer, or find information in your library about the steps that a manufacturer follows to produce a final product. Make a bulletin board display comparing or contrasting the steps for producing manufactured houses with those for site-built houses.

REVIEW

A. Questions. Give a brief answer for each question.

1. What is the name for an assembly of pipes and fittings to be installed in a manufactured house?
2. What were the first building components to be prefabricated?
3. Give two advantages of manufactured construction over traditional construction.
4. What kind of construction uses factory-built panels for floors, walls, and roofs which are erected at the construction site?
5. What portion of the total cost of a typical house is spent on the shell?
6. What kind of manufactured construction is used when several identical units are put together for a motel?
7. What is the name of a module containing plumbing, wiring, and heating equipment?
8. Why are manufactured homes built in two or three sections instead of one large unit?
9. What construction must be completed at the site before a manufactured house is delivered?
10. How many connections must be made to hook up the utilities in a core unit? What are they?

CHAPTER 31

Construction in Space

OBJECTIVES

After completing this chapter, you should be able to:

▼ list the major events in the history of space exploration;

▼ explain why space exploration is important to the advancement of technology;

▼ list the basic components necessary for a manned space station; and

▼ describe the special requirements of materials for construction in space.

KEY TERMS

Sputnik

Explorer

Yuri Gagarian

Valentina Tereshkova

Alan Shepard

Mercury

Gemini

Apollo

Neil Armstrong

Skylab

space shuttle

NASA

space station

deployable

composite

neutral buoyancy tank

automated beam builder

History of Space Exploration

On October 4, 1957, the Soviet Union launched **Sputnik I**, and the space age began. Actually, it would be impossible to pinpoint the beginning of space exploration. We know, for instance, that in A.D. 1232 the Chinese used gunpowder-powered rockets. Much later, the German-made V-2 rockets were a dreaded sight over London during World War II. However, Sputnik I was the first artificial satellite to *orbit* the earth. Sputnik I was an aluminum sphere about 23 inches in diameter, weighing 184 pounds. Sputnik I stayed in orbit for 57 days, orbiting the earth every 96.2 minutes, before it reentered the earth's atmosphere. Like all of the early satellites, Sputnik I was burned up by the heat it generated—on reentry—due to friction with the earth's atmosphere.

The United States launched its first satellite, **Explorer I**, on January 31, 1958. Explorer I was a cylinder 6 inches in diameter, 80 inches long, and weighed 31 pounds. It stayed in orbit 112 days.

The first man and woman in space were Russian cosmonauts. First was **Yuri A. Gagarian**, who, in *Vostok I*, made one orbit

FIGURE 31–1 Alan Shepard was the first American in space aboard a Mercury spacecraft. *(Courtesy of NASA)*

around the earth on April 12, 1961, and then **Valentina V. Tereshkova** in *Vostok VI* on June 6, 1963.

Alan Shephard became the first American to fly in space on May 5, 1961. Navy Commander Shepard flew in a **Mercury** spacecraft, named *Freedom 7*, **Figure 31–1** Freedom 7 flew a suborbital flight, which means that it did not orbit the earth, but left the earth's atmosphere and then fell back to earth, landing in the ocean. The Freedom 7, like all early manned American space flights, ended by parachuting into the ocean. In 1962 John Glenn became the first American to orbit the earth, making three revolutions in a Mercury spacecraft.

The United States **Gemini** program was designed to develop the technology necessary to go to the moon. Between 1965 and 1966, there were ten Gemini flights, each carrying two astronauts. In December 1965, Gemini 6 and Gemini 7 rendezvoused (met) within a few feet of each other in space.

In 1967, both of the space powers suffered tragedies. A fire during a ground test killed three astronauts: Grissom, White, and Chaffee, aboard an Apollo spacecraft. Cosmonaut Komarov was killed when the parachute lines of his Soyuz spacecraft became tangled during landing.

On July 16, 1969, **Apollo 11** was launched with astronauts Edwin Aldrin, Jr., **Neil Armstrong**, and Michael Collins. On July 20 Neil Armstrong stepped onto the surface of the moon and said, "That's one small step for man, one giant leap for mankind," **Figure 31–2**.

Skylab was launched on May 25, 1973, as a space-based laboratory, **Figure 31–3**. Three flights, with crews of three American astronauts each, spent a total of 171 days conducting experiments in Skylab.

The **space shuttle** is another milestone in space exploration, **Figure 31–4**. The shuttle was designed to carry astronauts and cargo into space; remain there for a week or more; then, return to land on earth like an airplane.

FIGURE 31–2 Neil Armstrong first stepped onto the moon in 1969. *(Courtesy of NASA)*

FIGURE 31–3 Skylab was the first U.S. space station. *(Courtesy of NASA)*

FIGURE 31–4 The space shuttle is the first craft that can carry cargo into space, return to earth, and make another trip. *(Courtesy of NASA)*

The first shuttle flight was launched April 12, 1981, and lasted 54 hours. Designed specifically to carry materials to build and then supply the completed space station, the shuttle is capable of carrying construction materials, unmanned spacecraft, and fully equipped laboratories into space.

The best known tragedy and the biggest setback in the American space program came on January 28, 1986. It was in that year that the Challenger shuttle exploded shortly after launch. All seven astronauts aboard were killed, including Christa McAuliffe, who was the first teacher in space. After the Challenger accident, all shuttle launches were stopped until the cause could be determined and corrected. Shuttle launches were resumed in September 1988, almost three years after the Challenger disaster.

FIGURE 31–5 The fabric on the roof of this Bullock's department store was first developed for space suits. *(Courtesy of NASA)*

The Lure of Space

There is little doubt that the first interest in space exploration was partially due to curiosity and the excitement of the unknown. However, even our first ventures into space with Sputnik, Explorer, and Mercury had more practical goals. At first we entered space so that we could gather data about our universe. The first unmanned satellites radioed back information about radiation, density of the upper atmosphere, and measurements of meteorites.

As space exploration expanded, especially with manned flights, useful spinoffs of space technology developed. Many of the products we take for granted today were originally developed to support the various space programs. For example, much of the con-venience food we eat was perfected for the astronauts who had no way of preparing their meals. A great number of high performance metal alloys and synthetic materials were first made for the space program, **Figure 31–5**.

Today we are more aware than ever of the industrial and technical uses and potential of space. Communications satellites already allow us to communicate almost instantaneously, with just about everyone in the world. In the low gravity of a space environment nearly pure crystals, which improve the performance of our electronic devices, **Figure 31–6**, are grown. Unobstructed by the screening effects of the atmosphere, we can collect great amounts of solar energy. It might some day be used on earth. Right now, though, only spacecraft and satellites are using it. It will probably be possible to produce medicines and other pharmaceutical products in space that would be impossible to develop on earth. The possibilities are only limited by one's imagination.

FIGURE 31–6 These very pure crystals were grown aboard Skylab. Crystals such as these are used in microelectronics. *(Courtesy of NASA)*

Manned Space Stations

In his State of the Union message of January 25, 1984, President Reagan announced a new plan for America's space program. He directed **NASA** (The National Aeronautics and Space Administration) to develop a permanently manned space station and to do it within a decade. The first stage of the U.S. Space Station *Freedom* is scheduled to be assembled in space in 1995.

With a comparatively small investment (about one-ninth the investment in Project Apollo and one-third the investment in the space shuttle), the space station offers great possibilities. A **space station** will not only increase our knowledge of our planet and the universe we live in, it will fulfill two other important goals: 1) it will stimulate more spinoffs of technology, and 2) it will help us develop industrial and biological technologies.

What Is a Space Station?

The space station, as envisioned by NASA, will be a permanent, multipurpose facility in orbit, **Figure 31–7**. It will serve as:

▼ a laboratory to conduct basic research;

▼ an observatory to look down at the earth or peer out into the sky;

▼ a garage to fix and service other spacecraft;

▼ a manufacturing plant to make exotic metal alloys, super-pure pharmaceuticals, or perfect crystals;

▼ an assembly plant to build structures too large to fit in the shuttle's cargo bay; and

▼ a storage warehouse to keep spare parts or even entire replacement satellites.

The space station concept provides for both manned and unmanned elements. The manned facility, as well as an unmanned free-flying platform, will be placed in a low earth orbit of about 250 miles (400 kilometers).

Space Station *Freedom*

Many designs were proposed and rejected before an extensive design-concept period resulted in the current plan for the space station. Engineers identified and evaluated alternative systems and first developed a configuration called the "power tower." Next came the "dual keel" design.

FIGURE 31–7 Proposed manned space station. *(Courtesy of NASA)*

The final design for Space Station *Freedom* is based on a single truss, **Figure 31–8**. The following are some facts about Space Station *Freedom*:

▼ Space Station *Freedom* will be assembled in space, **Figure 31–9**. The shuttle will carry all of the parts inside its large cargo bay.

▼ It will take 17 Shuttle flights over a period of four years to build Space Station *Freedom*. The first parts of *Freedom* will be launched in late 1995.

▼ It will take about one year to get Space Station *Freedom* ready for people to come aboard.

▼ The full length of Space Station *Freedom* will be 108 meters (353 feet). That is the length of a football field, including the end zones.

▼ Space Station *Freedom* will weigh 281,339 kilograms (296 tons), about as much as a fully loaded jumbo 747 airplane.

▼ Space Station *Freedom* will be in orbit about 400 kilometers (250 miles) above Earth (about the distance between New York City, and Washington, DC).

▼ Space Station *Freedom* will orbit the earth about every 90 minutes at a speed of approximately 18,000 miles per hour. At that speed you could fly from New York City to Los Angeles in 9 minutes.

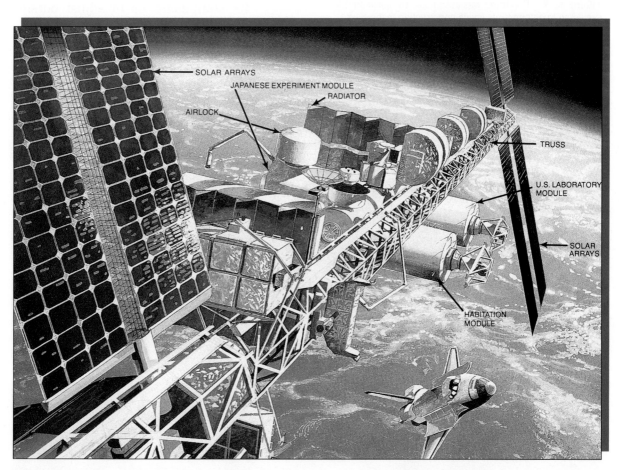

FIGURE 31–8 The design for Space Station Freedom has all major components mounted on a long truss.

FIGURE 31–9 This tower was assembled by space shuttle crewmen to demonstrate space construction techniques. *(Courtesy of NASA)*

▼ The truss of Space Station *Freedom* is made of aluminum.

▼ Two of the modules at the middle of *Freedom* were built by the United States: the Laboratory Module, **Figure 31–10** where the crew will work and perform experiments; and the Habitation Module, where the crew will eat, sleep, exercise, or relax after work. They are each 27.4 feet ong and 14.5 feet in diameter.

▼ Other countries are building parts for Space Station *Freedom*:

— Two more laboratories are being built by Japan and Europe.

— Canada is building a Mobile Servicing System, a robot on a movable platform, that will be used to build *Freedom*, to fix equipment, or to adjust outside experiments.

FIGURE 31–10 The laboratory module of Freedom

- Space Station *Freedom* will be powered by solar arrays attached to each end of the truss. They will rotate to catch all of the sun's rays and will provide up to 56,250 watts of power. That is as much power as five houses would need to use all of the owners' appliances (microwaves, refrigerators, air conditioners, hair dryers, etc.) at the same time.
- Four astronauts will live on Space Station *Freedom*. Most of the crew will be replaced by new crew arriving on Shuttle flights every three months. Some will stay on board for six months, or maybe a year, so that we can study the long-term effects of microgravity on people. This will help prepare us for future missions to the moon and Mars.
- There will only be one millionth the amount of gravity on Space Station *Freedom* as there is on Earth. The astronauts will float around inside *Freedom*.
- The kitchen on Space Station *Freedom* will be equipped with microwave and convection ovens. The astronauts will eat frozen, dehydrated, and vacuum-packed foods.
- The microgravity environment of space can cause medical problems for the astronauts. Besides causing motion sickness, which usually goes away after a couple of days, weightlessness causes muscles to weaken and bones to become brittle. In the Crew Health Care Facility, the astronauts will monitor their health, exercise to keep fit, and get treated if they are sick or injured.
- The astronauts will work 10 hours per day, 6 days a week on Space Station *Freedom*. They will sleep for 8 hours each night in sleeping bags that are held in place to keep the astronauts from floating around.
- An environment control and life support system (ECLSS) will provide the crew with a breathable atmosphere, supply water for drinking, bathing, and food preparation, remove contaminants from the air, and process biological wastes. The ECLSS system will be a closed one. This will permit oxygen to be recovered from the carbon dioxide expelled by the crew, and allow waste water to be recycled and reused. Only food and nitrogen will have to be periodically resupplied.

International Participation in the Space Station

In keeping with a long-term policy of international cooperation in space, President Reagan invited United States friends and allies to participate in the development of the space station. NASA has therefore signed agreements with Canada, Japan, and the European Space Agency that provide a framework for cooperation during the definition and preliminary design activities.

Canada is performing preliminary design on a mobile servicing center (a multipurpose structure equipped with manipulator arms that will be used to help assemble and maintain the space station). When the center is complete, it will help in the upkeep of instruments and experiments mounted on the station's framework.

Japan is conducting preliminary designs on an attached multipurpose research and development laboratory that will provide a shirtsleeve environment work space for station work crews. The Japanese experiment module will also include an exposed work deck, a scientific/equipment airlock, a local remote manipulator arm, and an experiment logistics module.

The European Space Agency (ESA) is doing preliminary design work on a permanently attached, pressurized laboratory module and a polar-orbiting platform.

Large Antennas

Now that we have a space shuttle to move large and bulky cargoes routinely into earth orbit, long-term planners and researchers in government, industry, and universities are shaping the work for a new era—the building of very large and complicated structures in space.

The shuttle orbiter's closed cargo bay carries up to 32 tons (29,500 kilograms, or three times the weight of a passenger bus) on each trip into space, **Figure 31–11** Freight costs are charged by both weight and length, so it will be wise to design space hardware that is both light and capable of being transported in short lengths. These large space-destined structures, that are too fragile to stand up under their own weight on earth, will now fold up in the cargo bay. When they reach their destination, they will deploy safely into their final shapes in the weightlessness of space.

Thoughts of outer space create exciting new possibilities for the engineering of space hardware, and pose a brand new set of challenges. What are the strongest, lightest, and most stable materials to use in space construction? How do you load the shuttle so as to build these colossal objects with the fewest trips into space? What are the best ways to assemble them once the materials are delivered to the orbiting sites? And the most obvious question: What kinds of structures will we build?

Deployable Antennas

The first large space antennas will be **deployables**. This means that the entire antenna will fold into a compact container on earth, go up on a shuttle trip, then open automatically in space in a single operation. The key, obviously, is to have the largest possible dish unfolding from the smallest and lightest possible package.

One type, the hoop-column or "maypole" antenna, would open up in orbit much as an umbrella does, **Figure 31–12** A cylinder no bigger than a school bus could be transformed within an hour into a gigantic antenna dish of two acres (100 meters across).

Depending on the length of the various strings that stretch the fabric taut inside its stiff outer hoop, this type of antenna can be designed for many shape variations. The bowl of the dish could be made flat, more hollowed out, or have four different surfaces, each focusing a beam in a different direction. Multibeam feeds could also allow one antenna to do the work of several by pointing signals toward different areas of the Earth's surface.

FIGURE 31–11 The shuttle's cargo bay. *(Photo by Dick Luria; courtesy Lockheed Corp.)*

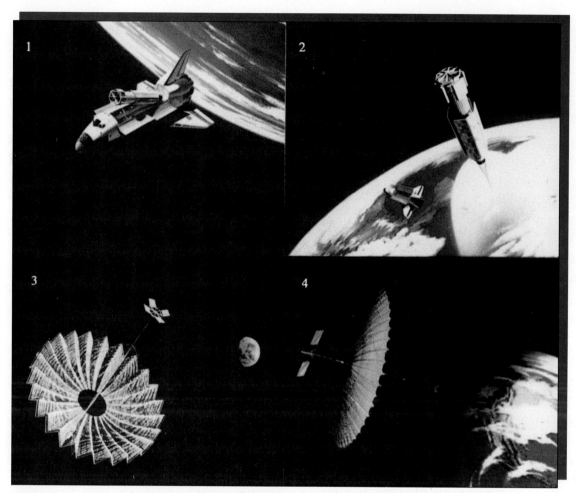

FIGURE 31–12 Deployment sequence for a 100-meter diameter hoop-column antenna. *(Courtesy of NASA)*

Materials

Whatever their shape, large space structures will put great demands on the materials from which they are made. Even though they'll be free from the weight stresses imposed by earth's gravity, there will be other strains from their tight packaging and from the hostile radiation and frigid temperatures of space.

Engineers will need to build with new materials for a new age. Materials will need to be—at the same time—light, super-strong, flexible or rigid (depending on the use), and thermally stable. For example, telescoping masts must be light, yet stay very stiff. Antenna

ribs need to be strong, but should be flexible enough to wrap around their hub.

And everything needs to remain fixed in position, equally well in the hot sun as in cold shadow. If a structure were to expand with heat, for example, it would ruin the precise shape of an antenna (some of which can be off no more than a few millimeters in a total diameter of 100 meters).

One substance that meets these hard demands quite well is the graphite-epoxy **composite** (a combination material) now used in lightweight tennis rackets, golf clubs, airplane parts, and in the space shuttle itself, **Figure 31–13** A three-meter-long hollow tube of this material can be lifted with one finger,

FIGURE 31–13 Engineers are experimenting with assembling possible structures for space use out of lightweight graphite-epoxy composite cones. *(Courtesy of NASA)*

yet for its purpose it is ten times stronger than steel.

There are other materials that will suit specific jobs. The hundreds of threads that pull and stretch a hoop-column antenna into shape might be made of a quartz filament, because quartz is very stable. The dishes themselves should be made of fabrics that fold like cloth before they are deployed. These would be metal meshes woven like nylon stockings or soft porch screening and coated with gold for reflectivity. A finer mesh will be used for dishes that deal in smaller wavelengths. For very small wavelengths there are ultra-thin membranes made of transparent films coated with metals that look and feel like sheets of Christmas tinsel.

Platforms: Outposts in the Sky

Suppose that five different groups want to fly scientific or remote-sensing instruments in earth orbit, all at about the same height. Instead of cluttering the sky with five individual satellites, why not build a huge platform to which all five can be attached? They could share the cost of power and communications systems, stability control, and cooling devices. Shuttle astronauts would need to make just one repair stop at a time instead of five, and could replace any of the original five devices with a new one as needed.

Multipurpose space platforms are now being designed. The possible uses for these platforms are almost as varied as their sizes and shapes. Any of the following equipment might be "plugged in" to a typical one: an astronomical telescope, a communications dish, or a sensor to trace air pollution or search for minerals.

Platforms will not be just simple rectangular slabs. There are birdlike configurations with cross arms to hold sensitive instruments apart from one another. There are modules, like large rafts, that would support several instrument pallets in a cluster. The only common element will be the central "bus" that houses the platform's power generator (attached to winglike solar collectors), and thermal and electronic systems.

What kinds of building blocks will we use on these floating construction sites? Ideally they should be basic, simple, and adaptable to many different kinds of structures. For example, twenty deployable boxes could be latched together to form a simple flat platform. Or, several of the raftlike modules could be snapped together to form a still larger surface.

All of these ideas have their roots in common household objects—in collapsible cardboard boxes, folding deck chairs, telescoping car radio antennas, accordion baby gates—in short, anything we have tried to make smaller and more portable. Masts for dish antennas will telescope into their full lengths from small cylinders. Latticed trusses will store as flat packages, unfold first into diamond shapes, then finally into tetrahedrons.

In each case, though, no matter how flexible their hinges when stored, the modules must be stiff when deployed, as would the hexagonal pieces for large antennas. Looking a bit like mini-trampolines when unfolded, these hexagons will be attached precisely and rigidly to form great reflecting surfaces many city blocks in area.

Not all of these potential building blocks will need to unfold, however. Some of them will store quite easily just as they are, like the light graphite-epoxy tubes that will stack inside one another and sit on racks in the cargo bay like arrows in a quiver. These tubes would then be attached to form struts—struts that can themselves be joined to build larger beams or trusses. Or they might be used to form a thin hoop for a space antenna. Highly adaptable to many creative shapes, these struts will be like giant tinkertoys for the practical construction of projects in space.

Eventually, no matter how cleverly the platforms and antennas are packed, they will be too large to unfold in a single deployable unit. At that point we will have to send up "erectables" in separate pieces, **Figure 31–14** Two such pieces (or a dozen, or even a hundred) can be loaded into the cargo bay on earth, lifted into space, unfolded, and finally assembled into a single gigantic structure in orbit.

Building in Orbit

Deployable antennas will, in a sense, build themselves. They will unfold with the push of a button. Erectables will not, so someone or something will have to put the separate pieces together.

Ongoing assembly projects, some of which will require several shuttle trips to deliver all

FIGURE 31–14 Segments of a modular structure are removed from the cargo bay of the shuttle by remote manipulator arms and are unfolded before being attached to the main structure. *(Courtesy of NASA)*

the pieces, will therefore mean having the first construction sites in space. And so a new type of work for the human race will be devised.

At various NASA Centers, researchers are now determining the most efficient ways to do the deployment and assembly jobs of the shuttle era. The testing is done underwater in a huge neutral buoyancy tank that simulates, as nearly as possible, the weightless conditions of space, **Figure 31–15** Here, technicians practice the

FIGURE 31–15 The neutral buoyancy tank allows divers to work in space suits underwater to simulate the weightless condition of Earth orbit. *(Courtesy of NASA)*

mechanical tasks necessary for the construction of these huge space structures.

Many factors are taken into account: safety and fatigue of the astronauts; speed in moving from one place to another; the requirement for simple tools and the need to restrain them so they don't float away; how much time the space shuttle loses lingering around the site when it could be returning to earth for another load.

In one method of assembly, astronauts tethered to the shuttle would simply move from beam to column to module, manually snapping, locking, or latching everything together. Their travel time could be shortened by wearing the jetpacks called manned maneuvering units (MMUs), models of which are used in neutral buoyancy tank simulations.

It is not yet certain, however, how manpower, machine operations, deployment, or assembly jobs to build these antennas and platforms will be combined. For some projects it might be more efficient to move astronauts around on a scaffold in a mobile work station instead of having them use the MMUs. The scaffold rests on a frame in the cargo bay and moves either up and down or right and left. As sections of the structure are finished they are moved away from the station so that the next part to be built is always in reach.

Astronauts might also stand in open cherry-pickers attached to the shuttle's 15-meter remote manipulator arm, and be moved from beam joint to beam joint like telephone linemen working on high wires. Even more sophisticated would be the closed cherry-pickers where workers inside a comfortable chamber would work with remote control arms.

Or, for repetitious or dangerous tasks, unmanned free-flying teleoperators—essentially programmed robots—could do the work with their mechanical arms. There might also be assembler devices to form three-dimensional structures from struts by following simple repeatable steps. Maneuverable

television (MTV) units would transmit pictures to technicians in the shuttle control room so that they could direct work by remote control.

These devices will most likely be used later in the shuttle era. In the meantime, astronauts will have to learn to erect structures the size of large stadiums in the peculiar world of low gravity. Seemingly easy tasks will become complicated—workers trying to turn ordinary bolts will be as likely to turn themselves as the bolts, thanks to the lack of leverage that comes with weightlessness.

These are precisely the problems studied during simulated assembly jobs in the neutral buoyancy tank. Those problems, in turn, influence the choice of technology, like using latches that snap firmly together with one quick motion instead of a series of twists and turns. The goal is to standardize hardware and assembly methods in order to get the jobs done as quickly and correctly as possible.

The Automated Beam Builder

After the deployables and after the erectables, the next logical step is to build large structures completely from scratch by

FIGURE 31–16 An automated beam builder fabricating triangular-shaped trussed beams from the cargo bay of the shuttle. *(Courtesy of Grumman Corp.)*

fabricating the building blocks in space. A machine for that very purpose has already been designed. Called the **automated beam builder**, it will sit at one end of the shuttle's cargo bay. Spools of ultra-light material, probably graphite-epoxy or metal matrix composites, would be loaded into the machine on earth and carried into orbit. Once at the space construction site the beam builder would heat, shape, and weld the material into 3-foot wide triangular beams that could be cut to any length, then latched together to build large structures, **Figure 31–16** By loading the cargo bay with extra spools, enough material could be carried in one trip to build thousands of yards of beams.

Now, with the beam builder, we will advance from the dreams of science fiction to practical blueprints for colossal structures that will dwarf the space shuttle flying around them. As the platforms grow in size, they will carry more science instruments and will grow even more gangly with their cross arms, dishes, and winglike solar panels, **Figure 31–17.**

FIGURE 31–17 Using space shuttles and advanced heavy lift launch vehicles to transport materials, equipment, and crew, kilometer-size structures may someday be erected. *(Courtesy of Boeing Aerospace Company)*

ACTIVITIES

Model Deployable Structure

One way of building large structures in space is to design them so that they can be built on earth, carried in the shuttle cargo bay, then unfolded (deployed) in space. In this activity you will design a structure that can be folded and unfolded. You are to choose the best materials for the job and decide on the procedures to use, so the Equipment and Materials list and the Procedure steps given here are only suggestions.

Equipment and Materials

Drinking straws
Strips of balsa wood
Stiff wire
String
Umbrella frame
Cardboard
Styrofoam
Glue
Wire brads
Pins

Procedure

1. Decide on the type of structure to be built such as a round antenna dish, a trussed beam, or a flat solar collector.
2. Draw several sketches of the structure.
3. Study each of your sketches to determine the major construction elements. Consider what materials could be used to make the model, where and how parts will be joined, how hinged parts will be hinged, etc.
4. Decide what scale you will use for the model, then put dimensions on your sketch. It may help to draw subassemblies or individual parts.
5. Cut and assemble the structure ready for loading into the shuttle. Check any moving parts to make sure they work easily.
6. Demonstrate to your teacher or your class how your structure will be deployed. Be prepared to discuss problems such as the effect of low gravity, packing the shuttle, and deploying the structure.

Applying Construction Across the Curriculum

Social Studies

Make a poster showing a timeline of the exploration of space.

Science

Explain how it is possible for a satellite, without any propulsion system of its own to remain in orbit around the Earth. Also explain the feeling of weightlessness aboard a space craft that is still within force of Earth's gravity.

REVIEW

A. **Matching.** Match the event in Column II with the person or space vehicle in Column I.

Column I

1. Yuri Gagarian
2. Sputnik I
3. Gemini 7
4. Explorer I
5. Alan Shepard
6. Roger Chaffee
7. Neil Armstrong
8. Mercury
9. Space shuttle
10. Skylab

Column II

a. First American space flight
b. First man in space
c. Shepard's space craft
d. Americans spent 171 days here
e. First man on the moon
f. First orbital space flight
g. Rendezvoused with another spacecraft
h. Killed in an Apollo ground test
i. First launched in 1981
j. First American in space

B. **Questions.** Give a brief answer for each of the following questions.

1. In what year was the first artificial satellite launched?
2. What can the space shuttle do that earlier space craft could not?
3. Name a spinoff product that was developed for space exploration and is now used on earth.
4. Name one reason why a manned space station would be useful.
5. List three utilities that must be provided by the systems in a manned space station.
6. Where will a manned space station most likely get electrical power?
7. How will a structure as large as the dual-keel space station be transported into space?
8. What are two characteristics that are important for materials to be used for construction in space?
9. What composite (combination of two materials) is being considered for rigid parts of space structures?
10. What is the name of the machine that automatically builds 3-foot wide beams of raw materials?

GLOSSARY

Abutment—The concrete structure against an embankment and upon which the end of a bridge is supported.

Aggregate—The solid particles that are bound together by the cement in concrete. Typical portland cement concrete contains sand as a fine aggregate and crushed stone as a coarse aggregate.

Air-entraining cement—A form of portland cement that traps tiny air bubbles as it cures. The entrained air acts as a cushion to help the concrete resist the forces of freezing and thawing.

Aluminum—A nonferrous (containing no iron) metal that is lightweight, light in color, and does not rust. Aluminum is widely used for architectural trim, roofing, siding, and the frames of windows and doors.

Ampere—The unit of measure of electrical current.

Apprenticeship—A training program in which the trainee, an apprentice, spends some time in class and some time working under the supervision and guidance of a journeyman.

Arbitrator—An outside expert in labor relations who is called upon to settle a labor dispute. Arbitration is usually the last step in a grievance procedure.

Architect's scale—A measuring device for making drawings that are dimensioned in feet, inches, and fractions of inches. Architect's scales are used for building construction drawings, but are not as widely used for civil construction drawings.

Armored cable—Electrical cable with metal sheathing.

Batter board—An arrangement of stakes and horizontal boards used to mark building lines. Batter boards are erected before any excavation for a foundation. They can be removed once the foundation is completed.

Beam—A horizontal structural member. Beams are heavier and not as closely spaced as joists.

Bearing wall—A structural wall or a wall that is intended to support weight from above.

Bid—A statement from a construction company that they will perform a job within a stated time period for a specified cost. A bid differs from a contract in that the bid does not contain all of the details of the contract. The bid is only signed by the company submitting the bid, not the one who wants the work done as in a contract.

British thermal unit (BTU)—A unit of measure of heat energy.

Building code—A law that controls the design and construction of buildings for the public good. Building codes cover such things as the design and installation of plumbing and electrical systems, the width and steepness of stairs, fire exits, and the number of people allowed in a public hall at one time.

Casing—The trim around an opening in a wall. Windows and doors are the most commonly trimmed openings.

Cathode ray tube (CRT)—A computer display device that looks like a television picture tube.

Circuit—A complete path for electric current from the source, through the load, and back to the original source.

Civil construction—Construction involving a large amount of earthwork, such as for roads, bridges, and airports.

Civil engineer's scale—A measuring device for making drawings that are dimensioned in feet and decimal parts of feet. Civil engineer's scales are used for civil construction projects, but not usually for building construction.

Clinch—To bend over the protruding point of a nail. Clinching is sometimes done on rough work for maximum holding power. Clinching is never done on finished work.

Cofferdam—A temporary barrier erected on a construction site to keep water out or to retain a high soil embankment.

Collar beam—A framing member fastened at mid height between a pair of rafters to prevent them from spreading apart. Collar beams are normally used on every second or third pair of rafters.

Collective bargaining—The procedure by which organized labor unions and managers negotiate employment contracts for all personnel in the bargaining unit (the union).

Commercial construction—Building construction for such business purposes as offices, stores, and hotels.

Common rafter—A rafter that extends from the ridge to the wall of the building.

Compressive strength—The ability of a material to resist being crushed.

Compressor—The part of a refrigeration system that raises the pressure and temperature of the refrigerant after it leaves the evaporator.

Computer-aided design (CAD)—The use of computer systems and special CAD software to design a project and provide the working drawings needed to produce the project as designed.

Concrete—A rock-like material formed by bonding aggregates of sand and stone together with portland cement.

Conduit—Pipe or tubing through which electrical cable is run.

Contour lines—The lines that show elevation and describe the contour of the land on a topographic drawing or site plan.

Contract—A binding agreement that specifies what each party promises to give to or do for the other. Construction contracts usually refer to specifications and drawings as contract documents.

Convection—The flow of heat from a warm mass to a cooler mass.

Cope—A shaped cut made on the end of a piece of molding so that it will fit around another piece to form a 90° angle or square corner.

Cornice—The construction with trim where the walls of a building meet the roof.

Corporation—The form of business ownership in which the owners hold shares of stock. A corporation is run by the board of directors and corporate officers, not the stockholders.

Cost-plus contract—A form of contract in which the payment is to be based on the actual cost of materials plus some percentage for labor and overhead. Because they do not control the total cost of the contract, cost-plus contracts are not widely used.

Course—One horizontal row of building material that is used in horizontal rows. Materials that are used in courses are bricks, concrete blocks, and shingles.

Crossuct—To saw a piece of lumber across the grain. Crosscutting is done with a crosscut saw.

CSI format—The Construction Specifications Institute's standard style for organizing specifications for building construction. The CSI format is widely used for commercial construction.

Curtain wall—A nonstructural wall panel, used only to close in the space between the structural members.

Cut—Earth that must be removed to create the desired finished grade.

Dead load—The load placed on a building by the weight of the materials used to construct the buildings.

Deck—A slang term used for a large horizontal area, such as a roof deck, bridge deck, or a floor.

Design temperature difference—The number of degrees difference between the expected outdoor temperature and the desired indoor temperature used for designing heating and air conditioning systems.

Detail drawing—A working drawing that shows more detail about construction than a standard plan or elevation view. Details are usually drawn at a larger scale than other working drawings.

Digitizer—A computer peripheral used with CAD systems. A digitizer is a flat surface to which drawings can be attached. Then, prints on the drawing can be digitized (entered into the computer) with a puck or stylus.

Direct cost—The costs involved in a project that are directly related to that project. Direct costs include materials, labor hired specifically for the project, and rental of special equipment.

DWV plumbing—Any plumbing in a system used for drainage, waste discharge, or venting the system to the atmosphere.

Effluent—Water and wastes discharged from a plumbing fixture.

Elevation—A working drawing that shows height. A set of working drawings for a building includes elevations of the building exterior and several detail elevations of the interior and the exterior.

Environmental impact statement—A report on the effect that a construction project will have on the environment. Environmental impact statements are usually required before major construction projects are allowed.

Evaporator—The part of a cooling system where heat is absorbed from the air into the refrigerant. The heat absorbed by the refrigerant causes it to boil, even though it is at a very low temperature.

Fascia—The vertical trim board that covers the ends of roof rafters. The fascia is part of the cornice.

Ferrous metal—Any metal that contains iron as its main ingredient. Ferrous metals are the various types of iron and steel.

Field notebook—A specially designed notebook in which civil engineers and land surveyors record data they collect in the field.

Fill—Earth that must be added to create the desired finished grade.

Finished grade—The contour of a site after all earthwork, including the spreading of topsoil, is finished.

Fixed-sum contract—A contract in which the payment is a specific amount of money.

Foamed plastics—Any plastics material that is formed by entrapping large amounts of air. Foamed plastics are excellent thermal insulators.

Footing—The bottom part of a foundation. The footing is the part that transmits all of the building loads to the soil beneath.

Foundation—The structural parts of a construction project that mainly serve to support the desired, useful structure or superstructure. The foundation may not be a necessary part of the design if it were not needed for support.

Frost line—The maximum depth to which the earth can be expected to freeze. In the southern most parts of the United States, there is no frost line. In the northern states and southern Canada, the frost line can be as deep as five feet.

Gable roof—A two-sided roof with a single pitch from two sides toward the ridge.

Gain—A depression, usually rectangular, cut out of a surface to accommodate something that is to be attached in that depression. This term is most often used to refer to the depression in which a door hinge is mounted.

Gambrel roof—A roof with steep pitch near the sidewalls and more gradual pitch near the ridge. A gambrel roof only slopes on two sides.

General partnership—The form of business ownership in which two or more persons share all investment, profit, loss, and management responsibilities.

Grade beam—A horizontal structural member at ground level.

Greenhouse effect—The effect of glass to trap heat from solar energy. When the sun's rays pass through glass and are reflected back to the glass from a surface, the reflected rays do not pass through the glass as easily.

Grievance—A formal complaint about something an employee feels is a violation of the union contract.

Gypsum—A rock-like mineral commonly used as the core of wallboard.

Hardboard—A forest product made by softening the natural adhesive that holds wood fibers together, then pressing the fibers back together in sheets and curing the natural adhesive with heat.

Hardwood—The wood from a deciduous tree. Some hardwoods are actually softer in texture than some softwoods. Hardwood is widely used for cabinets, furniture, and architectural trim.

Header—A short framing member across the end of other framing members. Headers are used to form rough openings.

Hip roof—A roof that slopes from all four sides at the same pitch toward the center.

Industrial construction—Construction of factories, power generation plants, and other uses involving heavy stationary machinery for industrial purposes. Much industrial construction includes some building construction and some civil construction.

Jack—A framing member that is less than full length because it intersects with a header, hip rafter, or valley rafter.

Joist—Horizontal framing members that make up most of the framing for a conventional frame floor or ceiling.

Journeyman—An experienced craftsperson. A journeyman has completed all of the training requirements for the job and is prepared to perform all aspects of that trade.

Kerf—The cut made by a saw.

Lift—One layer of soil in an embankment or other soil structure. Each lift is compacted before the next is placed.

Light construction—Building construction that involves the types of materials and construction practices normally found in the construction of a single-family home.

Limited partnership—The form of business ownership in which one or more of the partners is only responsible for part of the financial investment. Limited partners do not share in the management of the company and their financial liability cannot be more than their investment.

Live load—The variable load placed on a building, for example by people, furniture, and snow.

Mansard roof—A roof with very steep pitch at the outside and more gradual slope near the center. A mansard roof slopes on all four sides.

Miter—A 45° angle cut on the ends of two parts that are to be joined forming a 90° angle, or square corner. The joint formed by such cutting and joining is called a miter joint.

Mortar—A mixture of portland cement and lime (or masonry cement) and sand used for construction with bricks, concrete blocks, or stone.

Mouse—A hand-held computer peripheral that controls the location of a screen cursor and can be used to input data.

Natural grade—The contour of a site before construction begins.

Nonmetallic sheathed cable—Electrical cable which is held together by a plastic coating or sheathing.

Orthographic projection—A method of drawing three-dimensional objects on two-dimensional paper by showing views from separate directions as if they were projected onto a surface.

Overburden—The unusable soil layer above the structural grade soil or bedrock on a construction site. Overburden is generally removed before construction begins.

Overcurrent protection device—Circuit breaker or fuse that stops all current flow if there is too much current in the circuit.

Overhead—The costs of running a business that cannot be directly related to a particular project. Overhead includes office space, office staff, insurance, taxes, and personnel who are involved in more than one project at a time.

Parging—A coat of sand and cement that is plastered on the surface of masonry walls to prepare the wall for dampproofing.

Particleboard—A forest product made by gluing wood particles (very coarse sawdust or very fine wood chips) together into a sheet. Particleboard sheets are commonly used as underlayment on floors to be carpeted.

Percolation—The ability of soil to accept surface water.

PERT—Program Evaluation Review Technique, a system for plotting the time involved in each activity required for a project. PERT is used by managers to schedule work and delivery of materials, and to predict problem areas.

Pier—A vertical support made of concrete or masonry.

Pigment—The contents in paint that give the paint its color. Generally, the more pigment in paint, the better quality the paint is.

Pilaster—A concrete or masonry pier that is built as an integral part of a wall. Pilasters help a wall resist lateral (sideways) forces.

Pile—A wood or steel member driven into the earth as part of a foundation.

Pitch—The steepness of sloping construction, such as stairs or roofs. Pitch is measured as the ratio of the rise of a roof to its span (*fractional pitch*) or the number of units of rise in 12 units of run (*unit pitch*).

Plain sawing—Boards that are sawed from the log in such a way that the face of the boards shows the face of the wood grain are plain sawed. Sometimes this is called flat sawing.

Plan—Any construction drawing that shows the project as viewed from above. Plans do not show height. Typical plan views are site plans, foundation plans, floor plans, and roof or floor framing plans.

Plate—The horizontal framing members at the top and bottom of a wall frame.

Plot plan—A plan view of a building site showing where buildings are to be placed, the boundaries of the site, roads, drives, walkways, and other important information that would not be included on the building drawings.

Plotter—A computer output device used with a CAD system. The plotter holds one or several pens and a drawing sheet. Either the pen(s) or the paper is moved by the CAD program to produce a drawn copy of the data stored in the system.

Plywood—A forest product sheet material. Plywood is made by gluing an odd number of layers of wood veneer together. The most common sheet size is 4 feet by 8 feet, but other sizes are possible.

Portland cement—A powdery material made from limestone. Portland cement is mixed with water to form the bonding agent that holds the aggregates (sand and stone) together in concrete.

Proctor soil density—A system of measuring and recording the density of soil after compaction. It is named after Professor Proctor, who developed the system.

Quantity take-off—The method of construction estimating in which the estimator calculates the quantity of each material needed. Labor and related costs are usually based on the quantity of material to be used.

Quarter sawing—Boards that are sawed from the log in such a way that the saw cut is perpendicular to the growth rings of the tree.

Radiation—The movement of heat by heat rays. Radiation can take place in a vacuum.

Rail—The horizontal frame member in frame-and-panel construction. The horizontal parts in doors and window sashes are called rails.

Rake—The sloping cornice at the end of a pitched roof.

Reinforced plastics—Any of a number of materials made by reinforcing plastics with strands of another material which has very high tensile strength. The most widely used reinforcing material is fiberglass.

Ridge board—A single board forming the ridge of a roof and against which the tops of rafters rest.

Right of eminent domain—The right of a government to require a landowner to sell property to the government for public benefit.

Rip—To saw a piece of lumber in the direction of the grain or lengthwise. Ripping is done with a rip saw.

Rise—The vertical dimension of sloping construction, like stairs or roofs.

Run—The horizontal dimension of sloping construction, like stairs or roofs.

Running bond—The most common arrangement of masonry units. The joints between the units in one course are in line with the centers of the units in the courses immediately above and below.

R value—A measure of the ability of a building material to resist the flow of heat. The higher a material's R value, the better heat insulator the material is.

Sand cone—The device used for measuring the amount of dry sand placed in a test hole during a soil compaction test. This method of testing soil density, known as the *sand-cone method*, is often used to do a Proctor analysis.

Scale—The ratio of reduction or enlargement used on a drawing. Most construction projects are too large to be drawn full size, so they are drawn to scale.

Screed—When the concrete is placed for flatwork, like slabs and roadways, it is leveled with the top of the forms using a straightedge called a screed. A screed may be a specially made steel straightedge or a straight, strong piece of lumber.

Section view—A working drawing that shows a cross section of a part of a project. Most sets of working drawings include one or two section views of an entire building and several detail section views.

Set (nailing)—When the heads of finishing nails are to be concealed, they are driven slightly into the surface of the wood using a nail set. This is called setting the nail head.

Set (saws)—To prevent a saw blade from binding in the kerf, the teeth are bent outward slightly. The bend in the teeth is called the set. Setting a saw causes the kerf to be slightly wider than the blade is thick.

Setback—The distance from the edge of a street or road to the front of a building. Most zoning laws specify the minimum allowable setback.

Shear wall—A wall designed for rigidity and intended to help the building resist wind loads and other horizontal forces. Shear walls are widely used in large buildings and those constructed in earthquake-prone areas.

Shed roof—A sloping roof with no ridge. A shed roof has one continuous slope across the entire span.

Shim—A thin piece of material used to make an adjustment in the *fit* of mating parts.

Sill—The major structural member upon which a building frame rests. The sill is the framing member in contact with the foundation.

Skeleton frame—A style of building construction in which the major structural parts are framing members.

Slump—The amount that freshly mixed concrete sags or runs as a fluid when formed in a cone. A slump test is done to measure the stiffness of fresh concrete.

Soffit—The covering on an overhead horizontal area. The most common applications of soffits are on overhanging roof cornices, above bath tub enclosures, and over kitchen cabinets.

Soil sieve—A small container with an accurately sized sieve mesh bottom. Soil sieves are used to determine the size of soil particles.

Solar collector—A device that collects the sun's energy for use in heating water or air.

Solar orientation—The positioning of a building on its site in such a way as to take advantage of the sun's energy and shading provided by other structures and trees.

Sole proprietorship—The form of business ownership in which one person owns and manages a business.

Span—The distance between vertical supports beneath a horizontal or sloping element, like a roof, beam, or joist.

Specifications—Written requirements for the grade of materials, quality of workmanship, and other information about a construction project which cannot easily be shown on the working drawings.

Square (measure)—The amount of roofing, siding, or flooring material needed to cover 100 square feet.

Steel—An alloy (mixture) of iron and other elements. Other than stainless steel, steels will rust and are usually grayer in color than aluminum.

Stile—The vertical parts in frame-and-panel construction. The vertical parts in doors and window sashes are called stiles.

Stucco—A portland-cement based exterior wall covering system. Stucco cement is applied in two or three coats over wire lath.

Subcontractor—An individual or company who contracts for part of a construction project. Parts of the prime work may be let out in *subcontracts*.

Subfloor—A layer of rough flooring, usually plywood, applied to the floor framing over which a finished floor will be applied.

Superstructure—All of the structural parts of a constructed object above the foundation. The superstructure is the desired or useful part of the object which, although it may support other parts, would still be a necessary part of the design.

Tack rag—A rag made slightly sticky with a mixture of varnish and turpentine. Used to remove dust from objects to be varnished.

Technology—The use of available resources, including tools, materials, personnel, time, money, and knowledge, to alter our environment or extend our capabilities.

Tensile strength—The ability of a material to resist being pulled apart.

Thermoplastics—A synthetic material that softens when heated and rehardens when cooled. Thermoplastics can be reshaped by applying heat. Examples of thermoplastics used in construction are PVC pipes, acrylic for windows, and polyethylene film used for vapor barriers.

Thermosetting plastics—Plastics which require heat for a chemical reaction to result in their final form. Thermosetting plastics cannot be softened by applying heat after the chemical reaction has taken place. Examples of thermosetting plastics used in construction are melamine used for counter tops and foamed urethane boards used for insulating sheathing.

Toenailing—A method of nailing in which nails are driven into the face or edge of one member at an angle. Toenailing often holds parts more securely than does straight face nailing.

Topographical drawing—A plan view of a construction site or a map with contour lines that show the contour of the land. Topographic drawings show other important features of the land, such as streams, wooded areas, buildings, and roads.

Traffic survey—A count of traffic intensity, direction, and times. A traffic survey is the first step in planning for highway construction.

Trap—A plumbing fitting or an arrangement of fittings that causes some water to remain or be trapped in a drain. A trap prevents sewer gas from entering a building.

Trimmer—A second framing member at the side of an opening. Trimmers make up for some of the strength lost by the missing members in the opening.

Truss—A rafter, joist, or beam that is built with lightweight members joined together in such a way that they produce a very rigid assembly.

Vehicle (paint)—The base material in paint or varnish that serves mainly as a method for handling the pigment and other active ingredients. Paint vehicles can be water, oil, toluene, alcohol, etc. The thinner used with a paint must be compatible with the vehicle.

Vertical contour interval—The difference in elevation represented by topographical contour lines which are side by side.

Volt—The unit of measure of electrical pressure.

Wall tie—A metal strap that is nailed to a frame wall or imbedded in a masonry wall, then imbedded in masonry veneer to the wall.

Watt—The unit of measure for electrical power.

One watt is the amount of power produced by one ampere of current at one volt of force.

Wind bracing—Diagonal framing members installed in a building frame to resist wind and other horizontal forces.

Word processor—A computer-software system used for writing and editing text or written material.

Zoning—Local law that controls the types of buildings permitted in a neighborhood. Some common zones are single-family residential, multiple-family residential, commercial, and industrial.

MATH REVIEWS

Math Review 1:

Fractions and Mixed Numbers—Meanings and Definitions

▼ A *fraction* is a value which shows the number of equal parts taken of a whole quantity. A fraction consists of a numerator and a denominator.

$\dfrac{7}{16}$ ← Numerator
← Denominator

▼ *Equivalent fractions* are fractions which have the same value. The value of a fraction is not changed by multiplying the numerator and denominator by the same number.

EXAMPLE Express $\dfrac{5}{8}$ as thirty-seconds.

Determine what number the denominator is multiplied by to get the desired denominator. $(32 \div 8 = 4)$

$\dfrac{5}{8} = \dfrac{?}{32}$

Multiply the numerator and denominator by 4.

$\dfrac{5}{8} \times \dfrac{4}{4} = \dfrac{20}{32}$

▼ The *lowest common denominator* of two or more fractions is the smallest denominator which is evenly divisible by each of the denominators of the fractions.

EXAMPLE 1 The lowest common denominator of $\dfrac{3}{4}, \dfrac{5}{8}$, and $\dfrac{13}{32}$ is 32, because 32 is the smallest number evenly divisible by 4, 8, and 32.

$32 \div 4 = 8$
$32 \div 8 = 4$
$32 \div 32 = 1$

EXAMPLE 2 The lowest common denominator of $\dfrac{2}{3}, \dfrac{1}{5}$, and $\dfrac{7}{10}$ is 30, because 30 is the smallest number evenly divisible by 3, 5, and 10.

$30 \div 3 = 10$
$30 \div 5 = 6$
$30 \div 10 = 3$

▼ *Factors* are numbers used in multiplying. For example, 3 and 5 are factors of 15.

$3 \times 5 = 15$

▼ A fraction is in its *lowest terms* when the numerator and the denominator do not contain a common factor.

EXAMPLE Express $\dfrac{12}{16}$ in lowest terms.

Determine the largest common factor in the numerator and denominator. The numerator and the denominator can be evenly divided by 4.

$\dfrac{12 \div 4}{16 \div 4} = \dfrac{3}{4}$

▼ A *mixed number* is a whole number plus a fraction.

$$6\frac{15}{16}$$

Whole Number ⌐   ⌐— Fraction

$$6 + \frac{15}{16} = 6\frac{15}{16}$$

▼ *Expressing fractions as mixed numbers.* In certain fractions, the numerator is larger than the denominator. To express the fraction as a mixed number, divide the numerator by the denominator. Express the fractional part in lowest terms.

EXAMPLE   Express $\frac{38}{16}$ as a mixed number.

$$\frac{38}{16} = 2\frac{6}{16}$$

Divide the numerator 38 by the denominator 16.

$$\frac{6 \div 2}{16 \div 2} = \frac{3}{8}$$

Express the fractional part $\frac{6}{16}$ in lowest terms.

$$\frac{38}{16} = 2\frac{3}{8}$$

Combine the whole number and fraction.

▼ *Expressing mixed numbers as fractions.* To express a mixed number as a fraction, multiply the whole number by the denominator of the fractional part. Add the numerator of the fractional part. The sum is the numerator of the fraction. The denominator is the same as the denominator of the original fractional part.

EXAMPLE   Express $7\frac{3}{4}$ as a fraction.

$$\frac{7 \times 4 + 3}{4} = \frac{31}{4}$$

or

$$\frac{7}{1} \times \frac{4}{4} = \frac{28}{4}$$

Multiply the whole number 7 by the denominator 4 of the fractional part ($7 \times 4 = 28$). Add the numerator 3 of the fractional part to 28. The sum 31 is the numerator of the fraction. The denominator 4 is the same as the denominator of the original fractional part.

$$\frac{28}{4} + \frac{3}{4} = \frac{31}{4}$$

Math Review 2:
Adding Fractions

▼ Fractions must have a common denominator in order to be added.

▼ To add fractions, express the fractions as equivalent fractions having the lowest common denominator. Add the numerators and write their sum over the lowest common denominator. Express the fraction in lowest terms.

$$\frac{3}{8} = \frac{3}{8} \times \frac{4}{4} = \frac{12}{32}$$

$$\frac{1}{4} = \frac{1}{4} \times \frac{8}{8} = \frac{8}{32}$$

$$\frac{3}{16} = \frac{3}{16} \times \frac{2}{2} = \frac{6}{32}$$

$$+\frac{1}{32} = \quad\quad\quad \frac{1}{32}$$

$$\frac{27}{32}$$

EXAMPLE   Add: $\dfrac{3}{8} + \dfrac{1}{4} + \dfrac{3}{16} + \dfrac{1}{32}$

Express the fractions as equivalent fractions with 32 as the denominator.

Add the numerators.

▼ After fractions are added, if the numerator is greater than the denominator, the fraction should be expressed as a mixed number.

$$\frac{1}{2} = \frac{1}{2} \times \frac{8}{8} = \frac{8}{16}$$

$$\frac{3}{4} = \frac{3}{4} \times \frac{4}{4} = \frac{12}{16}$$

$$\frac{15}{16} = \quad\quad\quad \frac{15}{16}$$

$$+\frac{11}{16} = \quad\quad\quad \frac{11}{16}$$

$$\frac{46}{16}$$

EXAMPLE   Add: $\dfrac{1}{2} + \dfrac{3}{4} + \dfrac{15}{16} + \dfrac{11}{16}$

Express the fractions as equivalent fractions with 16 as the denominator.

Add the numerators.

Express $\dfrac{46}{16}$ as a mixed number in lowest terms.

$$\frac{46}{16} = 2\frac{14}{16} = 2\frac{7}{8}$$

Math Review 3:
Adding Combinations of Fractions, Mixed Numbers, and Whole Numbers

▼ To add mixed numbers or combinations of fractions, mixed numbers, and whole numbers, express the fractional parts of the numbers as equivalent fractions having the lowest common denominator. Add the whole numbers. Add the fractions. Combine the whole number and the fraction and express in lowest terms.

$$3\frac{7}{8} = 3\frac{14}{16}$$

$$5\frac{1}{2} = 5\frac{8}{16}$$

$$+9\frac{3}{16} = 9\frac{3}{16}$$

EXAMPLE 1   Add: $3\dfrac{7}{8} + 5\dfrac{1}{2} + 9\dfrac{3}{16}$

Express the fractional parts as equivalent fractions with 16 as the common denominator. Add the whole

$$17\frac{25}{16} = 17 + 1\frac{9}{16} = 18\frac{9}{16}$$

numbers. Add the fractions. Combine the whole number and the fraction. Express the answer in lowest terms.

EXAMPLE 2 Add: $6\dfrac{3}{4} + \dfrac{9}{16} + 7\dfrac{21}{32} + 15$

Express the fractional parts as equivalent fractions with 32 as the common denominator. Add the whole numbers. Add the fractions. Combine the whole number and the fraction. Express the answer in lowest terms.

$$6\dfrac{3}{4} = 6\dfrac{24}{32}$$
$$\dfrac{9}{16} = \dfrac{18}{32}$$
$$7\dfrac{21}{32} = 7\dfrac{21}{32}$$
$$\underline{+15\ \ = 15}$$
$$28\dfrac{63}{32} = 28 + 1\dfrac{31}{32} = 29\dfrac{31}{32}$$

Math Review 4:
Subtracting Fractions from Fractions

▼ Fractions must have a common denominator in order to be subtracted.

▼ To subtract a fraction from a fraction, express the fractions as equivalent fractions having the lowest common denominator. Subtract the numerators. Write their difference over the common denominator.

EXAMPLE Subtract $\dfrac{3}{4}$ from $\dfrac{15}{16}$

Express the fractions as equivalent fractions with 16 as the common denominator. Subtract the numerator 12 from the numerator 15. Write the difference 3 over the common denominator 16.

$$\dfrac{15}{16} = \dfrac{15}{16}$$
$$\underline{-\dfrac{3}{4} = -\dfrac{12}{16}}$$
$$\dfrac{3}{16}$$

Math Review 5:
Subtracting Fractions and Mixed Numbers from Whole Numbers

▼ To subtract a fraction or a mixed number from a whole number, express the whole number as an equivalent mixed number. The fraction of the mixed number has the same denominator as the denominator of the fraction which is subtracted. Subtract the numerators of the fractions and write their difference over the common denominator. Subtract the whole numbers. Combine the whole number and fraction. Express the answer in lowest terms.

EXAMPLE 1   Subtract $\dfrac{3}{8}$ from 7

Express the whole number as an equivalent mixed number with the same denominator as the denominator of the fraction which is subtracted

$(7 = 6\dfrac{8}{8})$.

Subtract $\dfrac{3}{8}$ from $\dfrac{8}{8}$

Combine whole number and fraction.

$$
\begin{array}{r}
7 = 6\dfrac{8}{8} \\[2mm]
-\dfrac{3}{8} = \dfrac{3}{8} \\[2mm]
\hline
6\dfrac{5}{8}
\end{array}
$$

EXAMPLE 2   Subtract $5\dfrac{15}{32}$ from 12

Express the whole number as an equivalent mixed number with the same denominator as the denominator of fraction which is subtracted

$(12 = 11\dfrac{32}{32})$.

Subtract fractions.

Subtract whole numbers.

Combine whole number and fraction.

$$
\begin{array}{r}
12 \quad = 11\dfrac{32}{32} \\[2mm]
-\,5\dfrac{15}{32} = 5\dfrac{15}{32} \\[2mm]
\hline
6\dfrac{17}{32}
\end{array}
$$

Math Review 6:

Subtracting Fractions and Mixed Numbers from Mixed Numbers

▼ To subtract a fraction or a mixed number from a mixed number, the fractional part of each number must have the same denominator. Express fractions as equivalent fractions having a common denominator. When the fraction subtracted is larger than the fraction from which it is subtracted, one unit of the whole number is expressed as a fraction with the common denominator. Combine the whole number and fractions. Subtract fractions and subtract whole numbers.

EXAMPLE 1 Subtract $\dfrac{7}{8}$ from $4\dfrac{3}{16}$

Express the fractions as equivalent fractions with the common denominator 16. Since 14 is larger than 3, express one unit of $4\dfrac{3}{16}$ as a fraction and combine whole number and fraction ($4\dfrac{3}{16} = 3 + \dfrac{16}{16} + \dfrac{3}{16} = 3\dfrac{19}{16}$).

$$\begin{array}{r} 4\dfrac{3}{16} = 4\dfrac{3}{16} = 3\dfrac{19}{16} \\[2mm] -\quad \dfrac{7}{8} = \dfrac{14}{16} = \dfrac{14}{16} \\ \hline 3\dfrac{5}{16} \end{array}$$

Subtract.

EXAMPLE 2 Subtract $13\dfrac{1}{4}$ from $20\dfrac{15}{32}$

Express the fractions as equivalent fractions with the common denominator 32.

$$\begin{array}{r} 20\dfrac{15}{32} = 20\dfrac{15}{32} \\[2mm] 13\dfrac{1}{4} = 13\dfrac{8}{32} \\ \hline 7\dfrac{7}{32} \end{array}$$

Subtract fractions.

Subtract whole numbers.

Math Review 7:
Multiplying Fractions

▼ To multiply two or more fractions, multiply the numerators. Multiply the denominators. Write as a fraction with the product of the numerators over the product of the denominators. Express the answer in lowest terms.

EXAMPLE 1 Multiply $\dfrac{3}{4} \times \dfrac{5}{8}$

Multiply the numerators.

Multiply the denominators.

$$\dfrac{3}{4} \times \dfrac{5}{8} = \dfrac{15}{32}$$

Write as a fraction.

EXAMPLE 2   Multiply $\frac{1}{2} \times \frac{2}{3} \times \frac{4}{5}$

Multiply the numerators.

Multiply the denominators.

Write as a fraction and express answer in lowest terms.

$$\frac{1}{2} \times \frac{2}{3} \times \frac{4}{5} = \frac{8}{30} = \frac{4}{15}$$

Math Review 8:

Multiplying any Combination of Fractions, Mixed Numbers, and Whole Numbers

▼ To multiply any combination of fractions, mixed numbers, and whole numbers, write the mixed numbers as fractions. Write whole numbers over the denominator 1. Multiply numerators. Multiply denominators. Express the answer in lowest terms.

EXAMPLE 1   Multiply $3\frac{1}{4} \times \frac{3}{8}$

Write the mixed number $3\frac{1}{4}$ as the fraction $\frac{13}{4}$.

Multiply the numerators.

Multiply the denominators.

Express as a mixed number.

$$3\frac{1}{4} \times \frac{3}{8} = \frac{13}{4} \times \frac{3}{8} = \frac{39}{32} = 1\frac{7}{32}$$

EXAMPLE 2   Multiply $2\frac{1}{3} \times 4 \times \frac{4}{5}$

Write the mixed number $2\frac{1}{3}$ as the fraction $\frac{7}{3}$.

Write the whole number 4 over 1.

Multiply the numerators.

Multiply the numerators.

Express as a mixed number.

$$2\frac{1}{3} \times 4 \times \frac{4}{5} = \frac{7}{3} \times \frac{4}{1} \times \frac{4}{5} = \frac{112}{15}$$

$$\frac{112}{15} = 7\frac{7}{15}$$

Math Review 9:
Dividing Fractions

▼ Division is the inverse of multiplication. Dividing by 4 is the same as multiplying by $\frac{1}{4}$. Four is the inverse of $\frac{1}{4}$ and $\frac{1}{4}$ is the inverse of 4. The inverse of $\frac{5}{16}$ is $\frac{16}{5}$ w.

▼ To divide fractions, invert the divisor, change to the inverse operation and multiply. Express the answer in lowest terms.

EXAMPLE Divide $\frac{7}{8} \div \frac{2}{3}$

Invert the divisor $\frac{2}{3}$

$$\frac{7}{8} \div \frac{2}{3} = \frac{7}{8} \times \frac{3}{2} = \frac{21}{16} = 1\frac{5}{16}$$

$\frac{2}{3}$ inverted is $\frac{3}{2}$.

Change to the inverse operation and multiply.

Express as a mixed number.

Math Review 10:
Dividing any Combination of Fractions, Mixed Numbers, and Whole Numbers

▼ To divide any combination of fractions, mixed numbers, and whole numbers, write the mixed numbers as fractions. Write whole numbers over the denominator 1. Invert the divisor. Change to the inverse operation and multiply. Express the answer in lowest terms.

EXAMPLE 1 Divide: $6 \div \frac{7}{10}$

Write the whole number 6 over the denominator 1.

Invert the divisor $\frac{7}{10}$; $\frac{7}{10}$ inverted is $\frac{10}{7}$.

$$\frac{6}{1} \div \frac{7}{10} =$$

$$\frac{6}{1} \times \frac{10}{7} = \frac{60}{7} = 8\frac{4}{7}$$

Change to the inverse operation and multiply.

Express as a mixed number.

EXAMPLE 2   Divide: $\dfrac{3}{4} \div 2\dfrac{1}{5}$

Write the mixed number divisor $2\dfrac{1}{5}$ as the fraction $\dfrac{11}{5}$.

$$\dfrac{3}{4} \div \dfrac{11}{5} =$$

$$\dfrac{3}{4} \times \dfrac{5}{11} = \dfrac{15}{44}$$

Invert the divisor $\dfrac{11}{5}$; $\dfrac{11}{5}$ inverted is $\dfrac{5}{11}$.

Change to the inverse operation and multiply.

EXAMPLE 3   Divide: $4\dfrac{5}{8} \div 7$

Write the mixed number $4\dfrac{5}{8}$ as the fraction $\dfrac{37}{8}$.

$$\dfrac{37}{8} \div \dfrac{7}{1} =$$

$$\dfrac{37}{8} \times \dfrac{1}{7} = \dfrac{37}{56}$$

Write the whole number divisor 7 over the denominator 1.

Invert the divisor $\dfrac{7}{1}$; $\dfrac{7}{1}$ inverted is $\dfrac{1}{7}$.

Change to the inverse operation and multiply.

Math Review 11:
Rounding Decimal Fractions

▼   To round a decimal fraction, locate the digit in the number that gives the desired number of decimal places. Increase that digit by 1 if the digit which directly follows is 5 or more. Do not change the value of the digit if the digit which follows is less than 5. Drop all digits which follow.

EXAMPLE 1   Round 0.63861 to 3 decimal places.

Locate the digit in the third place (8). The fourth decimal-place digit, 6, is greater than 5 and increases the third decimal-place digit 8, to 9. Drop all digits which follow.

$$0.638\underline{6}1 \approx 0.639$$

EXAMPLE 2   Round 3.0746 to 2 decimal places.

Locate the digit in the second decimal place (7). The third decimal-place digit 4 is less than 5 and does not change the value of the second decimal-place digit 7. Drop all digits which follow.

$$3.07\underline{4}6 \approx 3.07$$

Math Review 12:
Adding Decimal Fractions

▼ To add decimal fractions, arrange the numbers so that the decimal points are directly under each other. The decimal point of a whole number is directly to the right of the last digit. Add each column as with whole numbers. Place the decimal point in the sum directly under the other decimal points.

> **EXAMPLE**   Add: 7.65 + 208.062 + 0.009 + 36 + 5.1037
>
> Arrange the numbers so that the decimal points are directly under each other.
>
> Add zeros so that all numbers have the same number of places to the right of the decimal point.
>
> $$\begin{array}{r} 7.6500 \\ 208.0620 \\ 0.0090 \\ 36.0000 \\ +\ \ \ 5.1037 \\ \hline 256.8247 \end{array}$$
>
> Add each column of numbers.
>
> Place the decimal point in the sum directly under the other decimal points.

Math Review 13:
Subtracting Decimal Fractions

▼ To subtract decimal fractions, arrange the numbers so that the decimal points are directly under each other. Subtract each column as with whole numbers. Place the decimal point in the difference directly under the other decimal points.

EXAMPLE Subtract: 87.4 − 42.125

Arrange the numbers so that the decimal points are directly under each other. Add zeros so that the numbers have the same number of places to the right of the decimal point.

$$\begin{array}{r} 87.400 \\ -\ 42.125 \\ \hline 45.275 \end{array}$$

Subtract each column of numbers.

Place the decimal point in the difference directly under the other decimal points.

Math Review 14:
Multiplying Decimal Fractions

▼ To multiply decimal fractions, multiply using the same procedure as with whole numbers. Count the number of decimal places in both the multiplier and the multiplicand. Begin counting from the last digit on the right of the product and place the decimal point the same number of places as there are in both the multiplicand and the multiplier.

EXAMPLE Multiply: 50.216 × 1.73

Multiply as with whole numbers.

Count the number of decimal places in the multiplier (2 places) and the multiplicand (3 places).

Beginning at the right of the product, place the decimal point the same number of places as there are in both the multiplicand and the multiplier (5 places).

┌Multiplicand
50.216 ← (3 places)
× 1.73 ← Multiplier
─────── └(2 places)
150648
351512
50216
───────
86.87368 (5 places)

▼ When multiplying certain decimal fractions, the product has a smaller number of digits than the number of decimal places required. For these products, add as many zeros to the left of the product as are necessary to give the required number of decimal places.

EXAMPLE   Multiply: 0.27×0.18

Multiply as with whole numbers.

The product must have 4 decimal places.

Add one zero to the left of the product.

```
  0.27   (2 places)
× 0.18   (2 places)
  216
   27
0.0486   (4 places)
```

Math Review 15:
Dividing Decimal Fractions

▼ To divide decimal fractions, use the same procedure as with whole numbers. Move the decimal point of the divisor as many places to the right as necessary to make the divisor a whole number. Move the decimal point of the dividend the same number of places to the right. Add zeros to the dividend if necessary. Place the decimal point in the answer directly above the decimal point in the dividend. Divide as with whole numbers. Zeros may be added to the dividend to give the number of decimal places required in the answer.

EXAMPLE 1   Divide: $0.6150 \div 0.75$

Move the decimal point 2 places to the right in the divisor.

Move the decimal point 2 places in the dividend.

Place the decimal point in the answer directly above the decimal point in the dividend.

Divide as with whole numbers.

```
                     0.82
Divisor → 0.75. ) 0 61.50 ← Dividend
                  60 0
                   1 50
                   1 50
```

EXAMPLE 2   Divide: 10.7 ÷ 4.375. Round the answer to 3 decimal places.

Move the decimal point 3 places to the right in the divisor.

Move the decimal point 3 places in the dividend, adding 2 zeros.

Place the decimal point in the answer directly above the decimal point in the dividend.

Add 4 zeros to the dividend. One more zero is added than the number of decimal places required in the answer.

Divide as with whole numbers.

```
                2.4457 ≈ 2.446
4 375. ) 10 700.0000
         8 750
         1   0
         1 750 0
           200 00
           175 00
            25 000
            21 875
             3 1250
             3 0625
               625
```

Math Review 16:

Expressing Common Fractions as Decimal Fractions

▼ A common fraction is an indicated division. A common fraction is expressed as a decimal fraction by dividing the numerator by the denominator.

$$\text{EXAMPLE} \quad \text{Express } \frac{5}{8} \text{ as a decimal fraction.} \qquad 8 \overline{)\, 5}$$

Write $\frac{5}{8}$ as an indicated division. $8 \overline{)\, 5.000}$

Place a decimal point after the 5 and add zeros to the right of the decimal point. $8 \overline{)\, 5.000}$

Place the decimal point for the answer directly above the decimal point in the dividend.

$$\begin{array}{r} 0.625 \\ 8 \overline{)\, 5.000} \end{array}$$

Divide.

▼ A common fraction which will not divide evenly is expressed as a repeating decimal.

EXAMPLE   Express $\frac{1}{3}$ as a decimal.

$3\overline{)1}$

Write $\frac{1}{3}$ as an indicated division.

$3\overline{)1.0000}$

Place a decimal point after the 1 and add zeros to the right of the decimal point.

$3\overline{)1.0000}$

Place a decimal point for the answer directly above the decimal point in the dividend.

$\begin{array}{r} . \\ 0.3333 \\ 3\overline{)1.0000} \end{array}$

Divide.

Math Review 17:

Expressing Decimal Fractions as Common Fractions

▼ To express a decimal fraction as a common fraction, write the number after the decimal point as the numerator of a common fraction. Write the denominator as 1 followed by as many zeros as there are digits to the right of the decimal point. Express the common fraction in lowest terms.

EXAMPLE 1   Express 0.9 as a common fraction.

Write 9 as the numerator.

$\frac{9}{10}$

Write the denominator as 1 followed by 1 zero. The denominator is 10.

EXAMPLE 2   Express 0.125 as a common fraction.

Write 125 as the numerator.

$\frac{125}{1000}$

Write the denominator as 1 followed by 3 zeros. The denominator is 1000.

Express the fraction in lowest terms.

$\frac{125}{1000} = \frac{1}{8}$

Math Review 18:

Expressing Inches as Feet and Inches

▼ There are 12 inches in 1 foot.

▼ To express inches as feet and inches, divide the given length in inches by 12 to obtain the number of whole feet. The remainder is the number of inches in addition to the number of whole feet. The answer is the number of whole feet plus the remainder in inches.

EXAMPLE 1 Express $176\frac{7}{16}$ inches as feet and inches.

Divide $176\frac{7}{16}$ inches by 12.

There are 14 feet plus a remainder of $8\frac{7}{16}$ inches.

$$
\begin{array}{r}
14 \quad \text{(feet)} \\
12\overline{)\,176\frac{7}{16}} \\
\underline{12} \\
56 \\
\underline{48} \\
8\frac{7}{16} \leftarrow \begin{array}{l}\text{Remainder}\\ \text{(inches)}\end{array}
\end{array}
$$

$14' - 8\frac{7}{16}''$

EXAMPLE 2 Express 54.2 inches as feet and inches.

Divide 54.2 inches by 12.

There are 4 feet plus a remainder of 6.2 inches.

$$
\begin{array}{r}
41 \quad \text{(feet)} \\
12\overline{)\,54.2} \\
\underline{48} \\
6.2 \leftarrow \begin{array}{l}\text{Remainder}\\ \text{(inches)}\end{array}
\end{array}
$$

4 feet 6.2 inches

Math Review 19:

Expressing Feet and Inches as Inches

▼ There are 12 inches in one foot.

▼ To express feet and inches as inches, multiply the number of feet in the given length by 12. To this product, add the number of inches in the given length.

EXAMPLE Express 7 feet $9\frac{3}{4}$ inches as inches.

Multiply 7 feet by 12. There are 84 inches in 7 feet.

Add $9\frac{3}{4}$ inches to 84 inches.

$7 \times 12 = 84$
7 feet = 84 inches
84 inches + $9\frac{3}{4}$ inches =
$93\frac{3}{4}$ inches

Math Review 20:

Expressing Inches as Decimal Fractions of a Foot

▼ An inch is $\frac{1}{12}$ of a foot. To express whole inches as a decimal part of a foot, divide the number of inches by 12.

> **EXAMPLE** Express 7 inches as a decimal fraction of a foot.
>
> Divide 7 by 12.

$7 \div 12 = 0.58$
0.58 feet

▼ To express a common fraction of an inch as a decimal fraction of a foot, express the common fraction as a decimal, then divide the decimal by 12.

> **EXAMPLE 1** Express $\frac{3}{4}$ inch as a decimal fraction of a foot.
>
> Express $\frac{3}{4}$ as a decimal.
>
> Divide the decimal by 12.

$3 \div 4 = 0.75$
$0.75 \div 12 = 0.06$
0.06 feet

> **EXAMPLE 2** Express $4\frac{3}{4}$ inches as a decimal fraction of a foot.
>
> Express $4\frac{3}{4}$ as a decimal.
>
> Divide the decimal inches by 12.

$4 + \frac{3}{4} = 4 + 0.75 = 4.75$
$4.75 \div 12 = 0.39$

0.39 feet

Math Review 21:

Expressing Decimal Fractions of a Foot as Inches

▼ To express a decimal part of a foot as decimal inches multiply by 12.

> **EXAMPLE** Express 0.62 feet as inches.
>
> Multiply 0.62 by 12.

$0.62 \times 12 = 7.44$
7.44 inches

▼ To express the decimal fraction of an inch as a common fraction, see Math Review 17.

Math Review 22:
Area Measure

▼ A surface is measured by determining the number of surface units contained in it. A surface is two dimensional. It has length and width, but no thickness. Both length and width must be expressed in the same unit of measure. Area is expressed in square units. For example, 5 feet × 8 feet equals 40 square feet.

▼ *Equivalent Units of Area Measure:*

1 square foot (sq ft) =

　12 inches × 12 inches = 144 square inches (sq in)

1 square yard (sq yd) =

　3 feet × 3 feet = 9 square feet (sq ft)

▼ To express a given unit of area as a larger unit of area, divide the given area by the number of square units contained in one of the larger units.

EXAMPLE 1　Express 648 square inches as square feet.

Since 144 sq in = 1 sq ft, divide 648 by 144.

$$648 \div 144 = 4.5$$
648 square inches = 4.5 square feet

EXAMPLE 2　Express 28.8 square feet as square yards.

Since 9 sq ft = 1 sq yd, divide 28.8 by 9.

$$28.8 \div 9 = 3.2$$
28.8 square feet = 3.2 square yards

▼ To express a given unit of area as a smaller unit of area, multiply the given area by the number of square units contained in one of the larger units.

EXAMPLE 1　Express 7.5 square feet as square inches.

Since 144 sq in = 1 sq ft, multiply 7.5 by 144.

$$7.5 \times 144 = 1080$$
7.5 square feet = 1080 square inches

EXAMPLE 2   Express 23 square yards as square feet.

Since 9 sq ft = 1 sq yd, multiply 23 by 9.

23 × 9 = 207
23 square yards = 207 square feet

▼ *Computing Areas of Common Geometric Figures:*

1. **Rectangle**   A rectangle is a four-sided plane figure with 4 right (90°) angles.

   The area of a rectangle is equal to the product of its length and its width.

   Area = length × width (A = l × w)

   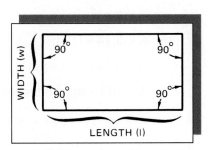

   EXAMPLE   Find the area of a rectangle 24 feet long and 13 feet wide.

   $A = l \times w$

   $A = 24 \text{ ft} \times 13 \text{ ft}$

   $A = 312 \text{ square feet}$

2. **Triangle**   A triangle is a plane figure with 3 sides and 3 angles.

   The area of a triangle is equal to one-half the product of its base and altitude.

   $A = \dfrac{1}{2} \text{ base} \times \text{altitude } (A = \dfrac{1}{2} \text{ b} \times \text{a})$

   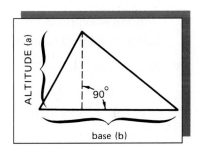

   EXAMPLE   Find the area of a triangle with a base of 16 feet and an altitude of 12 feet.

   $A = \dfrac{1}{2} \text{ b} \times \text{a}$

   $A = \dfrac{1}{2} \times 16 \text{ ft} \times 12 \text{ ft}$

   $A = 96 \text{ square feet}$

3. **Circle**   The area of a circle is equal to π times the square of its radius.

   Area = $\pi \times \text{radius}^2$ (A = $\pi \times r^2$)

NOTE: π (pronounced "pie") is approximately equal to 3.14. Radius squared (r^2) means r × r.

EXAMPLE   Find the area of a circle with a 15-inch radius.

$A = \pi \times r^2$

$A = 3.14 \times (15 \text{ in})^2$

$A = 3.14 \times 225 \text{ sq in}$

$A = 706.5 \text{ square inches}$

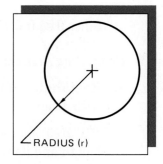

RADIUS (r)

Math Review 23:
Volume Measure

▼ A solid is measured by determining the number of cubic units contained in it. A solid is three dimensional; it has length, width, and thickness or height. Length, width, and thickness must be expressed in the same unit of measure. Volume is expressed in cubic units. For example, 3 feet × 5 feet × 10 feet = 150 cubic feet.

▼ *Equivalent Units of Volume Measure:*

1 cubic foot (cu ft) =

   12 in × 12 in × 12 in = 1728 cubic inches (cu in)

1 cubic yard (cu yd) =

   3 ft × 3ft × 3 ft = 27 cubic feet (cu ft)

▼ To express a given unit of volume as a larger unit of volume, divide the given volume by the number of cubic units contained in one of the larger units.

> **EXAMPLE 1**   Express 6048 cubic inches as cubic feet.
> Since 1728 cu in = 1 cu ft, divide 6048 by 1728.

6048 ÷ 1728 = 3.5
6048 cubic inches = 3.5 cubic feet

> **EXAMPLE 2**   Express 167.4 cubic feet as cubic yards.
> Since 27 cu ft = 1 cu yd, divide 167.4 by 27.

167.4 ÷ 27 = 6.2
167.4 cubic feet = 6.2 cubic yards

▼ To express a given unit of volume as a smaller unit of volume, multiply the given volume by the number of cubic units contained in one of the larger units.

EXAMPLE 1   Express 1.6 cubic feet as cubic inches.
Since 1728 cu in = 1 cu ft, multiply 1.6 by 1728.

1.6 × 1728 = 2764.8
1.6 cubic feet = 2764.8 cubic inches

EXAMPLE 2   Express 8.1 cubic yards as cubic feet.
Since 27 cu ft = 1 cu yd, multiply 8.1 by 27.

8.1 × 27 = 218.7
8.1 cubic yards = 218.7 cubic feet

▼ *Computing Volumes of Common Solids*

▼ A prism is a solid which has two identical faces called bases and parallel lateral edges. In a right prism, the lateral edges are perpendicular (at 90°) to the bases. The altitude or height (h) of a prism is the perpendicular distance between its two bases. Prisms are named according to the shapes of their bases.

The volume of any prism is equal to the product of the area of its base and altitude or height.

Volume = area of base × altitude
$(V = A_B \times h)$

Right Rectangular Prism.
   A right rectangular prism has rectangular bases.
   Volume = area of base × altitude
   $V = A_B \times h$

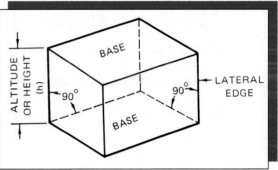

EXAMPLE   Find the volume of a rectangular prism with a base length of 20 feet, a base width of 14 feet and a height (altitude) of 8 feet.

$V = A_B \times h$

Compute the area of the base (AeB):

Area of base = length × width

$A_B = 20 \text{ ft} \times 14 \text{ ft}$

$A_B = 280 \text{ sq ft}$

Compute the volume of the prism:

$V = A_B \times h$

$V = 280 \text{ sq ft} \times 8 \text{ ft}$

$V = 2240 \text{ cu ft}$

Right Triangular Prism.

A right triangular prism has triangular bases.

Volume = area of base × altitude

$$V = A_B \times h$$

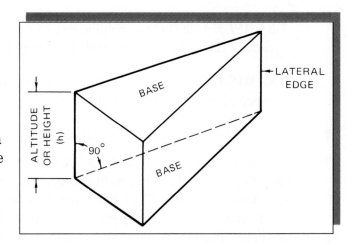

EXAMPLE   Find the volume of a triangular prism in which the base of the triangle is 5 feet, the altitude of the triangle is 3 feet, and the altitude (height) of the prism is 4 feet. Refer to the accompanying figure.

Volume = area of base × altitude

$$V = A_B \times h$$

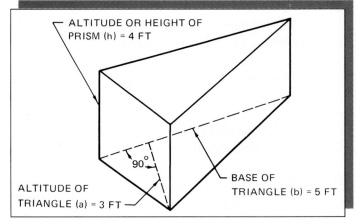

Compute the area of the base:

Area of base = $\frac{1}{2}$ base of triangle × altitude of triangle

$$A_B = \frac{1}{2} b \times a$$

$$A_B = \frac{1}{2} \times 5 \text{ ft} \times 3 \text{ ft}$$

$$A_B = 7.5 \text{ sq ft}$$

Compute the volume of the prism:

$$V = A_B \times h$$

$$V = 7.5 \text{ sq ft} \times 4 \text{ ft}$$

$$V = 30 \text{ cubic feet}$$

Right Circular Cylinder

A right circular cylinder has circular bases.

Volume = area of base × altitude

$$V = A_B \times h$$

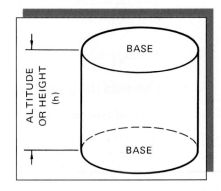

EXAMPLE   Find the volume of a circular cylinder 1 foot in diameter and 10 feet high.

NOTE: Radius = $\frac{1}{2}$ Diameter; Radius = $\frac{1}{2}$ × 1 ft = 0.5 ft.

$V = A_B \times h$

Compute the area of the base:

Area of base = π × radius squared

$A_B = 3.14 \times (0.5 \text{ ft})^2$

$A_B = 3.14 \times 0.5 \text{ ft} \times 0.5 \text{ ft}$

$A_B = 3.14 \times 0.25 \text{ sq ft}$

$A_B = 0.785 \text{ sq ft}$

Compute the volume of the cylinder:

$V = A_B \times h$

$V = 0.785 \text{ sq ft} \times 10 \text{ ft}$

$V = 7.85 \text{ cubic feet}$

Math Review 24:

Finding an Unknown Side of a Right Triangle, Given Two Sides

▼ If one of the angles of a triangle is a right (90°) angle, the figure is called a right triangle. The side opposite the right angle is called the hypotenuse. In the figure shown, c is opposite the right angle; c is the hypotenuse.

▼ In a right triangle, the square of the hypotenuse is equal to the sum of the squares of the other two sides:

$c^2 = a^2 + b^2$

If any two sides of a right triangle are known, the length of the third side can be determined by one of the following formulas:

$c = \sqrt{a^2 + b^2}$

$a = \sqrt{c^2 - b^2}$

$b = \sqrt{c^2 - a^2}$

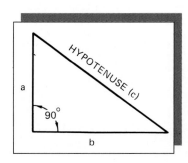

EXAMPLE 1   In the right triangle shown, a = 6 ft, b = 8 ft, find c.

$$c = \sqrt{a^2 + b^2}$$

$$c = \sqrt{6^2 + 8^2}$$

$$c = \sqrt{36 + 64}$$

$$c = \sqrt{100}$$

$$c = 10 \text{ feet}$$

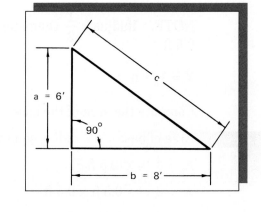

EXAMPLE 2   In the right triangle shown, c = 30 ft, b = 20 ft, find a.

$$a = \sqrt{c^2 - b^2}$$

$$a = \sqrt{30^2 - 20^2}$$

$$a = \sqrt{900 - 400}$$

$$a = \sqrt{500}$$

$$a = 22.36 \text{ feet (to 2 decimal places)}$$

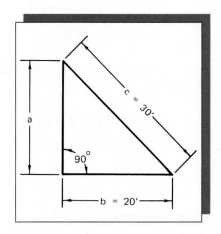

EXAMPLE 3   In the right triangle shown, c = 18 feet, a = 6 ft, find b.

$$b = \sqrt{c^2 - a^2}$$

$$b = \sqrt{18^2 - 6^2}$$

$$b = \sqrt{324 - 36}$$

$$b = \sqrt{288}$$

$$b = 16.97 \text{ feet (to 2 decimal places)}$$

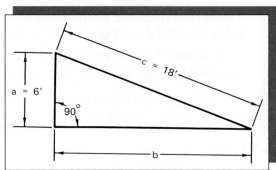

TECHNOLOGY EDUCATION STUDENT ASSOCIATIONS

Introduction

Technology education student associations help you work with others in your technology class, your school, and your community. The technology student organization is the group in your school that uses technology in a variety of interesting activities, projects, and contests.

Think of the other groups in your school. Some, called "clubs," serve a special interest or offer after-school activities. Associations such as the Student Council serve all students in the school. The technology education student association serves all students taking Technology Education or Industrial Technology in their school.

Technology education student associations should meet the standards for technology associations that have been established by an international association of teachers. These standards are helpful in making student associations serve the needs of students who must be technologically literate to live and work in our technological world.

Technology education student associations involve communication, construction, manufacturing, and transportation activities. Opportunities for creative thinking, problem solving, and decision making are also provided.

Learning how technology works can be greatly enhanced by the student association's activities.

Learning to Be Leaders

Every group or organization has a set of leaders, called officers. Most officers are elected by group members. In a corporation, for example, a board of directors may appoint (or employ) a president to run the company. Vice presidents, treasurers, and other managers or leaders are the top people who make things happen in the organization.

Every student should serve at some time as an officer in the student association. You can volunteer to be a candidate or be nominated by someone else. Experience as an elected officer will improve your ability to lead and to work with others. Leadership skill is especially valuable when you are attending college or working. Employers look for people who are willing to learn and are able to get along well with others.

Associations Start in Class

Many teachers recognize the benefit of student associations and involve all of their students in the associations. The students learn to lead and to use technology to make learning activities more significant. The officers and committees of an in-class student organization can help lead and manage activities during the class period. Each class may then want to take part in the school's technology student association activities.

The School Organization or Chapter

Most student groups in your school are organized to allow students to work together in activities related to a subject they take. A student association for technology education plans activities that relate to technology. By working together, the group accomplishes more than each member could accomplish individually.

The association forms committees to allow more of its members to lead activities. Each Committee plans and leads at least one activity relating to the school's technology education classes. The activity adds to students' understanding of technology and its impact on our world.

The student association in your school affiliates (joins together) with the state association so that all school chapters throughout your state are stronger and can share information. In the ame way, your state office affiliates with all other state offices to form a strong national association. The national association offers many services to its members states, schools, and students.

Advantages to Students and Their School

The technology education student association can help you continue your exploration in the field of construction technology. You learn about career options and opportunities. By working in groups, you learn leadership skills and work cooperatively with other students and adults.

Your school technology education program becomes better known because the student association attains recognition through the success of its members. You prepare for contests, follow the Achievement Programs, and travel to conferences in your state. Attending a national conference is an honor and a very educational experience.

Learning about Technology with the Technology Education Student Association

Your class and chapter officers can select several activities that will help students learn more about technology. Each activity allows students an opportunity to make technology work. Begin a brainstorming session, and find activities that are closely related to construction technology. Try to use technology in your student association activities. Refer to the activities in this book for activity ideas.

Contests to Motivate and Teach

Contests start in the classroom or laboratory, like many other activities of the technology education student association. The contest or project can be one of your class assignments. For example, all students might be required to design and build a bridge during the unit on highway and bridge construction. The winning bridges are then entered in the Technology Student Association (TSA) Bridge Contest for recognition at the state or national conferences.

Design a Bridge Contest

Statement of the Problem

Design a bridge to withstand a certain load in accordance with the following engineering specifications.

Engineering Specifications

1. Bridges are to be constructed from no more than thirty (30) feet of 3-32" square balsa wood and model airplane type of cement. While glue may be used to laminate the wood, it cannot be used to coat the entire piece. Paint may be used if desired. Contestants must supply their own materials. (Use of any other than the type of specificied material will disqualify the contestant.)

2. Bridge must be free standing and be of the following dimensions:
   a. total length—16"
   b. total height—10"
   c. total width—2"
   d. maximum height above water—5"
   e. the stream is 10" wide.

3. Bridges cannot be fastened to any common base. They will be tested by placing a standard 3-1/2" × 7-7/8" brick (approx. 3 lb) on the center of the span. The bridge should be able to support the total weight as specified below.

4. The event will be judged in accordance with the following parameters:
   a. Function—bridge must withstand the full load for at least 15 seconds.
   b. Form—bridge must be aesthetically attractive.
   c. Specifications—bridges must conform to all engineering specifications outlined above.

The Winner

The winning bridges will be placed on display in a suitable location in the school.

INDEX